George W. Bluman

Problem Book for First Year Calculus

With 384 Illustrations

George W Bluman
Department of Mathematics
University of British Columbia
Vancouver, B.C. V6T 1Y4
Canada

Editor
Paul R. Halmos
Department of Mathematics
Indiana University
Bloomington, IN 47405
USA

Illustration by Landy Lee

AMS Classification 26-01: 00A07

Library of Congress Cataloging-in-Publication Data
Bluman, Gearge W., 1943 -
Problem book for first year calculus.
(Problem books in mathematics)
1.Calculus—Problems, exercises, etc. I. Halmos,
Paul R. (Paul Richard), 1916- .II. Title.
III. Series
QA303.B674 1984 515'.076 83-25412
ISBN 0-387-90920-6 (U.S.)

ISBN 0-387-90920-6 Springer-Verlag New York Berlin Heidelberg Tokyo
ISBN 3-540-90920-6 Springer-Verlag Berlin Heidelberg New York Tokyo

Bluman: Problem Book for First Year Calculus

First Indian Reprint 2011

ISBN 978-81-8489-835-4

This edition is published by Springer (India) Private Limited, (a part of Springer Science+Business Media), Registered Office: 3rd Floor, Gandharva Mahavidyalaya, 212, Deen Dayal Upadhyaya Marg, New Delhi – 110 002, India

Printed in India by Rashtriya Printers, Delhi.

To

My wife Cynthia

My sons David and Benjamin

My parents Nathan and Susan

Preface for the Student

This book focuses on the application of one-variable calculus to problems connected with physics, engineering, business, economics, biology, and chemistry. It is intended for the student who wants to "learn by doing."

In the application of mathematics it is most important to represent and interpret visually the essentials of a given problem. Chapters I and II are concerned with problems in graphing and geometry.

Chapters III, IV, V, and VI, respectively, deal with problems pertaining to physics and engineering, business and economics, biology and chemistry, and numerical methods.

Chapters VII and VIII contain problems on the theory and techniques of calculus.

Each of the first six chapters begins with a discussion of background material necessary for doing the corresponding problems. Each chapter has a section of Solved Problems worked out in detail, often with alternative methods of solution, and a section of Supplementary Problems for which answers and occasional comments are given in Chapter IX.

In each chapter the order of problems approximates the order in which topics are encountered in most calculus courses. There are standard and difficult problems, the latter indicated by an asterisk (*). The student who masters the standard problems should have no trouble passing. The difficult problems offer a substantial challenge to the best of students.

Each chapter may be studied independently of the others. However, a student should work on Chapters I and II before proceeding to Chapters III, IV, and V.

An easy-to-use Index has each entry referenced to specific Solved and Supplementary Problems.

Preface for the Instructor

For a course in one-variable calculus this book can be used as

(1) the textbook for a course emphasizing problem-solving;
(2) a resource book for stimulating problems;
(3) a supplement to your textbook.

The book contains about 1000 problems, including over 300 problems solved in great detail. There are approximately 350 diagrams. In comparison with other calculus books the emphasis is more on applied problems. Moreover, they are collected together according to the field of application. Some of the challenging problems are open-ended and use actual data (e.g., extrapolation of population data, or the speed limit for optimizing traffic flow on a bridge). The applied problems stress the concepts of calculus and require few techniques.

Many of the problems have been drawn from homework exercises and recent examination papers of various Canadian universities.

A detailed index is included.

Acknowledgments

I am indebted to the goodwill of many Canadian universities in providing me with valuable problems. The Canadian Mathematical Society helped with the collection of the problems.

I want to thank my able student assistants Chun Lee, Christopher Lin, Marjorie Sayer, and, above all, Edmond Chow. (The University Youth Employment Program of the Province of British Columbia provided them with financial support).

I am grateful to the many colleagues and friends who offered encouragement and advice during the preparation of the manuscript. In particular I thank Professors Ed Barbeau, University of Toronto; David Parker and Tony Rayner, University of Nottingham; Robert Israel, Charles McDowell, Hugh Thurston, and, posthumously, Ron Riddell, University of British Columbia.

My wife, Cynthia, deserves special thanks for her patience through many trying moments.

I thank Miss Kelly McLaughlin and Mrs. Joyce Davis for their typing; Landy Lee for the final illustrations; and Philip Loewen for a reading of the proofs.

I also wish to thank Springer-Verlag New York for their cooperation in the final preparation and production of the manuscript. P. R. Halmos, the Editor of this series, provided some valuable suggestions for improvements.

Contents

Chapter I

Graphing

In general, problem solving is an art. It cannot be presented as a rigid set of rules which must be adhered to. However by studying examples from different points of view and through constant practice, a student can gain much insight into how to tackle fresh problems.

In the course of an applied problem a student is eventually confronted with a function to consider. Often particular properties of the function are important to its application (e.g. the domain of the function, its roots, its maximum value, the points where it is undefined, etc.). Since one is naturally attuned to see immediate insights from diagrams, one normally sketches the graph of a function as a first step in its analysis.

Usually the graph of a function $f(x)$ can be sketched by following a definite routine in which one considers successively $f(x)$, $f'(x)$, and $f''(x)$. The suggested procedure is as follows:

1. *Consider $f(x)$ itself*:

    (a) Find its domain.

    (b) Identify its symmetry and periodicity properties.

    (c) Note its behaviour as $x \to \pm\infty$ and near points where $f(x)$ is undefined.

    (d) If it is easy to do, locate the roots of $f(x)$ and identify regions of common sign, i.e., regions where $f(x) > 0$ and regions where $f(x) < 0$.

2. *Consider the first derivative of $f(x)$, namely $f'(x)$*:

    (a) Identify important points where $f'(x) = 0$ or $f'(x)$ is undefined.

    (b) Find regions of common sign of $f'(x)$ to determine where $f(x)$ is increasing or decreasing. Thus determine the nature of the important points of (a), i.e., determine whether such a point is a local maximum, local minimum, inflection point, cusp, etc.

(c) Use $f'(x)$ to check properties of $f(x)$ drawn from considering $f(x)$ itself. For example, between two roots of $f(x)$, $x_1 < x_2$, there must be a *critical point* X, $x_1 < X < x_2$, where $f'(X) = 0$ (*stationary point*) or $f'(X)$ fails to exist.

3. *Consider the second derivative of* $f(x)$, *namely* $f''(x)$ (*not always essential*):

(a) Identify important points where $f''(x) = 0$ or $f''(x)$ is undefined.

(b) Find regions of common sign of $f''(x)$ to determine the concavity of $f(x)$. Determine inflection points of $f(x)$.

(c) Use $f''(x)$ to check conclusions about $f(x)$ drawn from considering $f'(x)$. For example, if X is such that $f'(X) = 0$, $f''(X) < 0$, then $x = X$ corresponds to a local maximum of $f(x)$.

4. *Now sketch the graph of* $f(x)$.

Solved Problems

1. Identify, sketch and locate the centre of the curve

$$x^2 + y^2 + 2x + 3y = 0.$$

First observe that this equation describes a conic. (Note that this curve can be sketched without the use of calculus.) Completing squares in both variables:

$$x^2 + 2x + 1 + y^2 + 3y + \tfrac{9}{4} = 1 + \tfrac{9}{4} = \tfrac{13}{4}.$$

Then

$$(x+1)^2 + \left(y + \tfrac{3}{2}\right)^2 = \tfrac{13}{4}.$$

Thus the graph is a circle, centred at $(-1, -\tfrac{3}{2})$, with radius $\sqrt{13}/2$. To sketch: Note that the y-intercepts (that is, $x = 0$) are 0 and -3; the x-intercepts ($y = 0$) are 0 and -2 (Fig. I.1).

2. $f(x) = \sqrt{x}$, $g(x) = x^2 + 1$, $h(x) = (f \circ g)(x)$.

(a) Find the domain and range of h.
(b) Find $h'(x)$ and $h''(x)$.
(c) Where does h have a relative minimum? Relative maximum?
(d) Where is h concave up? Concave down?
(e) Sketch the graph of h.

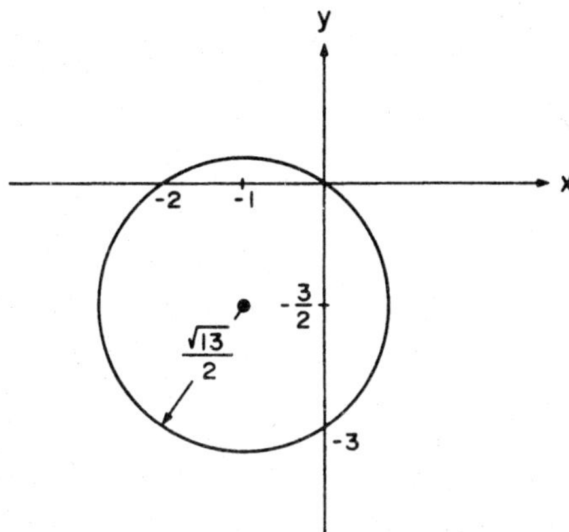

Figure I.1. Sketch of $x^2 + y^2 + 2x + 3y = 0$.

(a) Domain of h: $h(x)$ is defined for all real x. Range of h: $h(x) \geq 1$.

(b) Using the chain rule,

$$h'(x) = \frac{1}{2}(x^2+1)^{-1/2}(2x) = \frac{x}{\sqrt{x^2+1}};$$

$$h''(x) = \frac{\sqrt{x^2+1} - x\left(\dfrac{x}{\sqrt{x^2+1}}\right)}{x^2+1} = \frac{x^2+1-x^2}{(x^2+1)^{3/2}} = \frac{1}{(x^2+1)^{3/2}}$$

(or use $h'(x) = x(x^2+1)^{-1/2}$ and the product rule).

(c) $h'(x) = 0 \leftrightarrow \dfrac{x}{\sqrt{x^2+1}} = 0 \leftrightarrow x = 0$. Hence $h'(x)$ has the same sign for all $x < 0$ and the same sign for all $x > 0$. Moreover, since $h(x)$ is even (that is, $h(-x) = h(x)$), one concludes that $H(x) \equiv h'(x)$ is an odd function of x (i.e. $H(-x) = -H(x)$).

Take a sample point, say $x = 1$; one finds that $h'(1) > 0$. Therefore

$$\left.\begin{array}{r} h(x) \text{ is decreasing for } x < 0 \\ \text{increasing for } x > 0 \end{array}\right\} \leftrightarrow \text{relative minimum at } x = 0.$$

(d) $h''(x) = \dfrac{1}{(x^2+1)^{3/2}} > 0$ for all x. Hence $h(x)$ is concave up for all x.

(e) Graphing aids:

$h(1) = \sqrt{2}$. Remember that $h(x)$ is even.

As $|x|$ increases, x^2 dominates; that is, $h(x)$ behaves like $\sqrt{x^2} \equiv |x|$. Hence $y = |x|$ is an asymptote as $|x| \to \infty$. One can also recognize $h(x) = \sqrt{x^2+1}$ as the positive branch of the hyperbola $y^2 - x^2 = 1$, and graph accordingly (Fig. I.2).

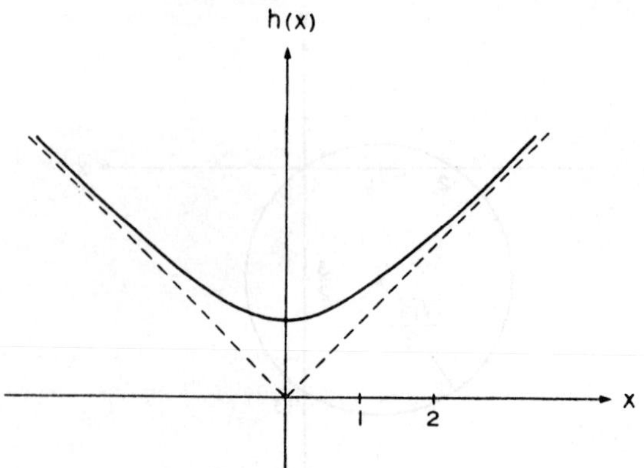

Figure I.2. Graph of $h(x)$.

3. Graph the function $f(x)=|x-1|+|x-2|$. Indicate those points (if any) at which $f(x)$ is (a) discontinuous, (b) non-differentiable.

$f(x)$ can be graphed without calculus. Since

$$|x-a|=\begin{cases} x-a & \text{if } x\geq a, \\ -(x-a) & \text{if } x\leq a, \end{cases}$$

$x=1$, $x=2$ are important points to consider.

For $x\leq 1$: $|x-1|=-(x-1)=-x+1,$

$|x-2|=-(x-2)=-x+2.$

Here $f(x)=-2x+3$.

For $1\leq x\leq 2$: $|x-1|=x-1,\quad |x-2|=-x+2.$

Here $f(x)=1$.

For $x\geq 2$: $|x-1|=x-1,\quad |x-2|=x-2.$

Here $f(x)=2x-3$.

(a) $f(x)$ is continuous everywhere.
(b) $f(x)$ is non-differentiable at $x=1$, $x=2$.

4. Sketch the graph of

$$f(x)=x^3-3x^2,$$

indicating the maxima, the minima, the inflection points, and the concavity.

$$f(x)=x^3-3x^2;$$

$$f'(x)=3x^2-6x;$$

$$f''(x)=6x-6.$$

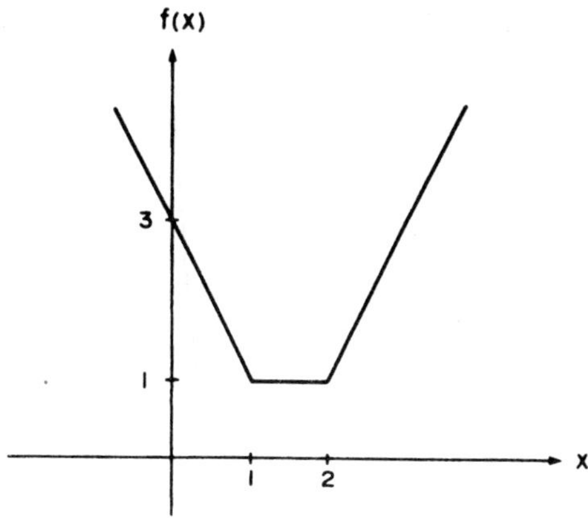

Figure I.3. Graph of $f(x) = |x-1| + |x-2|$.

The Function.

Domain and range of $f(x)$: all real x.
$f(x)$ behaves like x^3 as $|x| \to \infty$.
$f(x) = 0 \leftrightarrow x^2(x-3) = 0 \leftrightarrow x = 0,\ x = 3$.

The First Derivative.

$f'(x) = 0 \leftrightarrow 3x(x-2) = 0 \leftrightarrow x = 0,\ x = 2$. $[f(2) = -4]$.
Important points: $x = 0, 2$.

For $x < 0, f'(x) > 0 \leftrightarrow f(x)$ increasing
For $0 < x < 2, f'(x) < 0 \leftrightarrow f(x)$ decreasing
For $x > 2, f'(x) > 0 \leftrightarrow f(x)$ increasing

(first two lines) $\leftrightarrow$ local maximum at $x = 0$.
(last two lines) $\leftrightarrow$ local minimum at $x = 2$.

The Second Derivative.

$f''(x) = 0 \leftrightarrow 6(x-1) = 0 \leftrightarrow x = 1$. $[f(1) = -2]$.
Important point: $x = 1$.

For $x < 1, f''(x) < 0 \leftrightarrow f(x)$ concave down
For $x > 1, f''(x) > 0 \leftrightarrow f(x)$ concave up

$\leftrightarrow$ inflection point at $x = 1$.

5. For the function defined by $y = 3x^5 - 5x^3$ find:
   (a) All critical points, relative maxima, relative minima.
   (b) Where the function is increasing, decreasing.
   (c) Inflection points, if any.
   (d) Where the function is concave up, concave down.

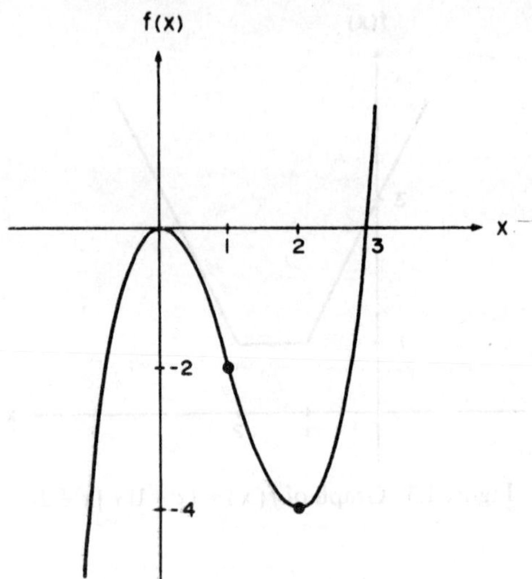

Figure I.4. Graph of $f(x) = x^3 - 3x^2$.

(e) Sketch the graph.

$$y = 3x^5 - 5x^3 = x^3(3x^2 - 5);$$

$$\frac{dy}{dx} = 15x^4 - 15x^2 = 15x^2(x^2 - 1);$$

$$\frac{d^2y}{dx^2} = 60x^3 - 30x = 30x(2x^2 - 1).$$

The Function.

Domain and range of y: all real x.
y behaves like $3x^5$ as $|x| \to \infty$.
$y = 0 \leftrightarrow x^3(3x^2 - 5) = 0 \leftrightarrow x = 0,\ x = \pm\sqrt{\frac{5}{3}} \simeq \pm 1.3$.
$y(-x) = -y(x)$. Hence y is an odd function of x.
Thus the graph for $x < 0$ can be obtained by reflection through the origin of the graph for $x > 0$. *Consequently only consider y for $x \geq 0$.*

The First Derivative.

$\frac{dy}{dx} = 0 \leftrightarrow x^3(x^2 - 1) = 0 \leftrightarrow x = 0,\ x = \pm 1.\ [y(\pm 1) = \mp 2]$.
Note that $\frac{dy}{dx}$ is even since y is odd.
Important points: $x = 0, \pm 1$.
For $0 < x < 1$, $\frac{dy}{dx} < 0 \Rightarrow$ for $-1 < x < 0$, $\frac{dy}{dx} < 0$.

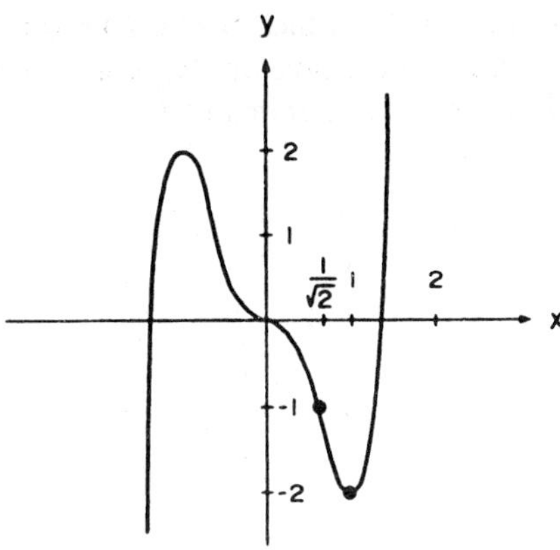

Figure I.5. Graph of $y = 3x^5 - 5x^3$.

For $x > 1$, $\dfrac{dy}{dx} > 0 \Rightarrow$ for $x < -1$, $\dfrac{dy}{dx} > 0$.
Hence $x = 1 \leftrightarrow$ relative minimum; $x = -1 \leftrightarrow$ relative maximum.

The Second Derivative.

$$\frac{d^2y}{dx^2} = 0 \leftrightarrow x(2x^2 - 1) = 0 \leftrightarrow x = 0,\ x = \pm\frac{1}{\sqrt{2}}.\ \left[y\left(\pm\frac{1}{\sqrt{2}}\right) = \mp\frac{7\sqrt{2}}{8}\right].$$

Note that $\dfrac{d^2y}{dx^2}$ is odd since $\dfrac{dy}{dx}$ is even.

Important points: $x = 0, \pm\dfrac{1}{\sqrt{2}}$.

For $0 < x < \dfrac{1}{\sqrt{2}}$, $\dfrac{d^2y}{dx^2} < 0 \Rightarrow$ for $-\dfrac{1}{\sqrt{2}} < x < 0$, $\dfrac{d^2y}{dx^2} > 0$.

For $x > \dfrac{1}{\sqrt{2}}$, $\dfrac{d^2y}{dx^2} > 0 \Rightarrow$ for $x < -\dfrac{1}{\sqrt{2}}$, $\dfrac{d^2y}{dx^2} < 0$.

(a) Critical points: $(0,0)$, $(\pm 1, \mp 2)$.
Relative maximum: $(-1, 2)$.
Relative minimum: $(1, -2)$.

(b) Increasing for $x < -1$ and $x > 1$.
Decreasing for $-1 < x < 1$.

(c) Inflection points: $(0,0)$, $\left(\pm\dfrac{1}{\sqrt{2}}, \mp\dfrac{7\sqrt{2}}{8}\right)$.

(d) Concave up for $\dfrac{-1}{\sqrt{2}} < x < 0$ and $x > \dfrac{1}{\sqrt{2}}$.
Concave down for $x < \dfrac{-1}{\sqrt{2}}$ and $0 < x < \dfrac{1}{\sqrt{2}}$.

6. Let $f(x) = \dfrac{1}{x^2} - \dfrac{1}{x}$.

(a) Determine the intervals on which $f(x)$ is (i) positive, (ii) negative, (iii) increasing, (iv) decreasing, (v) concave up, (vi) concave down.

(b) Find all local maxima and minima of f.

(c) Sketch the graph of f.

$$f(x) = \frac{1}{x^2} - \frac{1}{x} = \frac{1-x}{x^2};$$

$$f'(x) = \frac{-2}{x^3} + \frac{1}{x^2} = \frac{x-2}{x^3};$$

$$f''(x) = \frac{6}{x^4} - \frac{2}{x^3} = \frac{6-2x}{x^4}.$$

The Function.

$f(x)$ is undefined when $x = 0$. As $x \to 0$, $f(x) \sim \dfrac{1}{x^2} \to +\infty$.

As $x \to \pm\infty$, $f(x) \sim -\dfrac{1}{x} \to 0^{\mp}$.

$f(x) = 0$ when $x = 1$.

The First Derivative.

$f'(x)$ is undefined when $x = 0$. $f'(x) = 0$ when $x = 2$. $[f(2) = -\frac{1}{4}]$.

Important points: $x = 0, 2$.

For $x < 0$, $f'(x) > 0$.

For $0 < x < 2$, $f'(x) < 0$ }
For $x > 2$, $f'(x) > 0$ } $\leftrightarrow$ local minimum at $x = 2$.

The Second Derivative.

$f''(x)$ is undefined when $x = 0$. $f''(x) = 0$ when $x = 3$. $[f(3) = -\frac{2}{9}]$.

Important points: $x = 0, 3$.

For $x < 0$, $f''(x) > 0 \leftrightarrow f(x)$ concave up.

For $0 < x < 3$, $f''(x) > 0 \leftrightarrow f(x)$ concave up }
For $x > 3$, $f''(x) < 0 \leftrightarrow f(x)$ concave down } $\leftrightarrow x = 3$ is an inflection point.

Note that $f'(2) = 0$, $f''(2) > 0 \Rightarrow x = 2 \leftrightarrow$ local minimum.

(a) The function $f(x)$ is:

(i) positive on $-\infty < x < 0$, $0 < x < 1$.
(ii) negative on $1 < x < \infty$.
(iii) increasing on $-\infty < x < 0$, $2 < x < \infty$.
(iv) decreasing on $0 < x < 2$.
(v) concave up on $-\infty < x < 0$, $0 < x < 3$.
(vi) concave down on $3 < x < \infty$.

(b) $f(x)$ has a local minimum at $(2, -\frac{1}{4})$.

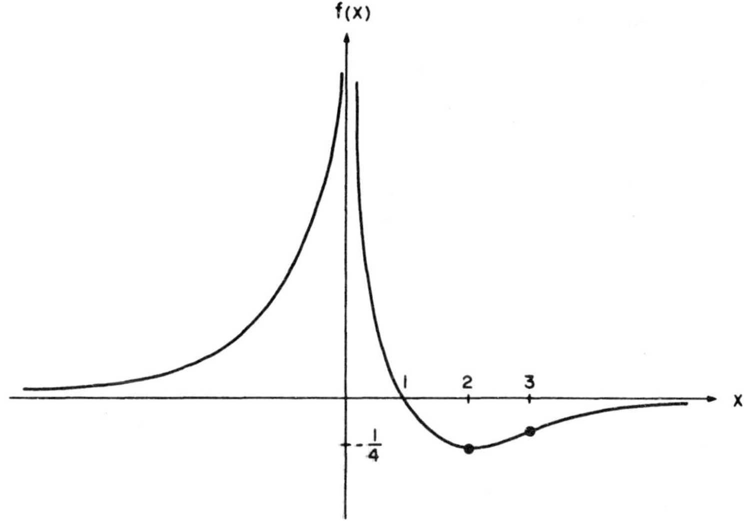

Figure I.6. Graph of $f(x)=\dfrac{1}{x^2}-\dfrac{1}{x}$.

7. Graph the function $f(x)=\dfrac{x^2+1}{x^2-1}$. Calculate the location of maxima and minima, intercepts with the coordinate axes, and the asymptotic behaviour for large $|x|$ and near any singular points that occur.

$$f(x)=\frac{x^2+1}{x^2-1}=\frac{(x^2-1)+2}{x^2-1}=1+\frac{2}{x^2-1};$$

$$\frac{1}{x^2-1}=\frac{1}{(x-1)(x+1)}=\frac{1/2}{x-1}-\frac{1/2}{x+1}.$$

Hence

$$f(x)=1+\frac{1}{x-1}-\frac{1}{x+1};$$

$$f'(x)=\frac{-1}{(x-1)^2}+\frac{1}{(x+1)^2}=-\frac{4x}{(x^2-1)^2};$$

$$f''(x)=\frac{2}{(x-1)^3}-\frac{2}{(x+1)^3}=\frac{4(3x^2+1)}{(x^2-1)^3}.$$

$f(-x)=f(x)$. Hence $f(x)$ has even symmetry. *Thus one need only graph $f(x)$ for $x\geq 0$* and reflect about the y-axis for $x<0$. Moreover $f'(x)$ has odd symmetry, i.e., $f'(-x)=-f'(x)$, and $f''(x)$ has even symmetry. $f(x)$ is undefined at $x=\pm 1$, i.e., $x=\pm 1$ are singular points of $f(x)$.

Asymptotic Behaviour.

As $x \to 1, f(x) = 1 + \dfrac{1}{x-1} - \dfrac{1}{x+1} \sim \dfrac{1}{x-1}$.
As $x \to -1, f(x) \sim \dfrac{-1}{x+1}$.
As $|x| \to \infty, f(x) = 1 + \dfrac{2}{x^2-1} \sim 1 + \dfrac{2}{x^2} \to 1^+$.

Critical Points.

$f'(x) = 0$ when $x = 0$. $[f(0) = -1]$.
$f''(0) < 0 \Rightarrow x = 0 \leftrightarrow$ local maximum.

Intercepts. $(0, -1)$.

Concavity.

$f''(x) \neq 0$ for any x, $f''(x)$ singular at $x = \pm 1$. Asymptotic behaviour of $f(x)$ near $x = \pm 1$, and as $x \to \pm\infty$, $\Rightarrow f(x)$ concave up on $-\infty < x < -1$, $1 < x < \infty$; $f(x)$ concave down on $-1 < x < 1$.

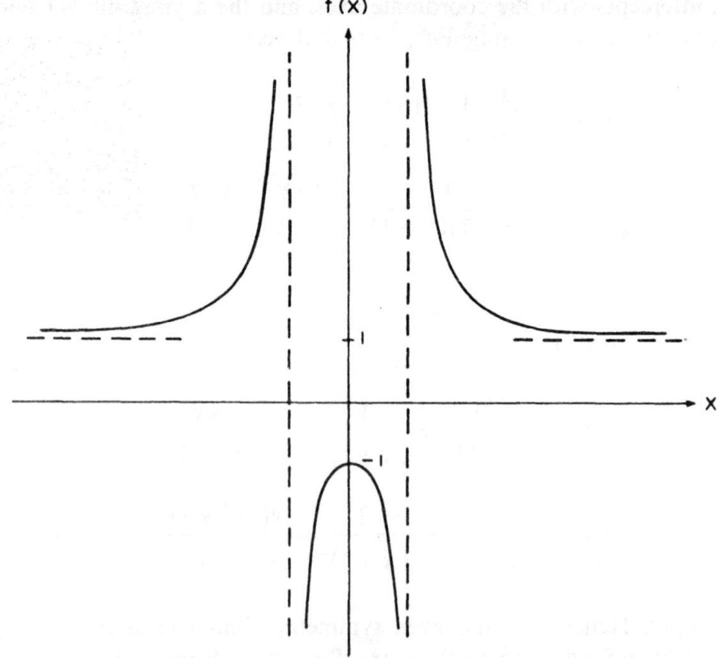

Figure I.7. Graph of $f(x) = \dfrac{x^2+1}{x^2-1}$.

8. Find maximum and minimum points, points of inflection, and sketch: $y = \frac{x^2}{8} - \frac{2}{x}$.

$$\frac{dy}{dx} = \frac{x}{4} + \frac{2}{x^2} = \frac{x^3+8}{4x^2};$$

$$\frac{d^2y}{dx^2} = \frac{1}{4} - \frac{4}{x^3} = \frac{x^3-16}{4x^3}.$$

The Function.

y is undefined at $x = 0$. As $x \to 0$, $y \sim \frac{-2}{x}$.

As $|x| \to \infty$, $y \sim \frac{x^2}{8}$.

$y = 0$ at $x = 2^{4/3}$.

The First Derivative.

$\frac{dy}{dx}$ is undefined at $x = 0$. $\frac{dy}{dx} = 0$ at $x = -2$.

$$\left.\begin{array}{l} \text{For } x < -2,\ \frac{dy}{dx} < 0 \\ \text{For } -2 < x < 0,\ \frac{dy}{dx} > 0 \end{array}\right\} \leftrightarrow \text{relative minimum at } x = -2.$$

For $x > 0$, $\frac{dy}{dx} > 0$.

The Second Derivative.

$\frac{d^2y}{dx^2}$ is undefined at $x = 0$. $\frac{d^2y}{dx^2} = 0$ at $x = 2^{4/3} \simeq 2.5$.

For $x < 0$, $\frac{d^2y}{dx^2} > 0$.

$$\left.\begin{array}{l} \text{For } 0 < x < 2^{4/3},\ \frac{d^2y}{dx^2} < 0 \\ \text{For } x > 2^{4/3},\ \frac{d^2y}{dx^2} > 0 \end{array}\right\} \leftrightarrow \text{inflection point at } x = 2^{4/3}.$$

Note that $\frac{d^2y}{dx^2} > 0$ at $x = -2$, checking out the fact that $x = -2 \leftrightarrow$ relative minimum.

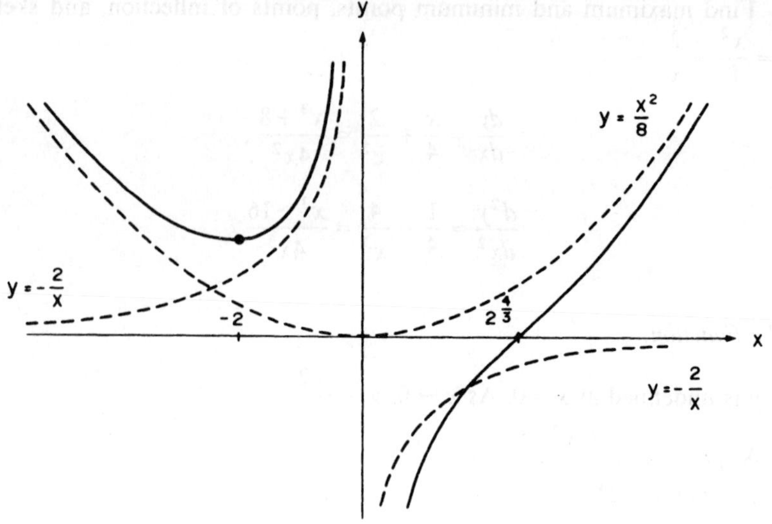

Figure I.8. Graph of $y = \frac{x^2}{8} - \frac{2}{x}$.

Graph.

Note that the graph can be obtained by adding the graphs of $y = \frac{x^2}{8}$ and $y = \frac{-2}{x}$ (see Fig. I.8).

9. Consider the function $x^2e^{-x}+1$.

(a) Calculate the stationary points and points of inflection. Using the sign of the second derivative, find whether the stationary points are relative maxima, minima or neither.

(b) Evaluate the limits:

$$\lim_{x \to \pm\infty} x^2e^{-x}+1.$$

(c) Obtain the numerical value of $x^2e^{-x}+1$ at every stationary point (making a rough estimate if necessary using $e \simeq 3$).

(d) Sketch the graph of the function.

$$f(x) = x^2e^{-x}+1;$$

$$f'(x) = 2xe^{-x} - x^2e^{-x} = x(2-x)e^{-x};$$

$$f''(x) = (x^2-4x+2)e^{-x} = (x-2+\sqrt{2})(x-2-\sqrt{2})e^{-x}.$$

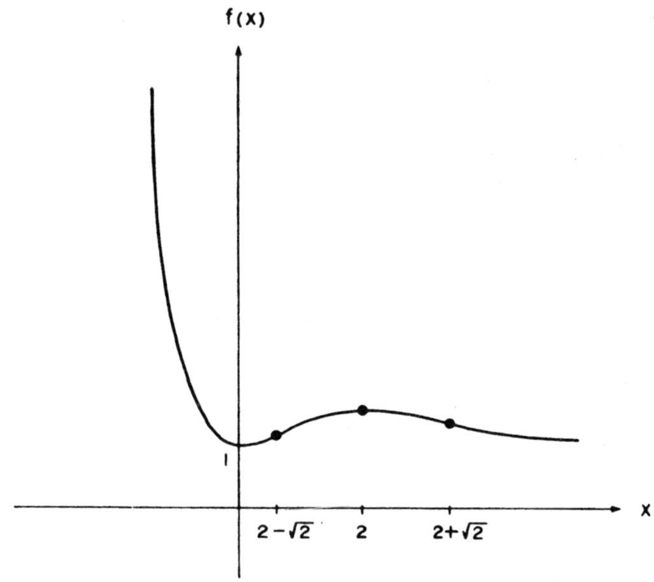

Figure I.9. Graph of $f(x)=x^2e^{-x}+1$.

(a) Stationary points: $f'(x)=0 \leftrightarrow x=0,\ x=2.$

$$f''(0)>0 \Rightarrow \text{relative minimum at } x=0.$$

$$f''(2)<0 \Rightarrow \text{relative maximum at } x=2.$$

Inflection points: $f''(x)=0 \leftrightarrow x=2\pm\sqrt{2}$.

(b) $$\lim_{x\to+\infty}(x^2e^{-x}+1)=\lim_{x\to+\infty}\left(\frac{x^2+e^x}{e^x}\right)=\lim_{x\to+\infty}\left(\frac{2x+e^x}{e^x}\right)$$

$$=\lim_{x\to+\infty}\left(\frac{2+e^x}{e^x}\right)=1^+, \text{ using l'Hôpital's rule.}$$

$$\lim_{x\to-\infty} f(x)=+\infty.$$

(c) $f(0)=1,\ f(2)=4e^{-2}+1\simeq 1.4.$

(d) The function is graphed in Figure I.9.

(*)**10.** Sketch the graph of $y=(x^2)^x$.

$$\log y = x\log(x^2);$$

$$\frac{1}{y}\frac{dy}{dx}=\log(x^2)+\frac{x(2x)}{x^2} \Rightarrow \frac{dy}{dx}=(x^2)^x(\log(x^2)+2);$$

$$\frac{d^2y}{dx^2}=(x^2)^x\left[(\log(x^2)+2)^2+\frac{2}{x}\right].$$

The Function.

$$y > 0 \text{ for all } x.$$

$$\lim_{x\to 0^\pm} \log y = \lim_{x\to 0^\pm} \frac{\log(x^2)}{\frac{1}{x}} = \lim_{x\to 0^\pm} \frac{\frac{2}{x}}{\frac{-1}{x^2}} = \lim_{x\to 0^\pm} -2x = 0^\mp$$

$$\Rightarrow \lim_{x\to 0^\pm} y = 1^\mp;$$

$$\lim_{x\to +\infty} y = +\infty; \quad \lim_{x\to -\infty} (x^2)^x = 0^+.$$

The First Derivative.

$$\frac{dy}{dx} = 0 \leftrightarrow \log(x^2) = -2 \leftrightarrow x = \pm\frac{1}{e}.$$

$$\left[y\left(\frac{1}{e}\right) \simeq 0.5, y\left(\frac{-1}{e}\right) \simeq 2.1.\right]$$

$\frac{dy}{dx}$ is singular at $x = 0$. Important points: $x = \frac{-1}{e}, 0, \frac{1}{e}$.

For $-\infty < x < \frac{-1}{e}$, $\frac{dy}{dx} > 0$.

For $\frac{-1}{e} < x < 0$, $\frac{dy}{dx} < 0$.

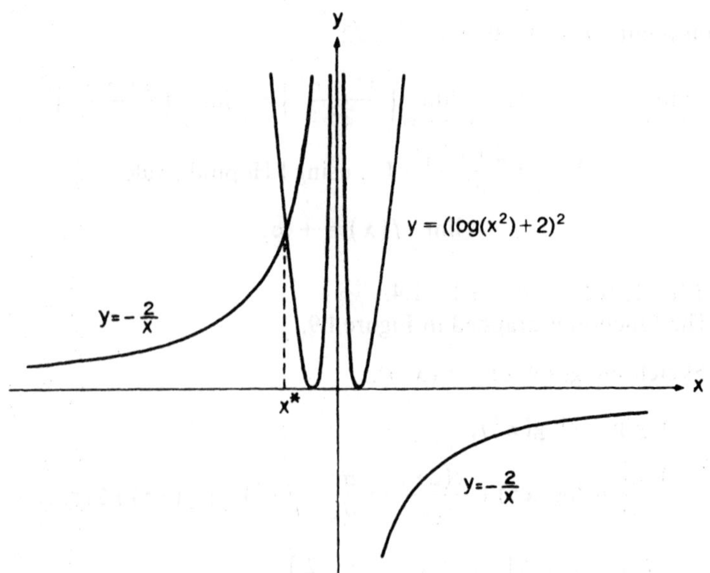

Figure I.10*a*. Graph of $y = \frac{-2}{x}$ vs. $y = (\log(x^2)+2)^2$.

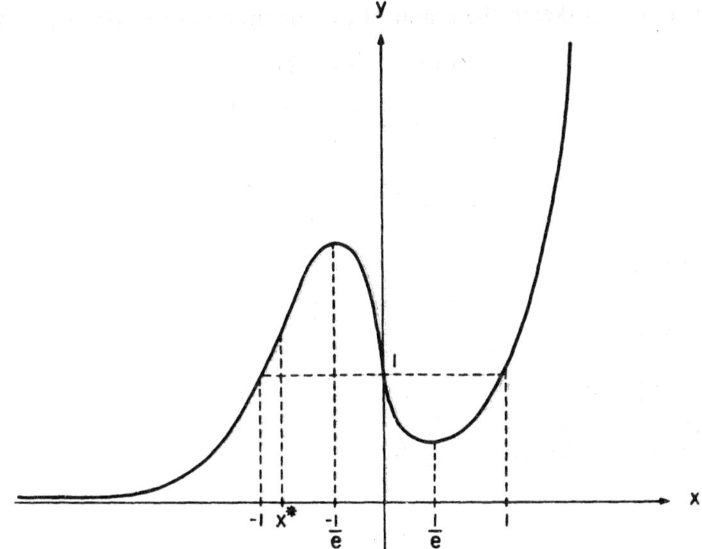

Figure I.10*b*. Graph of $y = (x^2)^x$.

For $0 < x < \frac{1}{e}$, $\frac{dy}{dx} < 0$.

For $\frac{1}{e} < x < \infty$, $\frac{dy}{dx} > 0$.

The Second Derivative.

$$\frac{d^2y}{dx^2} = 0 \leftrightarrow \left(\log(x^2)+2\right)^2 = \frac{-2}{x}. \tag{1}$$

Graph $y = \frac{-2}{x}$ vs. $y = (\log(x^2)+2)^2$ to find the nature of the roots of equation (1) (Fig. I.10*a*). Since the graphs intersect once, $\frac{d^2y}{dx^2} = 0$ at $x = x^*$.

$x^* \simeq -0.8$ from Newton's method to find roots. (See Chapter VI, Numerical Methods for information on Newton's method.)

$\frac{d^2y}{dx^2}$ is undefined (singular) at $x = 0$.

Important points: $x = x^*, 0$.

For $-\infty < x < x^*$, $\frac{d^2y}{dx^2} > 0 \leftrightarrow y$ concave up.

For $x^* < x < 0$, $\frac{d^2y}{dx^2} < 0 \leftrightarrow y$ concave down.

For $0 < x < \infty$, $\frac{d^2y}{dx^2} > 0 \leftrightarrow y$ concave up (Fig. I.10*b*).

11. Analyze and sketch the graph of the function $f(x)=|x|^{2/3}(x-2)^2$.

$$f(x)=x^{2/3}(x-2)^2;$$

$$f'(x)=\frac{4(2x^2-5x+2)}{3x^{1/3}};$$

$$f''(x)=\frac{8(5x^2-5x-1)}{9x^{4/3}}.$$

The Function.

Assume that $x^{1/3}=\begin{cases}|x|^{1/3} & \text{if } x>0,\\ -|x|^{1/3} & \text{if } x<0.\end{cases}$

$f(x)\geq 0$ for all x;
$f(x)=0\leftrightarrow x=0,2.$
As $|x|\to\infty$, $f(x)\sim|x|^{8/3}$.

The First Derivative.

$$f'(x)=0\leftrightarrow\frac{4(x-2)(2x-1)}{3x^{1/3}}=0\leftrightarrow x=2,\tfrac{1}{2};\ \begin{bmatrix}f(2)=0\\ f(\frac{1}{2})\simeq 1.4\end{bmatrix}.$$

Note $f'(x)$ is undefined at $x=0$. But $f(0)=0$; hence $f(x)$ has a cusp at $x=0$.

For $-\infty<x<0, f'(x)<0$
For $0<x<\frac{1}{2}, f'(x)>0$
For $\frac{1}{2}<x<2, f'(x)<0$
For $2<x<\infty, f'(x)>0$

$\leftrightarrow$ local minimum at $x=0$.
$\leftrightarrow$ local maximum at $x=\frac{1}{2}$.
$\leftrightarrow$ local minimum at $x=2$.

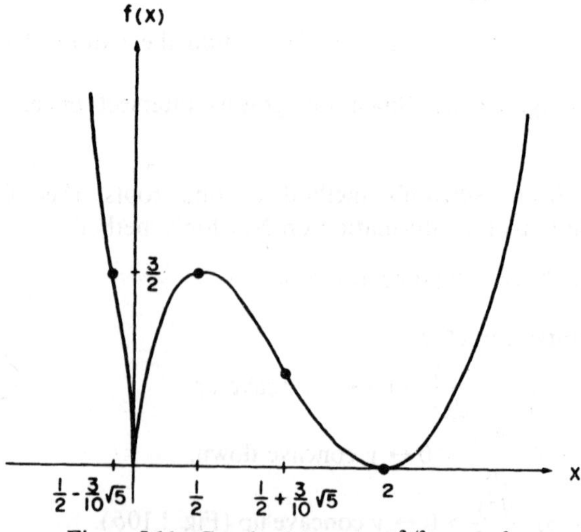

Figure I.11. Graph of $f(x)=|x|^{2/3}(x-2)^2$.

The Second Derivative.

$$f''(x)=0 \leftrightarrow \frac{8}{9}\frac{(5x^2-5x-1)}{|x|^{4/3}}=0 \leftrightarrow x=\frac{1}{2}\pm\frac{3}{10}\sqrt{5} \Rightarrow x_1 \simeq -0.17,$$

$x_2 \simeq 1.17.$

$$\left.\begin{array}{ll}\text{For } -\infty<x<x_1, f''(x) & >0\\ \text{For } x_1<x<0, f''(x) & <0\end{array}\right\} \leftrightarrow \text{inflection point at } x=x_1.$$

$$\left.\begin{array}{ll}\text{For } 0<x<x_2, f''(x) & <0\\ \text{For } x_2<x<\infty, f''(x) & >0\end{array}\right\} \leftrightarrow \text{inflection point at } x=x_2.$$

12. For the function $y=4x+2-5\log(1+x^2)$:

(a) Find all critical points.
(b) Locate the maxima and minima by using an appropriate test.
(c) Discuss the concavity of y and locate any points of inflection.
(d) Draw a sketch of the curve.

$$y=4x+2-5\log(1+x^2);$$

$$y'=\frac{2(2x^2-5x+2)}{1+x^2};$$

$$y''=\frac{-10(1-x^2)}{(1+x^2)^2}.$$

(a) $y'=0 \leftrightarrow (2x-1)(x-2)=0 \leftrightarrow x=\frac{1}{2},2;\ \left[\begin{array}{l} y(\frac{1}{2})=4-5\log\frac{5}{4}\simeq 2.9\\ y(2)=10-5\log 5\simeq 2.0\end{array}\right].$

(b) To find maxima:

Method 1

$$\begin{array}{ll}\text{For } -\infty<x<\frac{1}{2}, y' & >0\\ \text{For } \frac{1}{2}<x<2, y' & <0\\ \text{For } 2<x<\infty, y' & >0\end{array} \quad \begin{array}{l}\leftrightarrow \text{local maximum at } x=\frac{1}{2}.\\ \leftrightarrow \text{local minimum at } x=2.\end{array}$$

Method 2

$$y''(\tfrac{1}{2})=\frac{-10\left[1-(\frac{1}{2})^2\right]}{\left[1+(\frac{1}{2})^2\right]^2}<0 \Rightarrow \text{local maximum at } x=\tfrac{1}{2}.$$

$$y''(2)=\frac{-10[1-2^2]}{[1+2^2]^2}>0 \Rightarrow \text{local minimum at } x=2.$$

(c) $y''=0 \leftrightarrow 1-x^2=0 \leftrightarrow x=\pm 1;\ \begin{bmatrix} y(1)=6-5\log 2 \simeq 2.5 \\ y(-1)=-2-5\log 2 \simeq -5.5 \end{bmatrix}.$

$$\left.\begin{array}{l} \text{For } -\infty < x < -1,\ y'' > 0 \leftrightarrow y \text{ concave up} \\ \text{For } -1 < x < 1,\ y'' < 0 \leftrightarrow y \text{ concave down} \\ \text{For } 1 < x < \infty,\ y'' > 0 \leftrightarrow y \text{ concave up} \end{array}\right\} \begin{array}{l} \leftrightarrow \text{inflection point at } x=-1. \\ \leftrightarrow \text{inflection point at } x=1. \end{array}$$

(d) As $|x| \to \infty, y \to 4x^-$.

(*)13. Sketch the graph of the equation $y=\log(2+\sin x)$, and label all maxima and minima, all inflection points, and all asymptotes.

$$y=\log(2+\sin x);$$

$$\frac{dy}{dx}=\frac{\cos x}{2+\sin x};$$

$$\frac{d^2y}{dx^2}=\frac{-(2\sin x+1)}{(2+\sin x)^2}.$$

The Function.

Since $1 \le 2+\sin x \le 3$, one has $0 \le y \le \log 3$ for all x. Furthermore, $\sin(x+2\pi)=\sin x$. Hence $y(x+2\pi)=y(x) \Rightarrow y$ has period 2π. Therefore consider $0 \le x \le 2\pi$ as a representative period.

$$y=0 \leftrightarrow 2+\sin x=1 \leftrightarrow x=\frac{3\pi}{2}.$$

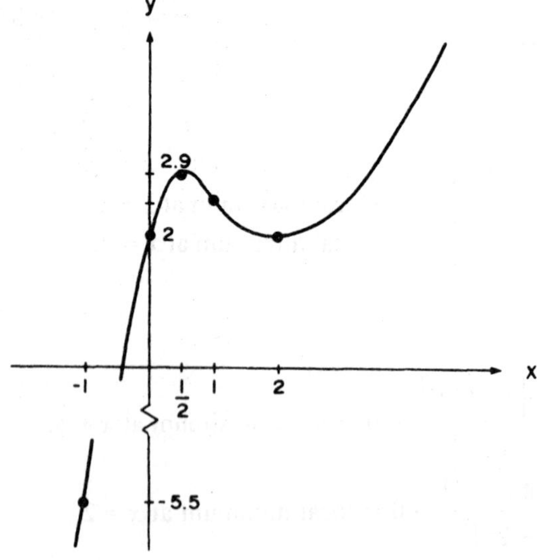

Figure I.12. Graph of $y=4x+2-5\log(1+x^2)$.

The First Derivative.

Note that $\frac{dy}{dx}$ is defined everywhere; thus the only points of interest are those where $\frac{dy}{dx} = 0$.

$$\frac{dy}{dx} = 0 \leftrightarrow \cos x = 0 \leftrightarrow x = \frac{\pi}{2}, \frac{3\pi}{2}; \quad \begin{bmatrix} y\left(\frac{\pi}{2}\right) = \log 3 \\ y\left(\frac{3\pi}{2}\right) = 0 \end{bmatrix}.$$

$$\left.\begin{array}{l} \text{For } 0 < x < \frac{\pi}{2},\ \frac{dy}{dx} > 0 \\ \text{For } \frac{\pi}{2} < x < \frac{3\pi}{2},\ \frac{dy}{dx} < 0 \end{array}\right\} \leftrightarrow \text{relative maximum at } x = \frac{\pi}{2}.$$

$$\left.\begin{array}{l} \text{For } \frac{\pi}{2} < x < \frac{3\pi}{2},\ \frac{dy}{dx} < 0 \\ \text{For } \frac{3\pi}{2} < x < 2\pi,\ \frac{dy}{dx} > 0 \end{array}\right\} \leftrightarrow \text{relative minimum at } x = \frac{3\pi}{2}.$$

The Second Derivative.

$$\frac{d^2y}{dx^2} = 0 \leftrightarrow 2\sin x + 1 = 0 \Rightarrow x = \frac{7\pi}{6}, \frac{11\pi}{6}; \quad \begin{bmatrix} y\left(\frac{7\pi}{6}\right) = \log\frac{3}{2} \\ y\left(\frac{11\pi}{6}\right) = \log\frac{3}{2} \end{bmatrix}.$$

$$\left.\begin{array}{l} \text{For } 0 < x < \frac{7\pi}{6},\ \frac{d^2y}{dx^2} < 0 \leftrightarrow y \text{ concave down.} \\ \text{For } \frac{7\pi}{6} < x < \frac{11\pi}{6},\ \frac{d^2y}{dx^2} > 0 \leftrightarrow y \text{ concave up.} \end{array}\right\} \leftrightarrow \text{inflection point at } x = \frac{7\pi}{6}.$$

$$\left.\begin{array}{l} \text{For } \frac{7\pi}{6} < x < \frac{11\pi}{6},\ \frac{d^2y}{dx^2} > 0 \leftrightarrow y \text{ concave up.} \\ \text{For } \frac{11\pi}{6} < x < 2\pi,\ \frac{d^2y}{dx^2} < 0 \leftrightarrow y \text{ concave down.} \end{array}\right\} \leftrightarrow \text{inflection point at } x = \frac{11\pi}{6}.$$

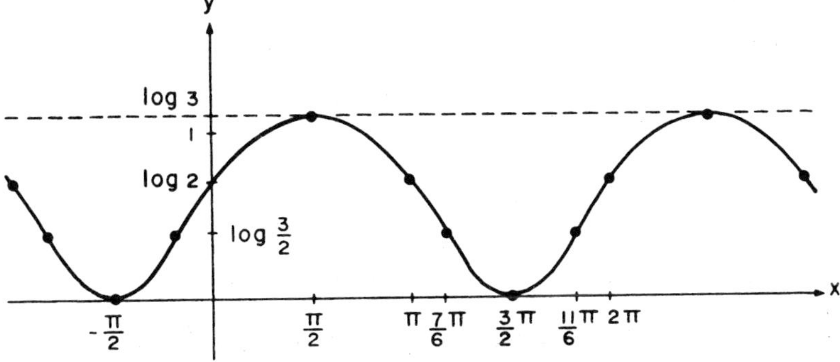

Figure I.13. Graph of $y = \log(2 + \sin x)$.

Graph.

Note that there are no asymptotes.

As a check, note that $\sin\left(\frac{\pi}{2}+x\right)=\sin\left(\frac{\pi}{2}-x\right)$

$$\Rightarrow y\left(\frac{\pi}{2}+x\right)=y\left(\frac{\pi}{2}-x\right) \Rightarrow \text{even symmetry about } x=\frac{\pi}{2}.$$

(*)14. Sketch the graph of $y=\dfrac{x}{\sin x}$, $0<x<\dfrac{\pi}{2}$.

$$y=\frac{x}{\sin x};$$

$$y'=\frac{\sin x - x\cos x}{\sin^2 x};$$

$$y''=\frac{x+x\cos^2 x-2\sin x\cos x}{\sin^3 x}.$$

The Function.

Using l'Hôpital's rule,

$\lim_{x\to 0^+}\dfrac{x}{\sin x}=\lim_{x\to 0^+}\dfrac{1}{\cos x}=1;\quad y>0$ for $0<x<\pi$.

The function "blows up" wherever $\dfrac{1}{y}=0$, i.e. wherever $\dfrac{\sin x}{x}=0$

$$\leftrightarrow x=\pi, 2\pi, 3\pi, \ldots\text{: all of these points lie outside } \left(0,\frac{\pi}{2}\right).$$

The First Derivative.

$$\lim_{x\to 0^+} y'=\lim_{x\to 0^+}\frac{\sin x - x\cos x}{\sin^2 x}=0 \text{ (l'Hôpital's rule).}$$

$y'=0 \leftrightarrow \sin x = x\cos x \leftrightarrow \tan x = x;\ x=?$

By comparing the graphs of $y=x$ and $y=\tan x$, note that $\tan x = x$ once in $\left(0,\frac{\pi}{2}\right)$: at $x=0^+ \leftrightarrow y'\to 0$ as $x\to 0^+$.

For $0<x<\frac{\pi}{2}$, $\tan x > x \leftrightarrow \sin x - x\cos x>0 \leftrightarrow y'>0$.

The Second Derivative.

$y''=0\leftrightarrow x=??$

Notice that $y''\left(\frac{\pi}{2}\right)=\frac{\pi}{2}$, $y''\left(\frac{\pi}{4}\right)=\frac{\sqrt{2}}{4}(3\pi-8)>0$.

Perhaps $y''>0$ for $0<x<\frac{\pi}{2}$; therefore try to show that

$$x+x\cos^2 x-2\sin x\cos x>0 \text{ for } 0<x<\frac{\pi}{2}.$$

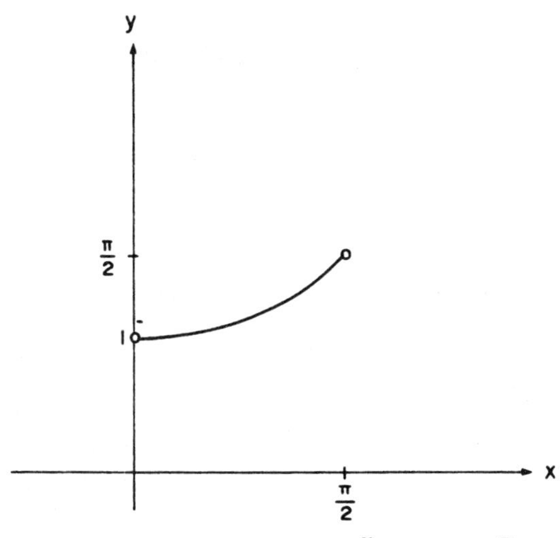

Figure I.14. Graph of $y = \dfrac{x}{\sin x}, 0 < x < \dfrac{\pi}{2}$.

Proof. $x > \sin x$ for $x > 0$. Hence for $x > 0$,

$$\begin{aligned} x(1+\cos^2 x) - 2\sin x\cos x &> \sin x(1+\cos^2 x) - 2\sin x \cos x \\ &= \sin x[1+\cos^2 x - 2\cos x] \\ &= \sin x[1-\cos x]^2 \\ &> 0 \text{ for } 0 < x < \frac{\pi}{2}. \end{aligned}$$

So y is concave up for $0 < x < \dfrac{\pi}{2}$. □

Graph.

$$y'\left(\frac{\pi}{2}\right) = 1, y\left(\frac{\pi}{2}\right) = \frac{\pi}{2}.$$

(*)15. Graph the inequality $2xy \le |x+y| \le x^2 + y^2$.

Case I. $\boxed{x + y \ge 0}$. Here $|x+y| = x + y$. Thus the inequality becomes $2xy \le x + y \le x^2 + y^2$.

Graph $\leftrightarrow \{(x, y)\colon x^2 - x + y^2 - y \ge 0\} \cap \{(x, y)\colon 2xy - y \le x\}$
$\cap \{(x, y)\colon x + y \ge 0\}$.

Completing squares, $x^2 - x + y^2 - y \ge 0$

$\leftrightarrow (x - \frac{1}{2})^2 + (y - \frac{1}{2})^2 \ge \frac{1}{2}$

$\leftrightarrow$ region beyond and including circle of radius $\dfrac{1}{\sqrt{2}}$, centred at $(\frac{1}{2}, \frac{1}{2})$.

$$2xy - y \le x \leftrightarrow y(2x-1) \le x$$
$$\leftrightarrow y \le \frac{x}{2x-1} \text{ if } x > \frac{1}{2}, y \ge \frac{x}{2x-1} \text{ if } x < \frac{1}{2};$$

$$y = \frac{x}{2x-1} = \frac{x}{2(x-\frac{1}{2})} = \frac{(x-\frac{1}{2})+\frac{1}{2}}{2(x-\frac{1}{2})} = \frac{1}{2} + \frac{1}{4x-2}$$
$\leftrightarrow$ hyperbola $y - \dfrac{1}{2} = \dfrac{1}{4(x-\frac{1}{2})}$.

Case II. $\boxed{x + y \le 0}$. Here $|x+y| = -x-y$. Thus the inequality becomes

$$2xy \le -x-y \le x^2+y^2.$$

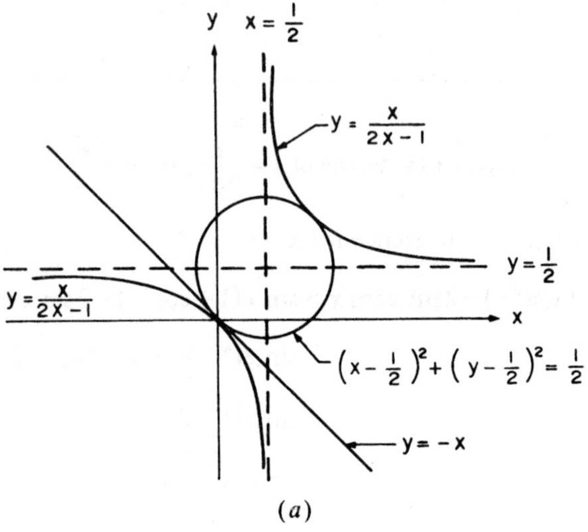

(*a*)

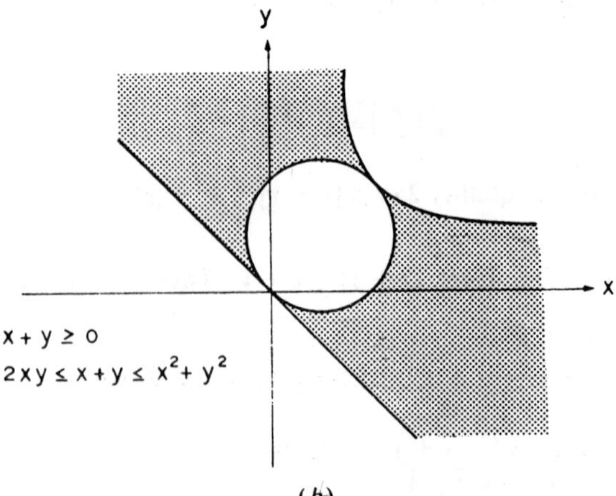

(*b*)

Figure I.15*a*. Sketch of boundary curves for *Case I*. *b*. Graph for *Case I* (shaded region).

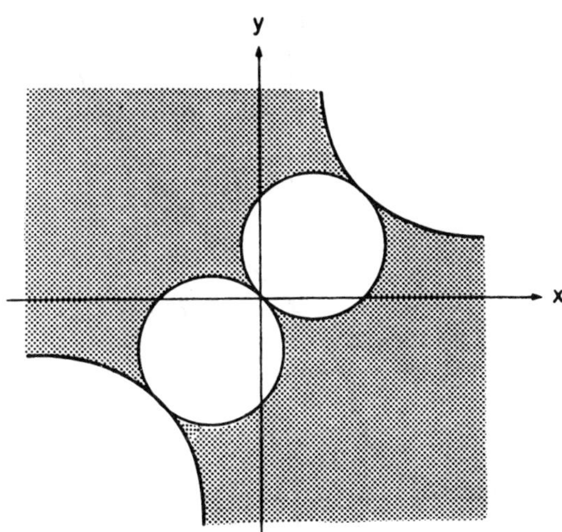

Figure I.15*c*. Graph of inequality $2xy \le |x+y| \le x^2+y^2$ (shaded region).

Graph $\leftrightarrow \{(x, y)\colon x^2+x+y^2+y \ge 0\} \cap \{(x, y)\colon 2xy+y \le -x\}$
$\cap \{(x, y)\colon x+y \le 0\}$
$\leftrightarrow$ reflection of *Case I* about $y=-x$ (i.e. $x \to -x$, $y \to -y$ in *Case I* leads to *Case II*).

(*)**16.** Draw the graph of a continuous function f with the following properties:

(a) $f'(x) > 0$ for all x;
(b) $f''(x) > 0$ for $|x| > 1$;
(c) $f''(x) < 0$ for $|x| < 1$;
(d) $\lim_{x \to -\infty} f(x) = 0$;
(e) $\lim_{x \to \infty} \dfrac{f(x)}{x} = 2$.

Consider each item in turn in extracting relevant information:

(a) $f'(x) > 0$ for all $x \leftrightarrow f(x)$ is always increasing.

(b) $f''(x) > 0$ for $|x| > 1 \leftrightarrow f(x)$ is concave up for $-\infty < x < -1$ and $1 < x < \infty$.

(c) $f''(x) < 0$ for $|x| < 1 \leftrightarrow f(x)$ is concave down for $-1 < x < 1$.

(d) $\lim_{x \to -\infty} f(x) = 0 \quad \leftrightarrow \quad y=0$ is a horizontal asymptote for large negative values of x.

(e) $\lim_{x \to \infty} \dfrac{f(x)}{x} = 2 \quad \leftrightarrow \quad y=2x$ is an asymptote for large positive x.

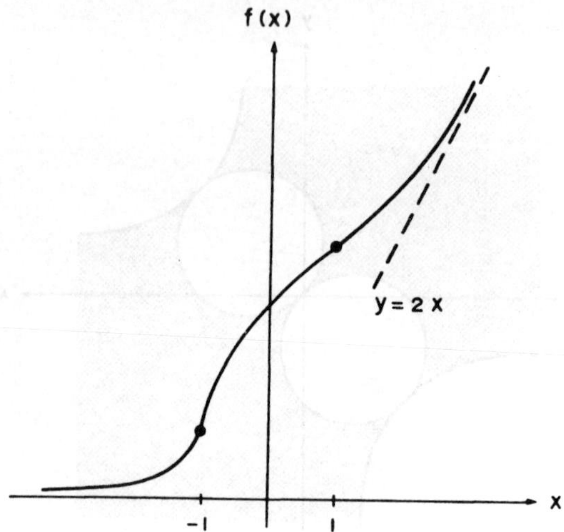

Figure I.16. Graph of a representative $f(x)$.

As the above information is synthesized into one graph, keep in mind that $f(x)$ is continuous (Fig. I.16).

(*)17. The graph of a function $f(x)$ looks like Figure I.17*a*. Sketch the graphs of the following functions:

(a) $f'(x)$.
(b) $\int_{-1}^{x} f(t)\,dt$.
(c) $1 + f(x^2)$.
(d) $(1-x)f(x)$.

(a) Graph $f'(x)$ (Fig. I.17*b*) from observing $f(x)$ directly:

$-\infty < x < -1$: $f'(x)$ is negative, becoming less negative as x increases, $f'(-2) \simeq -2$.
$x = -1$: $f'(-1) = 0$.
$-1 < x < 1$: $f'(x)$ is positive, reaching a maximum of about 1 at $x = 0$, becoming less positive as x approaches 1.
$x = 1$: $f'(1) = 0$.
$1 < x < 2$: $f'(x)$ is negative, becoming most negative at $x = 2$, where $f'(2^-) = -\tan 60° = -\sqrt{3}$.
$x = 2$: $f'(2)$ is undefined.
$2 < x < 3$: $f'(x) > 0$, most positive at $x = 2^+$, where $f'(2^+) = \sqrt{3}$.
$x = 3$: $f'(3) = 0$.
$3 < x < 5$: $f'(x) > 0$, most positive near $x = 4$, where $f'(4) \simeq 1$.
$x = 5$: $f'(5) = 0$.
$5 < x < \infty$: $f'(x) < 0$, becoming more negative as x increases.

(b) Consider $h(x) = \int_{-1}^{x} f(t)\,dt$:

One can interpret $h(x)$ as the area "under" the graph of $f(x)$ (including "negative" area), between -1 and x, $h(-1)=0$.

As x moves from -1 to 0, $h(x)$ is decreasing since area is being subtracted.

At $x=0$, no area change takes place. ($h(x)$ should level off here.)

As x increases (to $x=6$), $h(x)$ increases since the added area is positive. It hits zero near $x=1$, since the negative area from -1 to 0 is balanced by the positive area from 0 to 1.

At $x=6$, area stops being added, so $h(x)$ begins to decrease.

After $x=6$, $h(x)$ continues to decrease, hitting zero near $x=7$.

Moving to the left from $x=-1$, area is *added*, levels off at $x=-2$, then

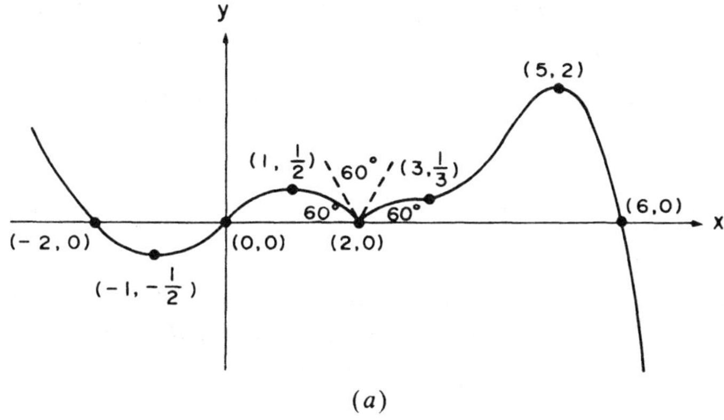

(*a*)

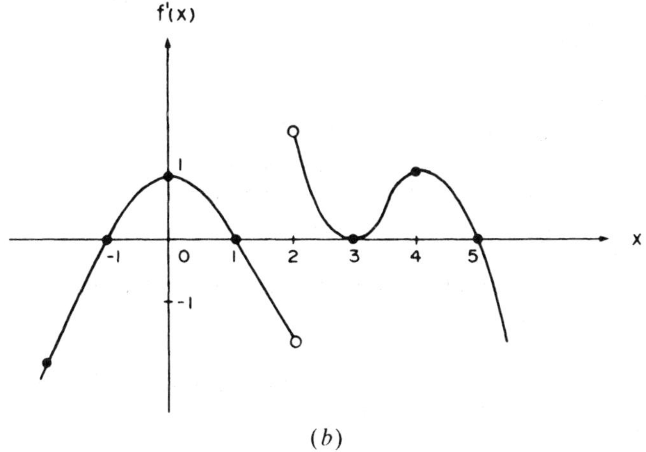

(*b*)

Figure I.17*a*. Graph of $f(x)$. *b*. Representative graph of $f'(x)$.

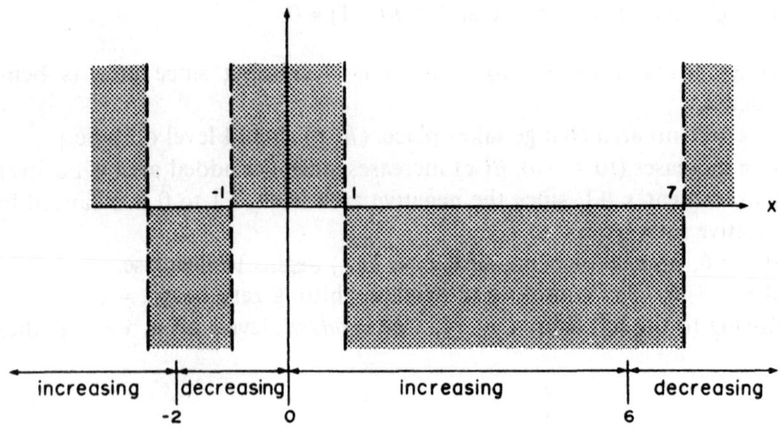

Figure I.17c. Sign and direction of $h(x) = \int_{-1}^{x} f(t)\, dt$.

begins to be subtracted. $h(x)$ hits zero near $x = -2.5$ as it continues to decrease.

From the above information one can construct Fig. 1.17c showing the sign and "direction" of $h(x)$. (The graph of $h(x)$ is *excluded* from the shaded regions.) But notice that $h'(x)$ is identically $f(x)$:

$$\frac{d}{dx} h(x) = \frac{d}{dx} \int_{-1}^{x} f(t)\, dt \equiv f(x),$$

i.e., $f(x)$ is the derivative of the integral $h(x)$. Proceeding from this fact, make the following observations.

Slope of $h(x)$.

$-\infty < x < -2$: $h(x)$ is increasing (since $f(x) > 0$). $h'(-3) \simeq +3$.

$x = -2$: $h'(-2) = 0 \leftrightarrow$ zero slope for $h(x)$.

$-2 < x < 0$: $h(x)$ is decreasing. In this interval the slope has its minimum value of $-\frac{1}{2}$ at $x = -1$.

$x = 0$: $h'(0) = 0$.

$0 < x < 2$: $h(x)$ is increasing. In this interval the slope has its maximum value of $\frac{1}{2}$ at $x = 1$.

$x = 2$: $h'(2) = 0$.

$2 < x < 6$: $h(x)$ is increasing. In this interval the slope has its maximum value of 2 at $x = 5$.

$x = 6$: $h'(6) = 0$.

$6 < x < \infty$: $h(x)$ is decreasing.

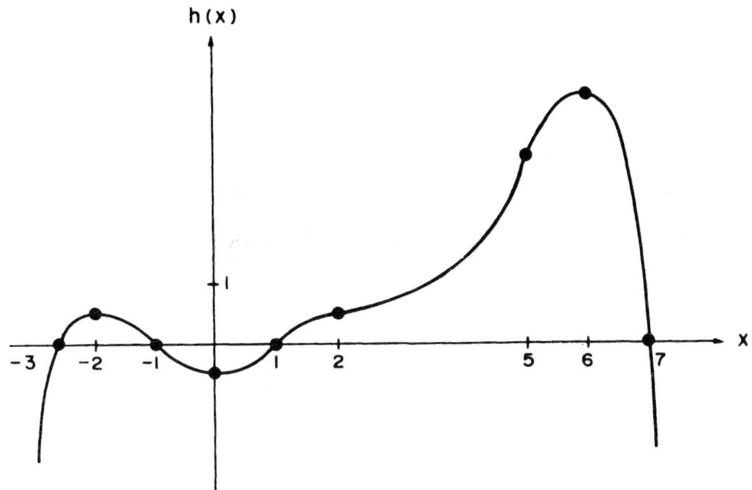

Figure I.17*d*. Graph of $h(x)=\int_{-1}^{x} f(t)\,dt$.

Concavity of $h(x)$.

$-\infty<x<-1$: $h''(x)\equiv f'(x)<0 \leftrightarrow h(x)$ concave down.

$x=-1$: $h''(-1)=0 \leftrightarrow$ inflection point for $h(x)$.

$-1<x<1$: $h''(x)>0 \leftrightarrow h(x)$ concave up.

$x=1$: inflection point for $h(x)$.

$1<x<2$: $h(x)$ is concave down.

$x=2$: inflection point for $h(x)$.

$2<x<5$: $h(x)$ is concave up.

$x=5$: inflection point for $h(x)$.

$5<x<\infty$: $h(x)$ is concave down.

Finally, sketch $h(x)$ (Figure I.17*d*).

(c) To graph $1+f(x^2)$ it is convenient to graph $g(x)\equiv f(x^2)$ and then suitably translate.

For the graph of $g(x)$ only the domain $x\geq 0$ of $f(x)$ is used.

(*) If $x>1$, then every point on the graph of $f(x)$ has a corresponding point on the graph of $g(x)$ *closer* to the line $x=1$.
(Note that $x>1$ in $f(x)\leftrightarrow x>1$ in $g(x)$.)

(**) If $0<x<1$, then every point on the graph of $f(x)$ has a corresponding point on the graph of $g(x)$ *further* away from the line $x=0$.
(Note that $0<x<1$ in $f(x)\leftrightarrow 0<x<1$ in $g(x)$.)

Remember that $g(-x)=f(x^2)$ as well: thus $g(x)$ has even symmetry about the y-axis.

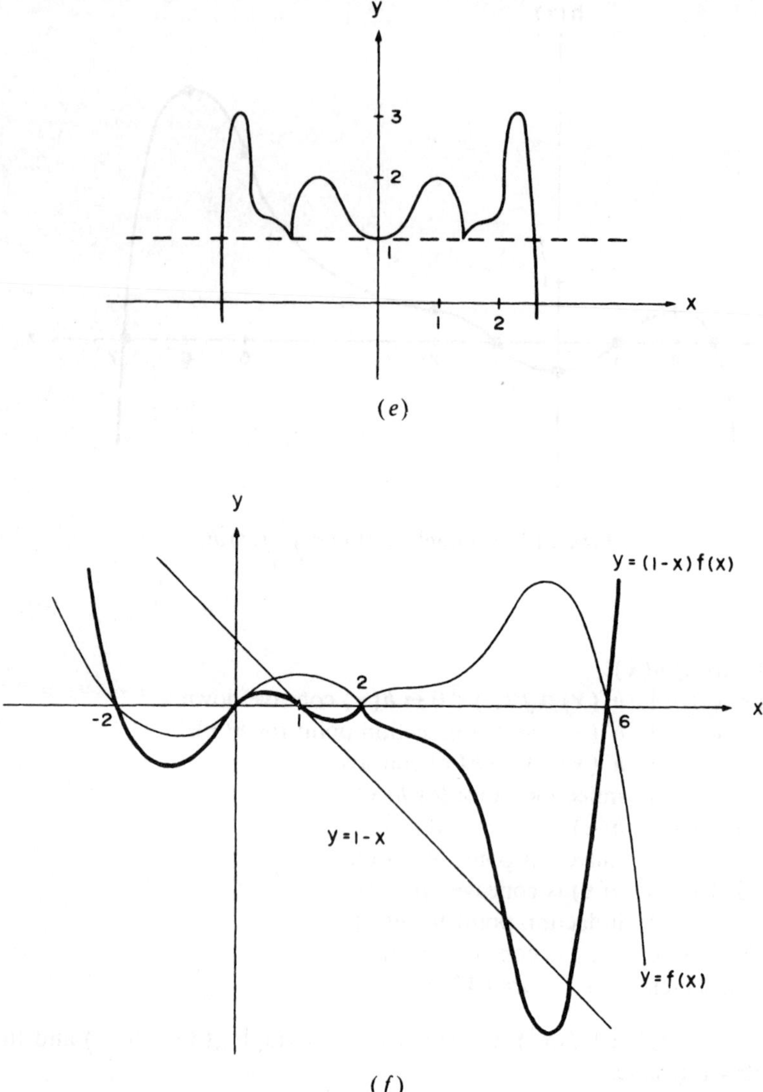

Figure I.17*e*. Graph of $y = 1 + f(x^2)$. *f*. Graph of $y = (1 - x)f(x)$.

If may help to plot points.

First graph on a "pseudo" x-axis; to add one to every value of $g(x)$, simply shift the x-axis DOWN by one, and rescale the y-axis.

(d) The function $(1 - x)f(x)$ can be seen as the product of the functions $y = 1 - x$ and $y = f(x)$ (Fig. 1.17*f*). Hence one can graphically multiply the two, i.e., multiply directly from the graphs of each.

Say $h(x) = (1 - x)f(x)$. Then $h(-2) = h(0) = h(1) = h(2) = h(6) = 0$ both from the graphs and from the equation for $h(x)$.

An alternative method is to plug values for x into the equation for $h(x)$; this may be used as a check.

An advantage of separately considering $y = 1 - x$, $y = f(x)$, is that it can be immediately seen on which intervals $h(x)$ is positive or negative.

Supplementary Problems

1. Sketch and discuss the graph of the equation

$$4x^2 + y^2 + 24x - 2y + 21 = 0.$$

2. Sketch the graph of $g(x) = x - |x|$.

3. Sketch the graph of $y = \sqrt{|2x - 1| - 2}$.

4. Sketch the curve $y = x^3 - 3x^2 - 9x + 8$. Label the axes and all significant features carefully.

5. Sketch the graph of $y = 2x^3 + 3x^2 - 12x - 4$. Consider maxima, minima, inflection points, etc.

6. For the function $f(x) = \dfrac{x^4}{4} - x^3 + 2$, find the intervals where $f(x)$ is increasing and where $f(x)$ is decreasing. Find all relative maximum and relative minimum points of $f(x)$. Find the intervals where $f(x)$ is concave up, intervals where $f(x)$ is concave down, and find all inflection points. Sketch the graph of $f(x)$.

7. Draw the graph $y = x^5 - x^3 - 2x$. Show all critical points and inflection points.

8. For the curve $y = (x + 1)(x - 1)^2$:

(a) Locate all local maxima and minima of the curve.

(b) Locate all points of inflection.

(c) Determine the values of x for which the curve is (i) increasing, (ii) decreasing, (iii) concave up, (iv) concave down.

(d) Sketch the curve.

9. Sketch a graph of the function $f(x) = (x - 1)^4(x - 6)$. Determine the relative extrema (if any), when the graph is increasing and decreasing, the concavity and the points of inflection, etc.

(*)10. Use calculus to identify all significant features of the curve:

$$y = x(x-1)(x-2)(x-3).$$

Sketch the curve; label carefully.

11. Sketch the graph of the function $y = \dfrac{x}{(x+1)^2}$. Check for maxima, minima, inflection points, asymptotes, convexity (also called concavity or curvature), and intercepts.

12. Sketch the graph of $y = \dfrac{x-1}{x+1}$. Consider maxima, minima, inflection points, etc.

13. Find all singular points and zeros of the function $\dfrac{x^2-1}{x(x^2+1)}$. Determine the approximate behaviour of the function near the singular points and for large $|x|$. Use this information to help you make a sketch of the function. (Do not use derivatives to help you make the sketch.)

14. Consider the function $y = \dfrac{x^2+x-5}{x-2}$.

(a) Obtain the two asymptotes to this curve.
(b) Locate any maxima or minima of the function.
(c) Obtain the points of intersection of the curve with both the x and y axes.
(d) Sketch a reasonable graph of the function that exhibits these pieces of information.

15. (a) Find $\lim\limits_{x \to -\infty} xe^x$.

(b) Make a careful sketch of the graph of the function $f(x) = xe^x$, indicating where the function is increasing or decreasing, concave up or down, and locating any maxima, minima, points of inflection and asymptotes.

16. Consider the function $f(x) = xe^{-x^2}$.

(a) Determine intervals where the function is increasing and where the function is decreasing. Locate all relative maxima and minima.
(b) Determine where the graph of the function is concave upward and where it is concave downward. Locate all points of inflection.
(c) Sketch the function.

17. Find the minimum value of $10\cosh x + 6\sinh x$, and draw a rough sketch of the graph of this function.

18. For

$$f(x)=\sqrt[3]{\frac{x^2}{(x-6)^2}} \quad \text{or} \quad \frac{x^{2/3}}{(x-6)^{2/3}}, f'(x)=\frac{-4}{x^{1/3}(x-6)^{5/3}},$$

$$f''(x)=\frac{8(x-1)}{x^{4/3}(x-6)^{8/3}}.$$

(a) Using limits, find all horizontal and vertical asymptotes, if any.

(b) Find the coordinates of all relative extrema and points of inflection, if any. Justify.

(c) Sketch the graph of f.

19. Let

$$g(x)=\begin{cases} 2x^2\log|x|-5x^2 & x\neq 0, \\ 0 & x=0. \end{cases}$$

(a) Compute $g'(0)$ from the definition of the derivative.

(b) Evaluate $\lim_{x\to 0} g(x)$ and $\lim_{x\to +\infty} g(x)$.

(c) Find and classify all the critical points of g.

(d) Find all points of inflection of g.

(e) Sketch the graph of g, identifying the intervals where it is concave up and where it is concave down.

20. For $x>-1$, sketch the graph of $y=f(x)=3x^2+x+\log(1+x)$. Indicate the asymptotes, local maxima and minima, and the inflection points. Find all roots of $f(x)$ to 2 decimal places.

(*)21. Show that $y=\dfrac{\sin(x+a)}{\sin(x+b)}$ has no critical points for any distinct constants a, b. Sketch.

(*)22. Sketch the region defined by the inequality $|x+y|\leq \sin x$.

23. The graphs of two functions $f(x)$ and $g(x)$ appear in Figure S.23a. The graphs of functions $f(x)-g(x)$, $f(x+1)$, $f(x)+1$, $f(-x)$, $f'(x)$, and $\int_0^x f(t)\,dt$ are shown in Figure S.23b in the wrong order as graphs I, II, III, IV, V and VI. Determine which graph belongs to which function.

24. The function f has the following properties:

(1) $0\leq f(x)\leq 1$ for $0\leq x\leq 1$;

(2) $f(x)=x$ for $x=0, \frac{1}{2}, 1$;

(3) $f'(x)=0$ for $x=0, \frac{1}{2}$;

$f'(1)=1$;

$f'(x)>0$ for the other values of x between 0 and 1.

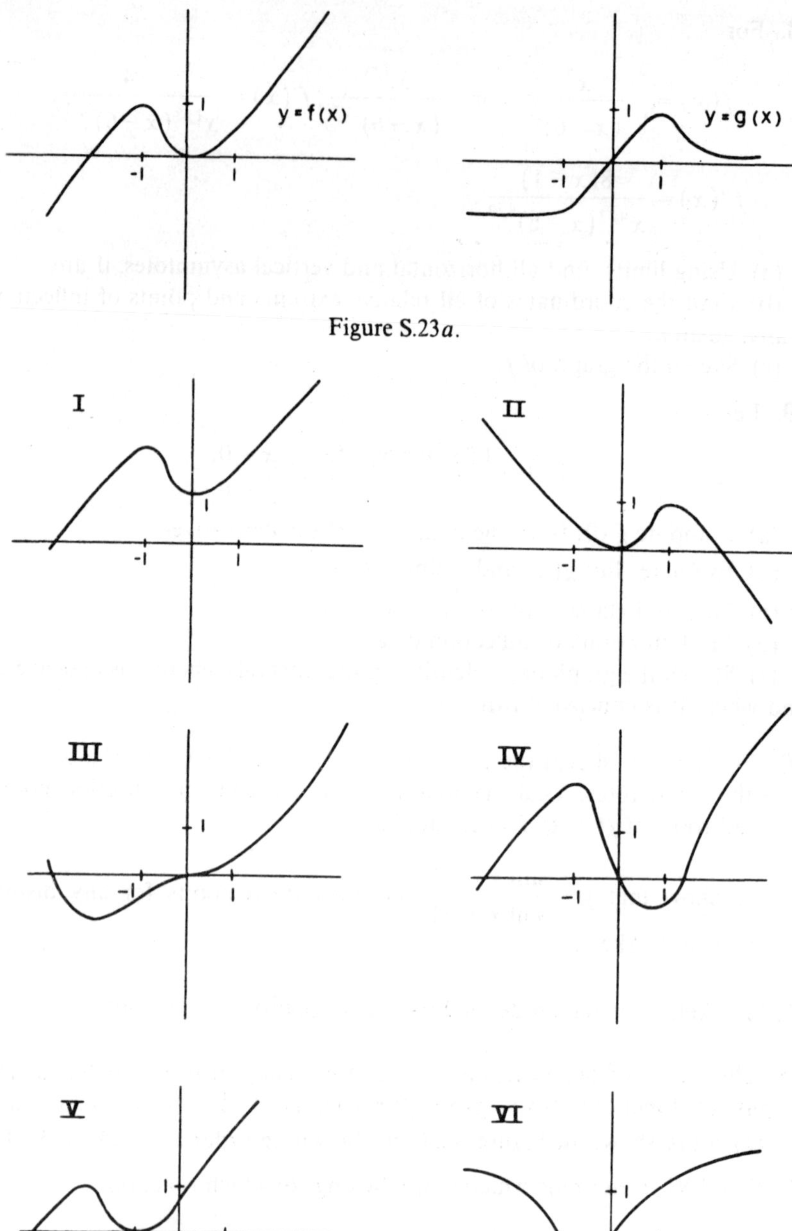

Figure S.23a.

Figure S.23b.

The function g has the following properties:
(1) g is defined for all real numbers.
(2) $g'(x)<0$ for $x<\frac{1}{2}$;
$g'(\frac{1}{2})=0$;
$g'(x)>0$ for $x>\frac{1}{2}$.
(a) Sketch a function which could be f.
(b) Is $g(f(x))$ increasing or decreasing when $x=\frac{1}{4}$? When $x=\frac{3}{4}$?
(c) For what values of x in the interval $[0,1]$ do you know $D_x(g(f(x)))=0$?

25. Sketch the graph of a function f satisfying all the following conditions.
(a) f is continuous everywhere.
(b) $f(-2)=1$, $f(0)=0$, $f(1)=2$.
(c) $f'(-2)=0$, $f'(0)$ does not exist, $f'(1)=0$, $f'(x)>0$ for $x<-2$, $0<x<1$, and $x>1$, and $f'(x)<0$ for $-2<x<0$.
(d) $f''(1)=0$, $f''(x)<0$ for $x<0$ and $0<x<1$, and $f''(x)>0$ for $x>1$.

26. Give examples by sketching a possible graph of:
(a) a continuous function on a closed interval which does not satisfy the conclusion of the Mean Value Theorem;
(b) a function which has a minimum at a point where the derivative is not zero;
(c) a function f such that $\lim_{x \to 2} f(x)=1$, while f is not continuous at $x=2$.

Chapter II

Geometry

In this chapter we consider three types of applied one-variable calculus problems which are basically geometrical in character: maxima and minima (extrema) associated with geometrical configurations, related rates, and volumes of solids with similar cross-sections. (Other geometrical problems are treated in Chapter III, Physics and Engineering: centre of mass, moment of inertia; in Chapter VIII, Techniques: tangent line, normal line, area of a region, volume of a solid of revolution, arc length, area of a surface of revolution, parametric curves, polar coordinates.)

Extremal and related rates problems involve applications of the differential calculus (derivatives) whereas volume problems involve the integral calculus (integrals). In all such problems the student should first construct a diagram including the essentials of the given problem.

1. Maxima and Minima

Here one tries to extremize some aspect of a geometrical figure such as (a) maximizing the area of a stated type of configuration inscribed in a given figure or (b) minimizing the length of all line segments from a given point to a given curve, i.e., finding the distance from a given point to a given curve.

The basic procedure in solving such problems is to:

(1) Set up a diagram including the essentials of the given problem.
(2) Find a formula (try to make it explicit) for the appropriate function $f(x)$ to be extremized, through relationships suggested from the diagram.

(3) Determine the domain of $f(x)$ from the data of the given problem. Usually its domain will be some bounded interval $a \le x \le b$.
(4) Find all extremal points where $f'(x)=0$, $a<x<b$. Usually there are a finite number of such points, i.e., $f'(x_i)=0$, $i=1,2,\ldots,n$.
(5) Now assume $f(x)$ has a finite number of extremal points and that $f(x)$ is defined on $a \le x \le b$. According to the problem, find the maximum (minimum) value of $f(x)$ for $a \le x \le b$. Its maximum (minimum) value is attained at one or more members of the set of points $S = \{a, x_1, x_2, \ldots, x_n, b\}$. This can be accomplished by:

(a) Computing the values of $f(x)$ at each point of S and then choosing the appropriate extremum.
(b) Graphing $f(x)$ (see Chapter I).

2. Related Rates

In such problems the relevant quantities depend *implicitly* on a single variable (usually time). One is interested in the rate of change (with time) of a particular quantity. The solution is obtained by *relating* its rate to the rates of other quantities. Usually it is most helpful to construct a diagram pertinent to the essentials of the given problem.

As an example, say one wishes to find the rate of change of area of a given circle, radius r, whose radius is changing with time t at the given rate $\frac{dr}{dt}$:

Let A be the area of the circle at time t, $A = \pi r^2$. Here r, and hence A, are defined implicitly as functions of t. The aiin is to find $\frac{dA}{dt}$. The related rates are:

$$\frac{dA}{dt} = \frac{d}{dt}(\pi r^2) = 2\pi r \frac{dr}{dt}.$$

3. Volumes of Solids with Known Cross-Sections

Say a solid can be described as having cross-sectional area $A(x)$ at x, $a \le x \le b$ (Fig. II.0). Then the volume V of such a solid is:

$$V = \int_a^b A(x)\,dx.$$

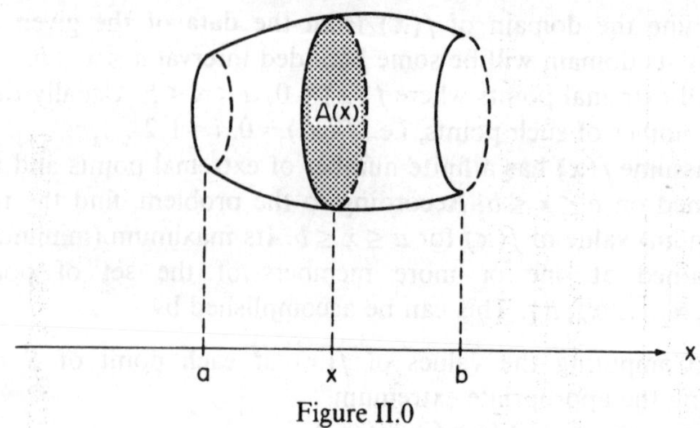

Figure II.0

Solved Problems

1. An open topped box is to be constructed by cutting equal squares from each corner of a 3 metre by 8 metre rectangular sheet of aluminum and folding up the sides. Find the volume of the largest such box.

Let x be the length dimension of the removed squares. Then the diagrams in Fig. II.1a describe the problem.

Let $V(x)$ be the volume of the constructed box $=(8-2x)(3-2x)x=4x^3-22x^2+24x$. $D=$ domain of $V(x)$: $0 \le x \le \frac{3}{2}$. (The dimension of each side must be nonnegative.)

$$\frac{dV}{dx}=12x^2-44x+24$$

$$=4(x-3)(3x-2);$$

$$\frac{dV}{dx}=0 \Rightarrow x=\tfrac{2}{3}. \qquad (x=3 \quad \text{is not in domain } D.)$$

$$\frac{d^2V}{dx^2}=24x-44<0 \quad \text{if} \quad 0 \le x \le \tfrac{3}{2}$$

$$\Rightarrow V(x) \quad \text{is concave down for} \quad 0 \le x \le \tfrac{3}{2}.$$

Hence $V(\frac{2}{3})=\dfrac{200}{27}\text{ m}^3$ is the maximum value of V in domain D (Fig. II.1b),

i.e., the largest such box has volume $\boxed{\dfrac{200}{27}\text{ m}^3}$.

2. Cylindrical soup cans are to be manufactured to contain a given volume V. There is no waste in cutting the metal for the sides of the can, but the circular endpieces will be cut from a square, with the corners wasted. Find the ratio of height to radius for the most economical can.

Let r be the radius of the top and bottom of each manufactured soup can (Fig. II.2a), h be the height of each can, and A be the area of the total used

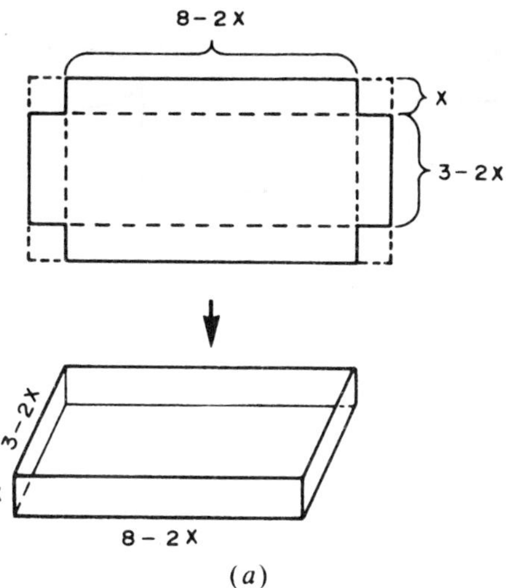

(a)

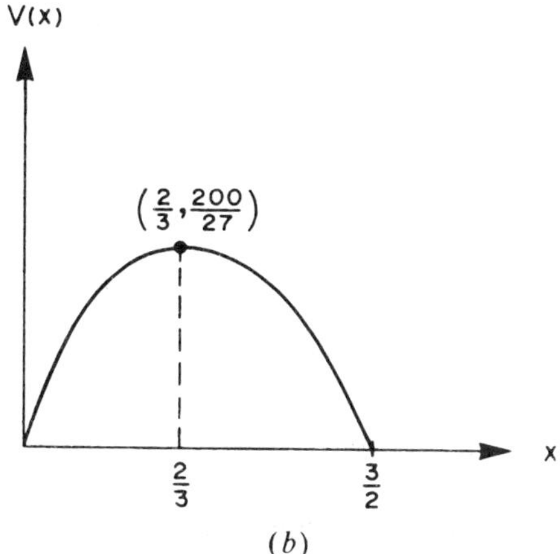

(b)

Figures II.1*a* and *b*.

material including waste for each can. Then

$$\begin{aligned} V &= \text{volume of the can} = \text{constant} \\ &= \pi r^2 h, \qquad h = h(r), \\ &\Rightarrow h = \frac{V}{\pi r^2}. \end{aligned}$$

$$\begin{aligned} A = A(r) &= (2r)^2 + (2r)^2 + 2\pi r h \\ &= 8r^2 + \frac{2V}{r}; \end{aligned}$$

$$D = \text{domain of } A(r): r > 0.$$

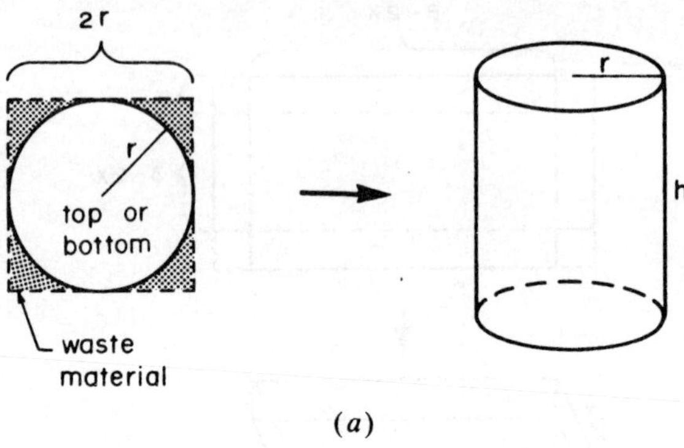

(a)

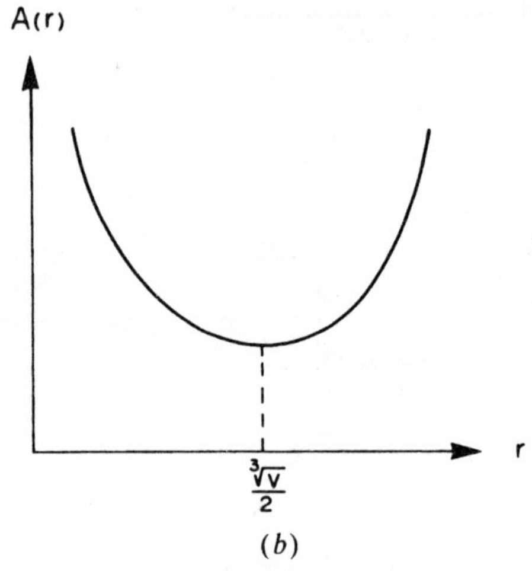

(b)

Figures II.2*a* and *b*.

The aim is to find the value of r minimizing $A(r)$ over the domain D:

$$\frac{dA}{dr} = \frac{-2V}{r^2} + 16r = 0 \leftrightarrow r^3 = \frac{V}{8};$$

$$\frac{d^2A}{dr^2} = \frac{4V}{r^3} + 16 > 0 \quad \text{if} \quad r > 0.$$

Hence $r = \dfrac{\sqrt[3]{V}}{2}$ minimizes $A(r)$ over D (Fig. II.2*b*);

$$h = \frac{V}{\pi r^2} \leftrightarrow \frac{h}{r} = \frac{V}{\pi r^3} = \frac{8}{\pi},$$

i.e., for the most economical can the ratio of height to radius is $\boxed{\dfrac{8}{\pi}}$.

3. A rectangle R is to be placed inside a triangle T so that one side of R coincides with part of one side of T (Fig. II.3a). If each side of T has length 2, and if R has the largest area consistent with these conditions, find the dimensions of R.

Let the dimensions of the rectangle be x by y. Then the area of the rectangle is $\alpha = xy$. Drop a perpendicular from A bisecting BC at O, and construct similar triangles AOC and DEC. The length of AO is $\sqrt{2^2 - 1^2} = \sqrt{3}$. Then, from these similar triangles,

$$\frac{\sqrt{3}}{1} = \frac{y}{1 - \frac{x}{2}}, \qquad \text{i.e.,} \qquad y = \sqrt{3}\left(1 - \frac{x}{2}\right).$$

Hence the aim is to maximize $\alpha(x) = \sqrt{3}\,x\left(1 - \frac{x}{2}\right)$, over the domain D: $0 \leq x \leq 2$.

$$\frac{d\alpha}{dx} = \sqrt{3}\,(1 - x) = 0 \leftrightarrow x = 1;$$

$$\frac{d^2\alpha}{dx^2} = -\sqrt{3} < 0 \quad \text{for all} \quad x.$$

Hence $x = 1 \leftrightarrow$ global maximum for $\alpha(x) \Rightarrow$ dimensions of the rectangle are $\boxed{1 \text{ by } \frac{\sqrt{3}}{2}}$. (Since $\alpha(x)$ is a parabola (Fig. II.3b), one can solve this problem without calculus.)

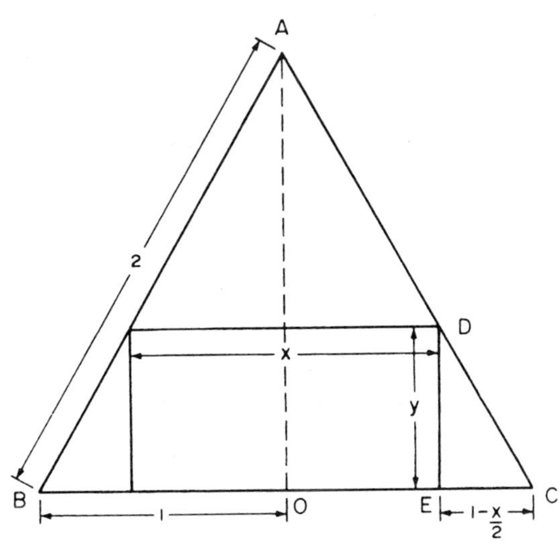

Figure II.3a

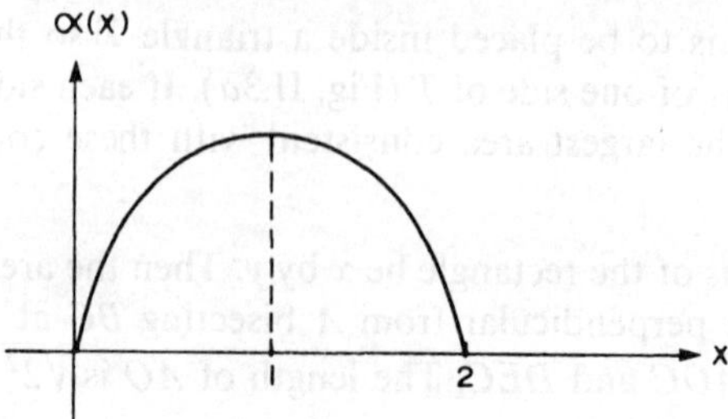

Figure II. 3*b*

(*)4. A circular disc of radius r is to be used in an evaporator. It is to rotate in a vertical plane (its axis of rotation being horizontal) and is to be submerged in a liquid in such a way as to maximize the exposed wetted area.

Show that the centre of the disc should be at a distance h above the surface of the liquid where $h = \dfrac{r}{\sqrt{1+\pi^2}}$.

The aim of this problem is to maximize the shaded area $= A =$ exposed wetted area (Fig. II.4*a*). $A = \pi r^2 - \pi h^2$ minus the area of the disc always immersed in the liquid.

Method 1

The immersed area = area of sector OPQ minus the area of ΔOPQ.
In terms of θ:

$$\text{area of sector } OPQ = \tfrac{1}{2}r^2(2\theta) = r^2\theta;$$

$$\text{area of } \Delta OPQ = ah = (r\sin\theta)(r\cos\theta) = \frac{r^2\sin 2\theta}{2}.$$

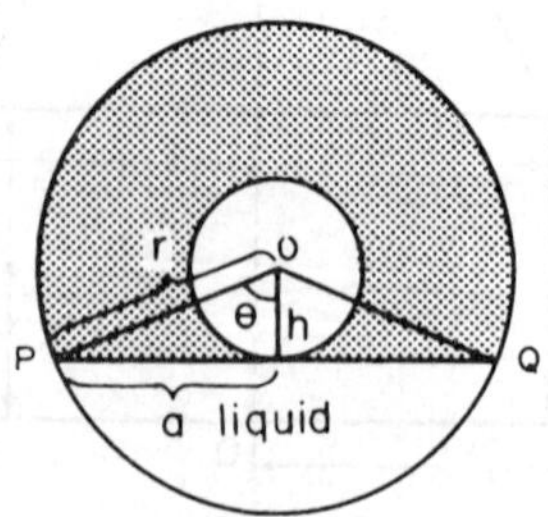

Figure II.4*a*

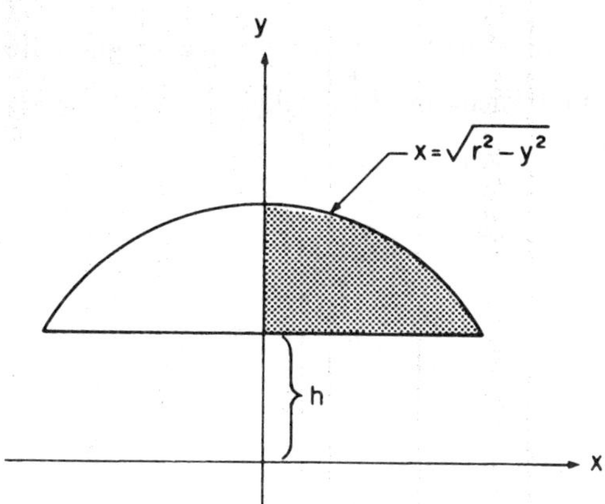

Figure II.4b

Hence

$$A = A(\theta) = \pi r^2(1-\cos^2\theta) - \left(r^2\theta - \frac{r^2\sin 2\theta}{2}\right)$$
$$= r^2\left[\pi\sin^2\theta + \frac{\sin 2\theta}{2} - \theta\right].$$

From Fig. II.4a, D = domain of $A(\theta)$: $0 \le \theta \le \frac{\pi}{2}$.

$$\frac{dA}{d\theta} = r^2[2\pi\sin\theta\cos\theta + \cos 2\theta - 1]$$
$$= r^2[2\pi\sin\theta\cos\theta - 2\sin^2\theta]$$
$$= 2r^2\sin\theta[\pi\cos\theta - \sin\theta].$$

Inside D, $\frac{dA}{d\theta} = 0 \leftrightarrow \pi\cos\theta = \sin\theta \leftrightarrow \tan\theta = \pi \Rightarrow \theta = \theta^* = \tan^{-1}\pi$. Note that $\frac{\pi}{3} < \theta^* < \frac{\pi}{2}$. If $\theta = \frac{\pi}{4}$, $\frac{dA}{d\theta} = r^2[\pi - 1] > 0$. So $\frac{dA}{d\theta} > 0$ for $0 < \theta < \theta^*$. If $\theta = \frac{\pi}{2}$, $\frac{dA}{d\theta} = -2r^2 < 0$. So $\frac{dA}{d\theta} < 0$ for $\theta^* < \theta < \frac{\pi}{2}$.

Hence $\theta = \theta^* = \tan^{-1}\pi \leftrightarrow$ the maximum exposed wetted area. In this case

$$h = r\cos\theta^* = \boxed{\frac{r}{\sqrt{\pi^2+1}}}.$$

Method 2

From integral calculus the wetted area is twice the shaded area of Fig. II.4b, or

$$2\int_h^r \sqrt{r^2 - y^2}\,dy.$$

Hence $A = A(h) = \pi r^2 - \pi h^2 - 2\int_h^r \sqrt{r^2 - x^2}\,dx$. Domain of $A(h)$: $0 \le h \le r$.
By the Fundamental Theorem of Calculus,

$$\frac{dA}{dh} = -2\pi h + 2\sqrt{r^2 - h^2};$$

$$\frac{dA}{dh} = 0 \leftrightarrow h = \frac{r}{\sqrt{1+\pi^2}}.$$

$$\text{At} \quad h = 0, \quad \frac{dA}{dh} > 0; \qquad \text{at} \quad h = r, \quad \frac{dA}{dh} < 0;$$

hence

$$\frac{dA}{dh} > 0 \quad \text{for} \quad 0 < h < \frac{r}{\sqrt{1+\pi^2}},$$

$$\frac{dA}{dh} < 0 \quad \text{for} \quad \frac{r}{\sqrt{1+\pi^2}} < h < r,$$

$$\Rightarrow h = \boxed{\frac{r}{\sqrt{1+\pi^2}}} \text{ for maximum exposed wetted area.}$$

5. The strength of a wooden beam of rectangular cross-section and fixed length is proportional to the product of its width and the square of its height. Find the dimensions of the strongest such beam that can be cut from a log of circular cross-section if the log has a diameter of four feet. Demonstrate that your result is a maximum.

Let w be the width of the beam, h be the height of the beam, and S be the strength of the beam (Fig. II.5). $S = kwh^2$, where k is a proportionality constant. $h^2 + w^2 = 4^2 = 16 \Rightarrow h^2 = 16 - w^2$. Hence the aim is to maximize $S = S(w) = kw(16 - w^2)$ over the domain D: $0 \le w \le 4$.

$$\frac{dS}{dw} = k(16 - 3w^2);$$

$$\frac{dS}{dw} = 0 \Rightarrow w = \frac{4}{\sqrt{3}};$$

$$\frac{d^2S}{dw^2} = -6kw \le 0 \quad \text{if} \quad w \ge 0.$$

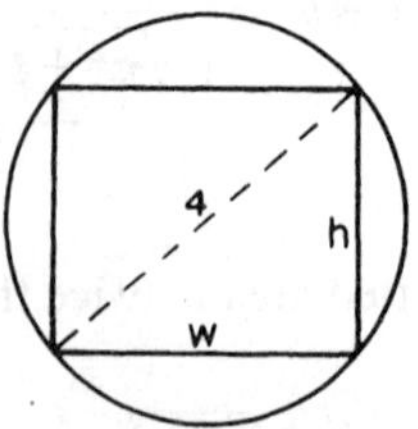

Figure II.5

Thus $w = \dfrac{4}{\sqrt{3}}$ corresponds to the maximum value of $S(w)$ over the domain D. Hence the

$$\text{strongest beam} \leftrightarrow w = \boxed{\frac{4}{\sqrt{3}} \text{ feet}},$$

$$h = \boxed{\frac{4\sqrt{6}}{3} \text{ feet}}.$$

6. A man wants to fence off a rectangular plot of 800 ft^2 which will be open to a straight river so that only 3 sides need to be fenced. What dimensions should the plot have in order to minimize the amount of fencing required?

Let x be the length of fencing parallel to the river, and let y be the length of fencing perpendicular to the river. The diagram corresponding to this is Fig. II.6a. The area enclosed by the fence is $xy = \text{constant} = 800$. Hence $y = \dfrac{800}{x}$. The total length of fencing required is

$$L(x) = x + 2y = x + \frac{1600}{x}.$$

D = domain of $L(x)$: $x > 0$.

$$L'(x) = 1 - \frac{1600}{x^2},$$

$$L''(x) = \frac{3200}{x^3} > 0 \quad \text{in domain } D.$$

The aim is to minimize $L(x)$ over D. Hence

$$L'(x) = 0 \leftrightarrow \frac{1600}{x^2} = 1 \leftrightarrow x = \pm 40.$$

Only $x = 40$ lies in D. Since $L(x)$ is concave up for all x in D, $x = 40$ corresponds to the absolute minimum value of $L(x)$ on D (Fig. II.6b). Thus

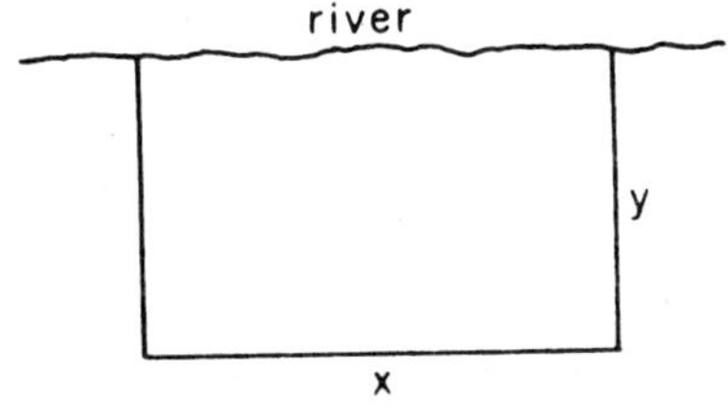

Figure II.6a

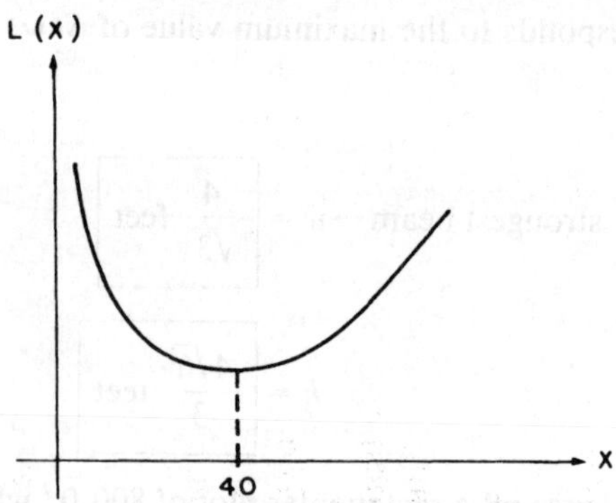

Figure II.6b

the dimensions must be

$$x = \boxed{40 \text{ feet}},$$

$$y = \frac{800}{x} = \boxed{20 \text{ feet}}.$$

7. **Three thousand metres of fencing are used to enclose a rectangular area and to divide it into 3 equal areas as shown in Fig. II.7a. Find the dimensions such that the total area enclosed is greatest.**

The total length of the fence is $6x + 4y = \text{constant} = 3000$. Hence

$$y = \frac{1500 - 3x}{2}.$$

The area enclosed by the fence is

$$A(x) = (3x)(y) = 3x\frac{(1500 - 3x)}{2} = \frac{9}{2}(500x - x^2).$$

D = domain of $A(x)$: $0 \le x \le 500$ since $y \ge 0$.

$$\frac{dA}{dx} = \frac{9}{2}(500 - 2x),$$

$$\frac{d^2A}{dx^2} = -9 < 0 \quad \text{for all} \quad x.$$

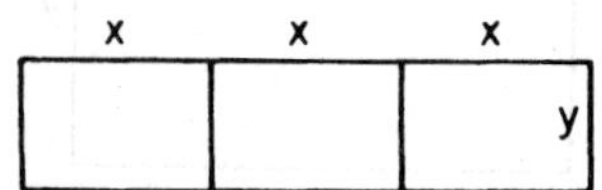

Figure II.7a

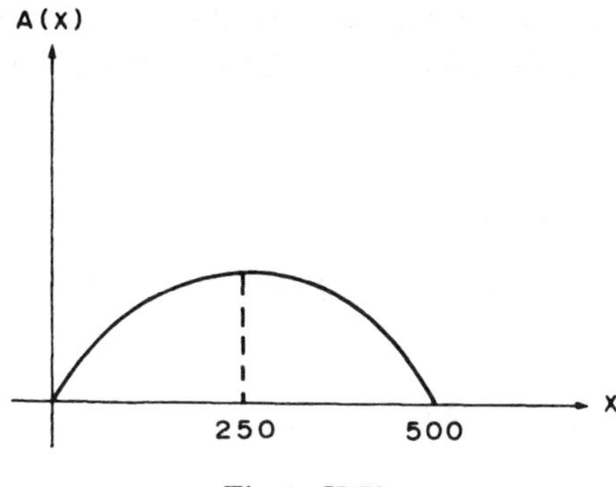

Figure II.7*b*

The aim is to maximize $A(x)$ on D:

$$\frac{dA}{dx} = 0 \leftrightarrow 2x = 500 \leftrightarrow x = 250.$$

Since $A(x)$ is concave down for all x, $x = 250$ corresponds to the absolute maximum value of $A(x)$ on D. Therefore in order to obtain maximum area,

$$x = \boxed{250 \text{ metres}},$$

$$y = \frac{1500 - 3x}{2} = \boxed{375 \text{ metres}}.$$

Observe (Fig. II.7*b*) that $A(x)$ is a parabola opening downwards, with vertex at $x = 250$:

$$\begin{aligned} A(x) &= \frac{-9}{2}(x^2 - 500x) \\ &= \frac{-9}{2}\left(x^2 - 500x + (250)^2\right) + \frac{9}{2}(250)^2 \\ &= \frac{-9}{2}(x - 250)^2 + \frac{9}{2}(250)^2. \end{aligned}$$

(Thus one can solve this problem without the use of calculus.)

8. A farmer has 100 metres of fencing to enclose a rectangular pasture. He may use part of the wall of his barn as one side of the pasture or all of the wall of his barn as one side.

(a) What is the maximum area A of the pasture he can fence in if the barn is 75 metres long?

(b) Same as (a) but for a barn 10 metres long.

Let the dimensions of the fence be x(parallel to wall) by y(perpendicular) (Fig. II.8a). Total length of fence is $x+2y = \text{constant} = 100$. Hence

$$y = \frac{100-x}{2}.$$

The area enclosed by the fence is

$$A(x) = (x)(y) = \tfrac{1}{2}x(100-x) = \tfrac{1}{2}(100x - x^2).$$

D = domain of $A(x)$: $0 \le x \le 75$.

$$A'(x) = \tfrac{1}{2}(100-2x);$$
$$A''(x) = -1.$$

The aim is to maximize $A(x)$:

$$A'(x) = 0 \leftrightarrow 2x = 100 \leftrightarrow x = 50.$$

Since $A''(x) < 0$ for all x, $x = 50$ corresponds to the maximum value of $A(x)$ over any domain containing this point. Hence the maximum area is

$$A(50) = \tfrac{1}{2}(50)(50) = \boxed{1250 \text{ m}^2}.$$

To solve (b), the problem is set up exactly as in (a), except in this case D: $0 \le x \le 10$. Hence $x = 50$ is not in D. Since no other critical points exist, the maximum value for $A(x)$ occurs at x equal to one of the endpoints, either 0 or 10 (Fig. II.8b).

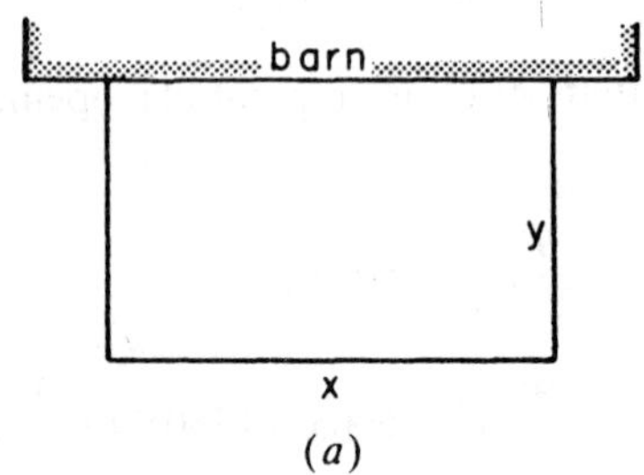

(a)

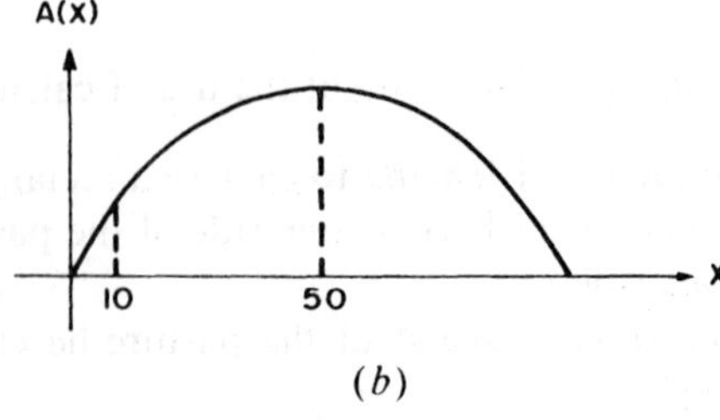

(b)

Figures II.8a and b.

By computing the slope of $A(x)$ at some point in D, one finds that $A'(x) > 0$ in D. Thus the maximum value of $A(x)$ occurs for $x = 10$. In this case

$$A(10) = \tfrac{1}{2}(10)(90) = \boxed{450 \text{ m}^2}.$$

Again note that $A(x)$ is a parabola; thus no calculus is necessary.

(*)9. A sheet of metal is 10 ft long and 4 ft wide. It is bent lengthwise down the middle to form a V-shaped trough 10 ft long. What should be the width across the top of the trough in order that it have maximum capacity?

Since the length of the trough is fixed, all that is necessary is that one maximize the cross-sectional area, which is that of an isosceles triangle (Fig. II.9a). Let the length of the base of the cross-sectional triangle (that is, the width across the top of the trough) be w, and the altitude of the triangle (i.e. the depth of the trough) be h (Fig. II.9b).

Method 1

Since the altitude to the base of an isosceles triangle also bisects it, one obtains the following relationship between h and w:

$$h^2 + \left(\frac{w}{2}\right)^2 = 2^2 = 4,$$

defining h implicitly as a function of w, $h(w) \geq 0$. Differentiating this relation with respect to w, one obtains

$$2h\frac{dh}{dw} + \frac{w}{2} = 0 \leftrightarrow \frac{dh}{dw} = -\frac{w}{4h}.$$

The aim is to maximize the area

$$A(w) = \tfrac{1}{2}(w)(h) = \tfrac{1}{2}wh.$$

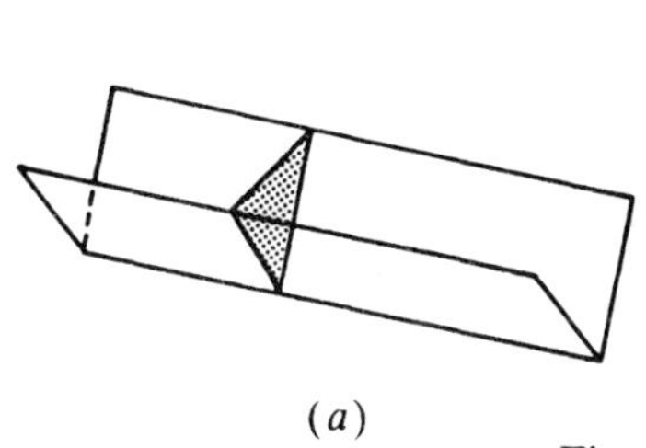

(a)

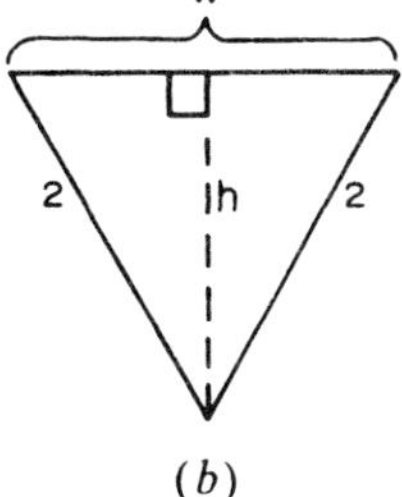

(b)

Figures II.9a and b.

D = domain of $A(w)$: $0 \leq w \leq 4$. Remembering that $h = h(w)$,

$$\frac{dA}{dw} = \frac{1}{2}h + \frac{1}{2}w\frac{dh}{dw}$$

$$= \frac{1}{2}\left(h - \frac{w^2}{4h}\right);$$

$$\frac{dA}{dw} = 0 \leftrightarrow h = \frac{w^2}{4h} \leftrightarrow h^2 = \frac{w^2}{4} = \left(\frac{w}{2}\right)^2.$$

Substituting $h^2 = \left(\frac{w}{2}\right)^2$ into

$$h^2 + \left(\frac{w}{2}\right)^2 = 4:$$

$$2\left(\frac{w}{2}\right)^2 = 4 \leftrightarrow w = \pm 2\sqrt{2}.$$

Only $w = 2\sqrt{2}$ is in D. This corresponds to maximum A, for

$$\text{if} \quad h^2 > \frac{w^2}{4} \quad (\text{i.e.} \quad w < 2\sqrt{2}), \qquad \text{then} \quad \frac{dA}{dw} > 0;$$

$$\text{if} \quad h^2 < \frac{w^2}{4} \quad (\text{i.e.} \quad w > 2\sqrt{2}), \qquad \text{then} \quad \frac{dA}{dw} < 0.$$

Thus the trough has maximum capacity when $\boxed{w = 2\sqrt{2} \text{ ft}}$.

Method 2

One can find an explicit formula for area A, by introducing the angle θ (Fig. II.9c). Hence

$$\frac{w}{2} = 2\sin\theta,$$

$$h = 2\cos\theta,$$

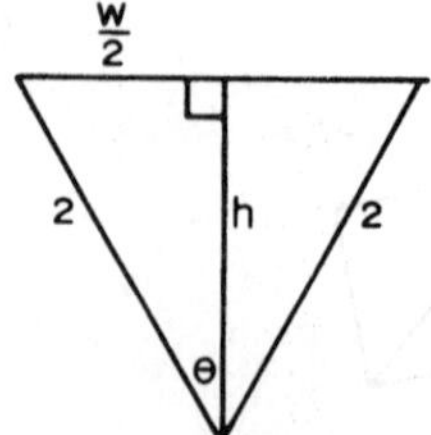

Figure II.9c

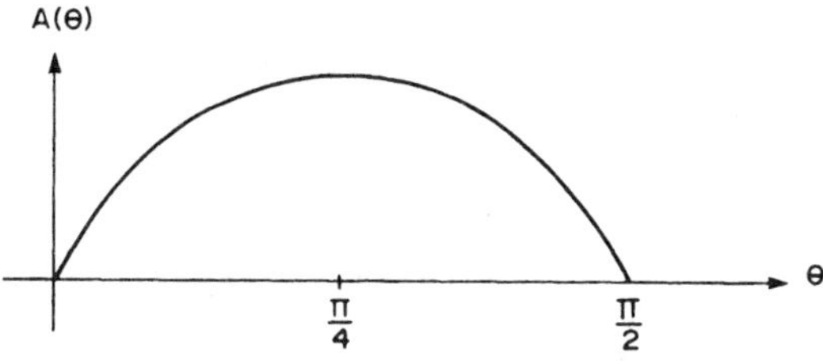

Figure II.9d

and $A(\theta) = \frac{1}{2}wh = 4\sin\theta\cos\theta = 2\sin 2\theta$;

$$\frac{dA}{d\theta} = 4\cos 2\theta;$$

$$\frac{d^2A}{d\theta^2} = -8\sin 2\theta.$$

D = domain of $A(\theta)$: $0 \le \theta \le \frac{\pi}{2}$. The aim is to maximize $A(\theta)$ over domain D:

$$\frac{dA}{d\theta} = 0 \leftrightarrow \cos 2\theta = 0 \leftrightarrow 2\theta = \frac{\pi}{2}, \frac{3\pi}{2}, \frac{5\pi}{2}, \ldots.$$

Only $\theta = \frac{\pi}{4}$ lies in D. Note that $\frac{d^2A}{d\theta^2} \le 0$ for all θ in D; hence $\theta = \frac{\pi}{4}$ corresponds to the maximum value for $A(\theta)$ (see Fig. II.9d) and in this case

$$w = 4\sin\theta = 4\sin\frac{\pi}{4} = \boxed{2\sqrt{2}\text{ ft}}.$$

10. A rectangular box-shaped house is to have a square floor. Three times as much heat per square metre is lost through the roof as through the walls; no heat is lost through the floor. What shape should the house be if it is to enclose 1500 cubic metres and minimize heat loss?

Let the house have height y and a floor width of x. Choose (arbitrarily) to make the heat loss an explicit function of x; therefore find an explicit formula for y in terms of x. The volume enclosed by the house is x^2y = constant = 1500. Hence

$$y = \frac{1500}{x^2}.$$

Let k be the constant amount of heat lost per unit wall area (heat units).

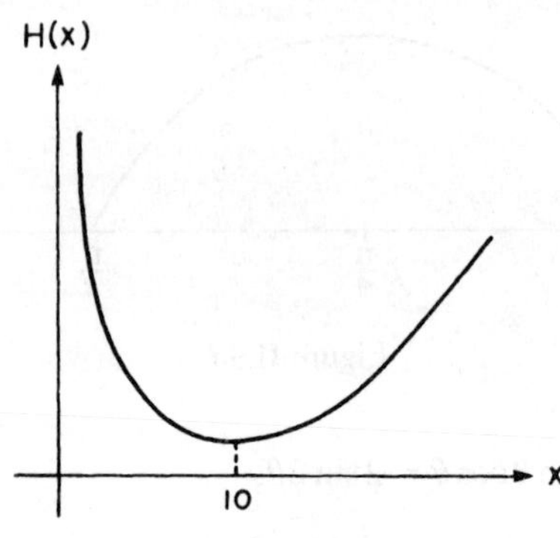

Figure II.10

Then the total heat loss is

$$\begin{aligned} H(x) &= k(\text{total area of walls}) + 3k(\text{total roof area}) \\ &= k(4xy + 3x^2) \\ &= k\left[3x^2 + (4x)\left(\frac{1500}{x^2}\right)\right] \\ &= 3k\left(x^2 + \frac{2000}{x}\right). \end{aligned}$$

D = domain of $H(x)$: $x > 0$. Aim: to minimize $H(x)$ on D;

$$\frac{dH}{dx} = 3k\left[2x - \frac{2000}{x^2}\right];$$

$$\frac{d^2H}{dx^2} = 3k\left[2 + \frac{4000}{x^3}\right];$$

$$\frac{dH}{dx} = 0 \leftrightarrow 2x = \frac{2000}{x^2} \leftrightarrow x^3 = 1000 \leftrightarrow x = 10.$$

Clearly $x = 10$ corresponds to the minimum value of $H(x)$ on D since $\frac{d^2H}{dx^2} > 0$ for all x in D (Fig. II.10). The desired dimensions are $\boxed{x = 10\text{m}}$, $\boxed{y = 15\text{m}}$.

11. One hundred feet directly overhead, a plane and a dirigible start flying due east. If the plane travels three times faster than the dirigible, what is the greatest angle of sight between the two vehicles from your position? Assume that the plane passes from sight after 6000 feet.

Let

$\alpha(t)$ = the angle between ground and the line of sight to the dirigible;
$\beta(t)$ = the angle between ground and the line of sight to the plane;
$\theta(t)$ = the angle of sight between the dirigible and the plane.

Let the speed of the dirigible be c = constant, and the altitude of plane and dirigible be a = constant (Fig. II.11). (Here $a = 100$.) Note that

$$\tan\alpha = \frac{a}{\text{length of } AD},$$

$$\tan\beta = \frac{a}{\text{length of } AP},$$

and that at time t, the length of AD is ct, and the length of AP is $3ct$, assuming $t = 0$ when both plane and dirigible are directly overhead. Then the angle of sight θ is

$$\theta(t) = \alpha - \beta = \tan^{-1}\frac{a}{ct} - \tan^{-1}\frac{a}{3ct}.$$

Domain D of $\theta(t)$: $0 \le t \le \frac{6000}{c}$. Aim: to maximize $\theta(t)$;

$$\theta'(t) = \frac{-ac}{c^2t^2 + a^2} + \frac{3ac}{9c^2t^2 + a^2};$$

$$\begin{aligned}\theta'(t) = 0 &\leftrightarrow ac(9c^2t^2 + a^2) = 3ac(c^2t^2 + a^2)\\ &\leftrightarrow 6c^2t^2 = 2a^2\\ &\leftrightarrow t = \pm\frac{a}{\sqrt{3}\,c}.\end{aligned}$$

Only $t = \frac{a}{\sqrt{3}\,c}$ lies in D. $\theta \ge 0$ for all t; $\theta(0) = 0$ (geometrically); $\lim_{t\to\infty} \theta = 0$; moreover there is only one critical point for $t \ge 0$.

Hence $t = \frac{a}{\sqrt{3}\,c}$ corresponds to the maximum value for $\theta(t)$ on D. The greatest angle of sight, corresponding to $t = \frac{a}{\sqrt{3}\,c}$, is

$$\begin{aligned}\theta &= \tan^{-1}\sqrt{3} - \tan^{-1}\frac{1}{\sqrt{3}}\\ &= \frac{\pi}{3} - \frac{\pi}{6} = \boxed{\frac{\pi}{6} \text{ radians}}.\end{aligned}$$

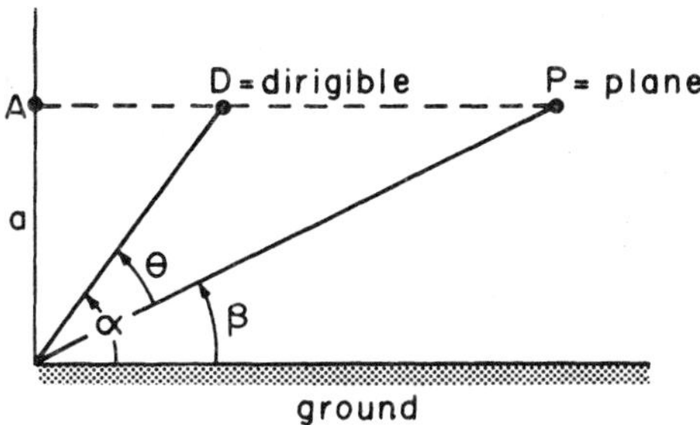

Figure II.11

12. A cable will be laid from a small island Q, which is 3 miles from the nearest point P on a straight shoreline (Fig. II.12a), to a town T, which is 12 miles down the shore from P. If cable costs \$40 per mile for laying under water and \$20 per mile for laying on land, at what distance from P should the cable leave the water in order to give the minimum cost?

Let the cable leave the water at the point R, a distance x from P (positive in the direction of T). Then the distances (cf. Fig. II.12b)

$$QR = \sqrt{9+x^2},$$
$$RT = 12-x.$$

The total cost $C(x)$ of the cable is

$$C(x) = 40\sqrt{9+x^2} + 20(12-x).$$

Domain of $C(x)$ is D: $0 \le x \le 12$. (It is clear that any x outside D will result in higher cost.) Aim: to minimize $C(x)$ over D;

$$\frac{dC}{dx} = \frac{40x}{\sqrt{9+x^2}} - 20;$$

$$\frac{d^2C}{dx^2} = \frac{360}{(9+x^2)^{3/2}};$$

$$\frac{dC}{dx} = 0 \leftrightarrow 40x = 20\sqrt{9+x^2} \Rightarrow 4x^2 = 9 + x^2 \leftrightarrow x^2 = 3 \leftrightarrow x = \pm\sqrt{3}.$$

Since $\dfrac{d^2C}{dx^2} > 0$ for all x, and only $x = \sqrt{3}$ is contained in D, one concludes that $x = \sqrt{3}$ gives the minimum value of $C(x)$ for all x in D. Hence the cable should leave the water a distance $\boxed{\sqrt{3}\text{ miles}}$ from P (Fig. II.12c).

13. A piece of wire 12 inches long is cut into 2 lengths, one of which is bent into a circle, the other into a square. Determine the ratio of the side length of the square to the radius of the circle for a minimum value of the sum of

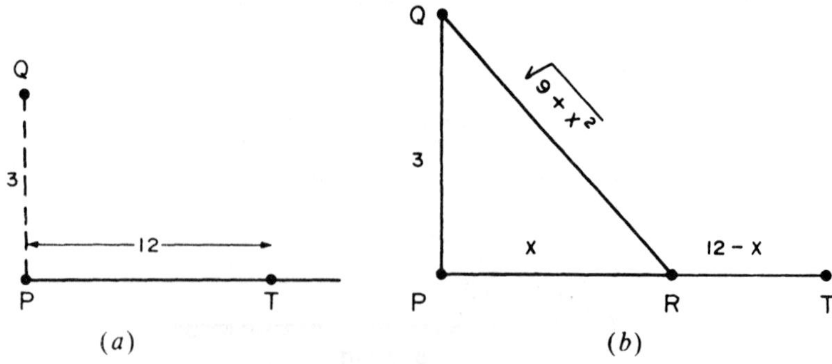

Figures II.12a and b.

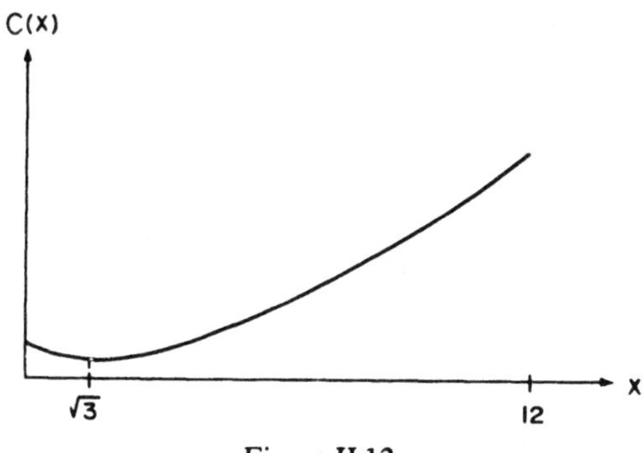

Figure II.12c

the areas of the circle and the square, and show that the minimum area is $\frac{36}{4+\pi}$ in^2.

Say the portion of the wire which is bent into a circle has length x; then the square must be formed from the remaining $12-x$ inches.

Therefore the radius of the circle is $\frac{x}{2\pi}$, and hence the area of the circle is $\pi\left(\frac{x}{2\pi}\right)^2=\frac{x^2}{4\pi}$.

The side length of the square is $\frac{12-x}{4}$; hence the area of the square is $\left(\frac{12-x}{4}\right)^2$.

The total area formed is

$$A(x)=\frac{x^2}{4\pi}+\frac{(12-x)^2}{16}.$$

The aim is to minimize $A(x)$ on the domain D: $0\le x\le 12$;

$$A'(x)=\frac{x}{2\pi}-\frac{(12-x)}{8};$$

$$A''(x)=\frac{1}{2\pi}+\frac{1}{8}>0 \quad \text{for all} \quad x;$$

$$A'(x)=0\leftrightarrow\frac{x}{2\pi}=\frac{12-x}{8}\leftrightarrow 4x=12\pi-\pi x\leftrightarrow x=x^*=\frac{12\pi}{4+\pi}.$$

Since $x^*=\frac{12\pi}{4+\pi}$ is in D, and $A''(x)>0$ for all x, one concludes that the minimum total area is obtained for this value of x.

The ratio of side length to radius is

$$\frac{\pi}{2}\left(\frac{12-x^*}{x^*}\right)=\frac{\pi}{2}\left[\frac{12}{x^*}-1\right]=\frac{\pi}{2}\left[\frac{4+\pi}{\pi}-\frac{\pi}{\pi}\right]=\boxed{2},$$

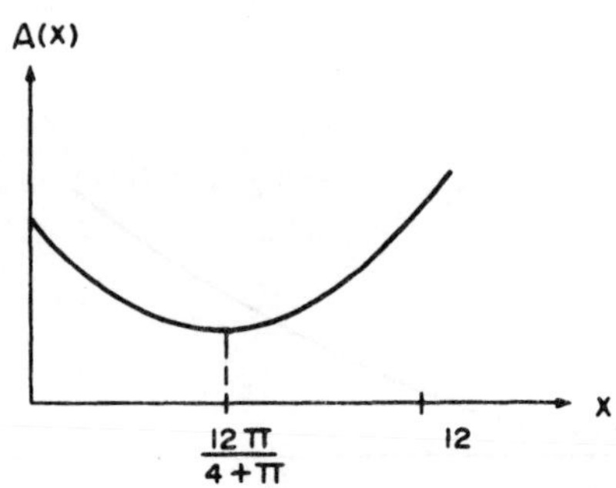

Figure II.13

and the minimum area is

$$A(x^*) = \frac{1}{4\pi}\frac{(12\pi)^2}{(4+\pi)^2} + \frac{1}{16}\left(\frac{48+12\pi-12\pi}{4+\pi}\right)^2$$
$$= \frac{36\pi+144}{(4+\pi)^2} = \frac{36}{4+\pi}\text{in}^2.$$

Note that $A(x)$ is a parabola with vertex at $x = \dfrac{12\pi}{4+\pi}$ (Fig. II.13).

(*)**14.** A man can swim at 20 feet/second and run at 25 feet/second. If he stands at point $A=(0,50)$ on the edge of a circular swimming pool of radius 50 feet (with its centre at the origin), find his optimum path from A to $B=(50,0)$. (Hint: Suppose that he begins by running and then swims the rest of the way.)

Say the man jumps into the pool at point P, and that the angle subtended by arc $\widehat{PB}$ is θ. Let the centre of the pool be O. The length of arc $\widehat{AP}$, i.e. the length of the path on ground, is

$$(\text{radius})\times(\text{angle subtended}) = 50\left(\frac{\pi}{2}-\theta\right).$$

Let segment $\overline{OQ}$ bisect segment $\overline{PB}$ (Fig. II.14a). Since triangle OPB is isosceles, $\angle OQP = \frac{\pi}{2}$; hence $\sin\frac{\theta}{2} = \frac{PQ}{50}$ and the length of the path in water is

$$PB = 2(PQ) = 100\sin\frac{\theta}{2}.$$

Therefore the total time T taken to travel from A to B is

$$T(\theta) = \frac{50\left(\frac{\pi}{2}-\theta\right)}{25} + \frac{100\sin\frac{\theta}{2}}{20} \qquad \text{(seconds)}$$
$$= \pi - 2\theta + 5\sin\frac{\theta}{2}.$$

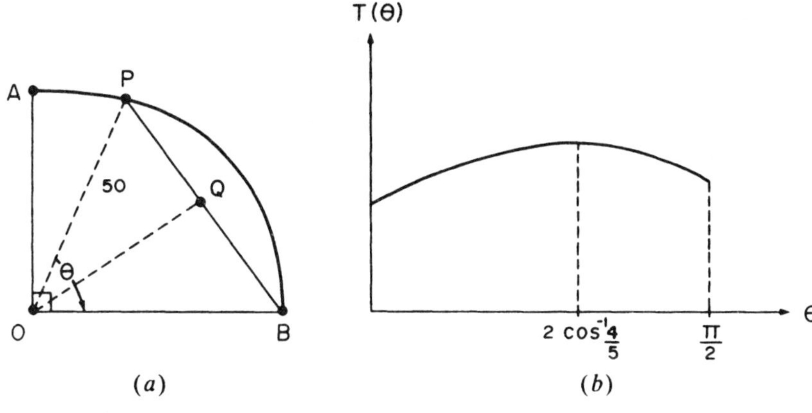

Figures II.14*a* and *b*.

$D=$ domain of $T(\theta)$: $0\le\theta\le\frac{\pi}{2}$. Aim: to minimize $T(\theta)$ over D;

$$\frac{dT}{d\theta}=-2+\frac{5}{2}\cos\frac{\theta}{2};$$
$$\frac{d^2T}{d\theta^2}=-\frac{5}{4}\sin\frac{\theta}{2}.$$

The derivative of $T(\theta)$ is zero where $\cos\frac{\theta}{2}=\frac{4}{5}$. The actual value of $\theta=2\cos^{-1}\frac{4}{5}$ is unimportant since $T(\theta)$ attains a *maximum* there! This is true because for $0<\theta<\frac{\pi}{2}$, $\sin\frac{\theta}{2}>0$; hence $\frac{d^2T}{d\theta^2}$ is negative at $\theta=2\cos^{-1}\frac{4}{5}$. Since there exists only one critical point in D, one must look for the minimum value of $T(\theta)$ at one of the endpoints of D: either $\theta=0$ or $\theta=\frac{\pi}{2}$.

$$\text{At}\quad\theta=0,\qquad T(0)=\pi\simeq 3.14.$$
$$\text{At}\quad\theta=\frac{\pi}{2},\qquad T\left(\frac{\pi}{2}\right)=\pi-\pi+5\sin\frac{\pi}{4}=\frac{5}{\sqrt{2}}\simeq 3.54.$$

Hence minimum time occurs for $\theta=0$; that is, the man runs all the way (Fig. II.14*b*).

(*)15. Find the length of the longest rod which can be carried horizontally (i.e. without tilting up or down) from a corridor 6m wide into one 3m wide.

To find the longest such rod, consider the situation when the rod just cuts the corridor at points A, B, C (Fig. II.15*a*).

Let l be the length of the rod $=AB+BC$

$$\Rightarrow l=l(\theta)=\frac{6}{\sin\theta}+\frac{3}{\cos\theta}.$$

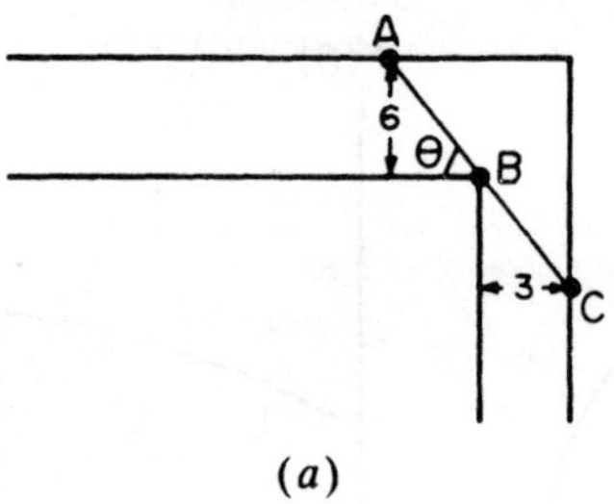

(a)

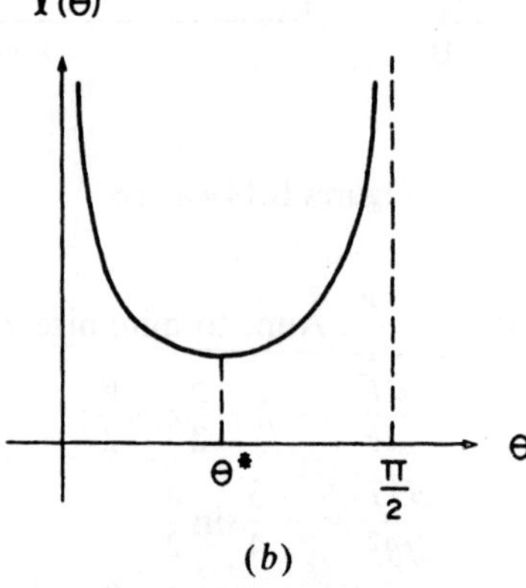

(b)

Figures II.15*a* and *b*.

D = domain of $l(\theta)$: $0 < \theta < \frac{\pi}{2}$. The solution to the problem corresponds to *minimizing* $l(\theta)$ over D (Fig. II.15*b*). (Any rod larger than this minimum length will not get through the corner.)

$$\frac{dl}{d\theta} = \frac{-6\cos\theta}{\sin^2\theta} + \frac{3\sin\theta}{\cos^2\theta} = \frac{3\sin^3\theta - 6\cos^3\theta}{\sin^2\theta\cos^2\theta}$$

$$= \frac{12(\sin^3\theta - 2\cos^3\theta)}{\sin^2 2\theta};$$

$$\frac{dl}{d\theta} = 0 \leftrightarrow \tan^3\theta = 2\left(0 < \theta < \frac{\pi}{2}\right)$$

$$\leftrightarrow \theta = \theta^* = \tan^{-1}(2^{1/3}).$$

As $\theta \to 0^+$, $l(\theta) \to +\infty$. As $\theta \to \left(\frac{\pi}{2}\right)^-$, $l(\theta) \to +\infty$. Hence $\theta = \theta^*$ corresponds to the minimum value of $l(\theta)$ over D. The longest rod has length $\boxed{3(2^{2/3}+1)^{3/2}\text{m}}$.

(*)16. The Statue of Liberty is 150 feet tall, and stands on a 150 foot pedestal. How far from the base should you stand to have the statue subtend the largest possible angle at your camera lens, assuming the camera is held 5 feet off the ground and the ground is flat?

Let θ be the angle subtended by the statue at the camera lens (Fig. II.16a). Let x be the distance at which the camera-holder stands from the base of the pedestal. From Fig. II.16a:

$$\theta = \alpha - \beta, \qquad \tan\alpha = \frac{295}{x}, \qquad \tan\beta = \frac{145}{x}.$$

Hence

$$\tan\theta = \tan(\alpha - \beta) = \frac{\tan\alpha - \tan\beta}{1 + \tan\alpha\tan\beta}$$

$$= \frac{\dfrac{150}{x}}{1 + \dfrac{(295)(145)}{x^2}} = \frac{150x}{x^2 + (295)(145)}, \qquad (1)$$

defines θ implicitly as a function of x. D = domain of $\theta(x)$: $0 \le x < \infty$. Differentiating equation (1) with respect to x yields

$$\frac{d\theta}{dx}\sec^2\theta = \frac{150}{x^2 + (295)(145)} - \frac{300x^2}{\left[x^2 + (295)(145)\right]^2}$$

$$= \frac{(150)\left[(295)(145) - x^2\right]}{\left[x^2 + (295)(145)\right]^2}.$$

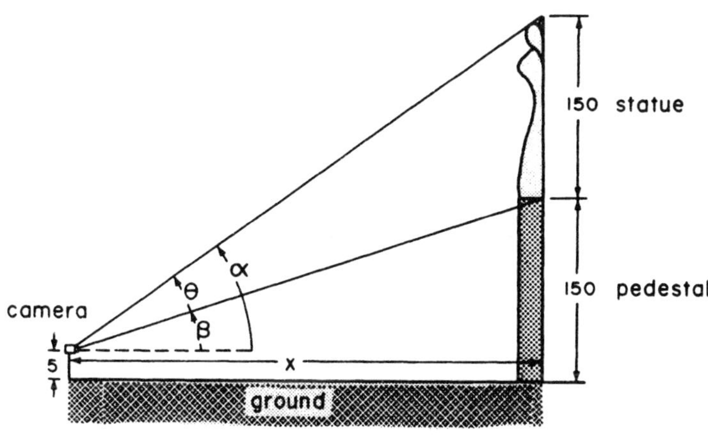

Figure II.16a

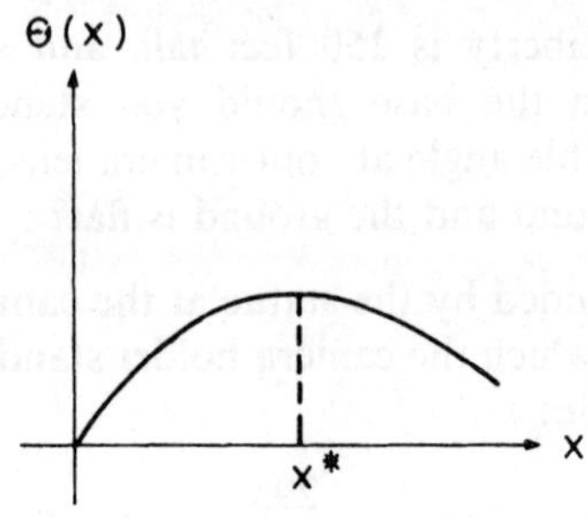

Figure II.16b

Hence

$$\frac{d\theta}{dx} = 0 \Rightarrow x = x^* = \sqrt{(295)(145)} \simeq 207.$$

$$\theta(0) = 0, \qquad \lim_{x \to +\infty} \theta(x) = 0, \qquad \theta(x^*) > 0$$

$$\Rightarrow x = x^* \leftrightarrow \text{maximum value of } \theta(x) \text{ in } D \text{ (Fig. II.16}b\text{)}.$$

The camera-holder should stand $\simeq$ **207 ft** from the base.

(*)17. Find the points Q (if any) on $y^2 = 4x$ for which the distance from the point P (4,0) to Q has a relative maximum, and those (if any) for which it has a relative minimum.

Let $S = \overline{PQ}$, $h = S^2$ (Fig. II.17a). (Extremizing $h \leftrightarrow$ extremizing S.) Then $h = (x-4)^2 + y^2$ where $y^2 = 4x$, $\Rightarrow h = h(x) = (x-4)^2 + 4x$. D = domain of $h(x)$: $0 \le x < \infty$. Aim: to find relative maxima and minima of $h(x)$

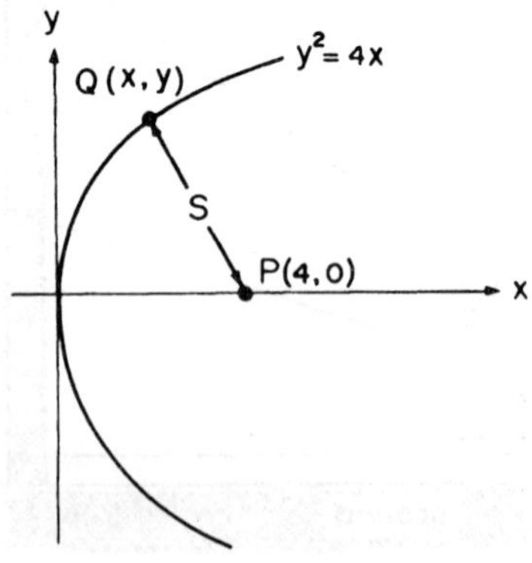

Figure II.17a

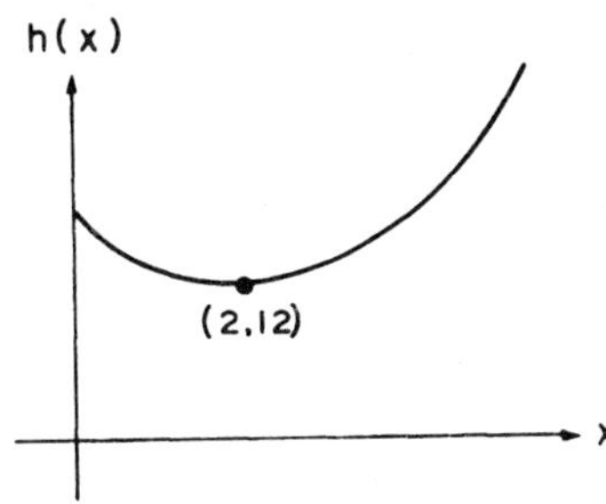

Figure II.17*b*

over D.

$$h(x) = x^2 - 8x + 16 + 4x = x^2 - 4x + 16$$

$$= (x-2)^2 + 12, \qquad \text{a parabola with vertex located at} \quad (2,12);$$

$$h'(x) = 0 \leftrightarrow 2x - 4 = 0 \leftrightarrow x = 2.$$

From the graph of $h(x)$ over D (Fig. II.17*b*), one sees that $x = 0 \leftrightarrow$ relative maximum, $x = 2 \leftrightarrow$ relative minimum. The corresponding Q points are:

$$\boxed{(0,0) \leftrightarrow \text{relative maximum}},$$

$$\boxed{(2, \pm 2\sqrt{2}) \leftrightarrow \text{relative minima}}.$$

(*)18. At what point on the parabola $y = 1 - x^2$ does the tangent have the property that it cuts from the first quadrant a triangle of minimum area? Find this minimum area.

Let $P_0 = (x_0, y_0) = (x_0, 1-(x_0)^2)$ be a point on the parabola lying in the first quadrant (Fig. II.18*a*). Let l be the tangent line to the parabola at P_0.

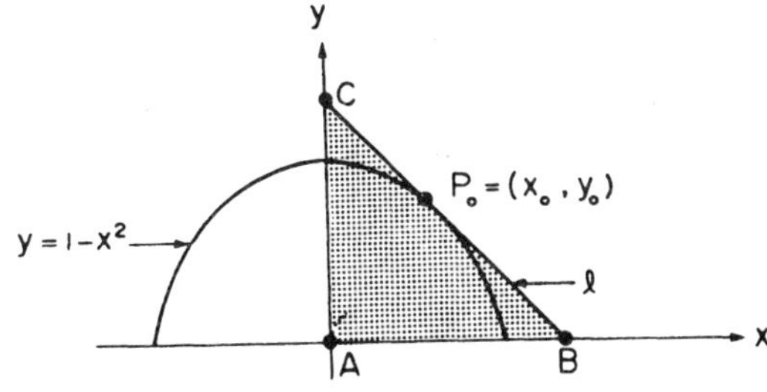

Figure II.18*a*

The slope of l is $-2x_0$. Hence the equation for l is:

$$\frac{y-y_0}{x-x_0}=-2x_0.$$

Let ABC be the triangle cut by l from the first quadrant. Then $A=(0,0)$,

$$B=\left(\frac{y_0+2(x_0)^2}{2x_0},0\right)=\left(\frac{1+(x_0)^2}{2x_0},0\right),$$

$$C=\left(0,\, y_0+2(x_0)^2\right)=\left(0,1+(x_0)^2\right).$$

Let $\alpha=\alpha(x_0)$ be the area of ΔABC. Then $\alpha(x_0)=\dfrac{\left[1+(x_0)^2\right]^2}{4x_0}$. $D=$ domain of $\alpha(x_0)$: $0<x_0<\infty$. The aim is to minimize $\alpha(x_0)$ over D:

$$\frac{d\alpha}{dx_0}=\frac{\left(12(x_0)^2-4\right)\left(1+(x_0)^2\right)}{16(x_0)^2},\qquad 0<x_0<\infty;$$

$$\frac{d\alpha}{dx_0}=0\leftrightarrow x_0=\frac{1}{\sqrt{3}};$$

$$\lim_{x_0\to 0^+}\alpha(x_0)=+\infty,\qquad \lim_{x_0\to +\infty}\alpha(x_0)=+\infty.$$

Hence $x_0=\dfrac{1}{\sqrt{3}}\leftrightarrow$ minimum (Fig. II.18b): the desired point is

$$(x_0,y_0)=\left(\frac{1}{\sqrt{3}},\frac{2}{3}\right).$$

The minimum area is $\alpha\left(\dfrac{1}{\sqrt{3}}\right)=\boxed{\dfrac{4\sqrt{3}}{9}}$.

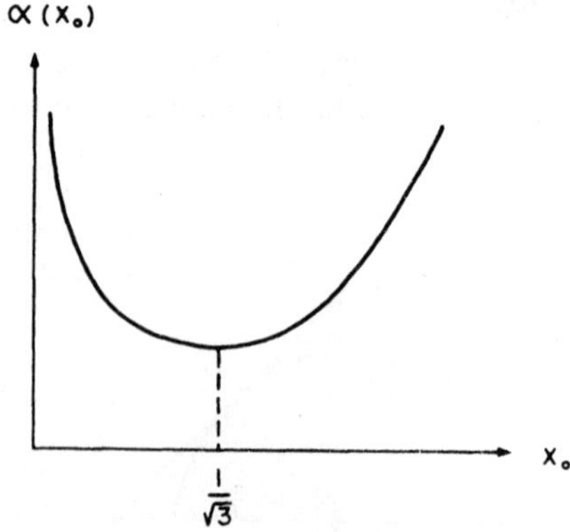

Figure II.18b

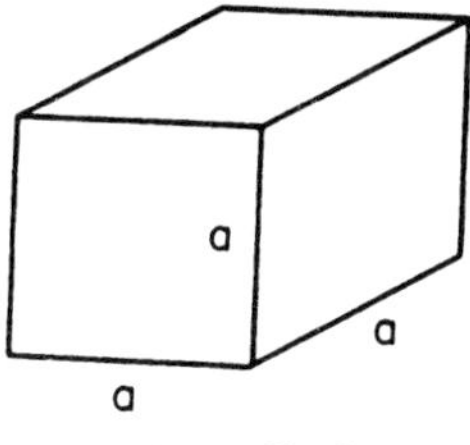

Figure II.19

19. The volume of a cube is increasing at a rate of 7 cubic inches per minute. How fast is the surface area increasing when the length of an edge is 12 inches?

Let a be the length of a side of the cube (Fig. II.19), V be the volume of the cube, S be the surface area of the cube. $S = 6a^2$, $V = a^3$, where a is an implicit function of t. Aim: to find $\dfrac{dS}{dt} = 12a\dfrac{da}{dt}$ when $a = 12$.

$$7 = \frac{dV}{dt} = 3a^2\frac{da}{dt} \Rightarrow \frac{da}{dt} = \frac{7}{3a^2} \qquad \text{and so} \qquad \frac{dS}{dt} = 12a\frac{da}{dt} = \frac{28}{a}.$$

Hence when $a = 12$, $\dfrac{dS}{dt} = \boxed{\dfrac{7}{3}\dfrac{(\text{inches})^2}{\text{minute}}}$.

20. A rope is fastened to the ground at a distance of eight feet from a vertical wall. A monkey climbs the wall holding onto the loose end of the rope. If the monkey holds the rope taut and climbs at the rate of four feet per second, at what rate is the length of the rope (between the monkey's hand and where it is fastened to the ground) increasing when the monkey is six feet up the wall?

At a given time t, the monkey is located at M, a distance x up the wall with y the length of the rope (Fig. II.20). Aim: to find $\dfrac{dy}{dt}$ when $x = 6$, given that $\dfrac{dx}{dt} = 4$. Since $y^2 = x^2 + 8^2 \Rightarrow y\dfrac{dy}{dt} = x\dfrac{dx}{dt}$. Hence $\dfrac{dy}{dt} = \dfrac{x}{y}\dfrac{dx}{dt}$. When $x = 6$, one has $y = 10$. Thus at the given instant $\boxed{\dfrac{dy}{dt} = \dfrac{12}{5}\text{ feet/second}}$.

21. A man on a dock is pulling in a boat at the rate of 50 feet/minute by means of a rope attached to the boat at water level. If the man's hands are 16 feet above the water level, how fast is the boat approaching the dock when the amount of rope out is 20 feet?

Let

$M \leftrightarrow$ location of man's hands,

$O \leftrightarrow$ location of dock (Fig. II.21).

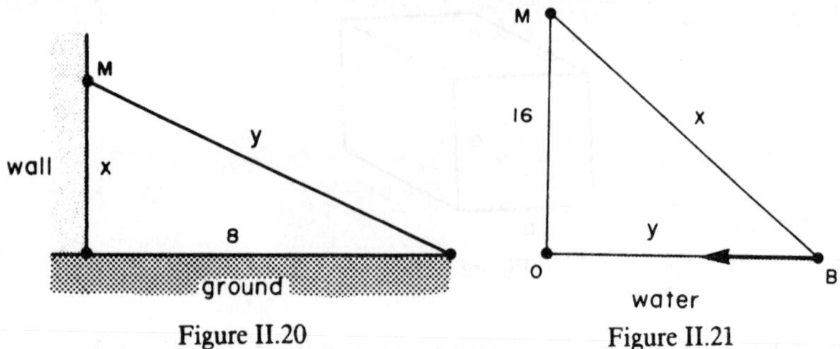

Figure II.20 Figure II.21

At time t, the boat is located at B, a distance y from the dock and x is the amount of rope which is out. Aim: to find $\frac{dy}{dt}$ when $x = 20$, given that $\frac{dx}{dt} = -50$. Since $x^2 = (16)^2 + y^2 \Rightarrow \frac{dy}{dt} = \frac{x}{y}\frac{dx}{dt}$. When $x = 20$, $y = 12 \Rightarrow \frac{dy}{dt} = -\frac{250}{3}$. Hence the boat is being *pulled in* at the rate of $\boxed{\frac{250}{3}\text{ ft/min}}$ when 20 feet of rope are out.

(*)22. Ship A is 15 miles east of O and moving west at 20 miles per hour; ship B is 60 miles south of O and moving north at 15 miles per hour.

(a) Are they approaching or separating after one hour and at what rate?
(b) When are they nearest one another?

Let $x(t) \leftrightarrow$ location of ship A at time t
$(x(t) > 0 \leftrightarrow$ ship A being east of $O)$.
Let $y(t) \leftrightarrow$ location of ship B at time t
$(y(t) < 0 \leftrightarrow$ ship B being south of $O)$.
Let $S(t) \leftrightarrow$ distance between the ships at time t (Fig. II.22a).

One is given that $\frac{dx}{dt} = -20$, $\frac{dy}{dt} = 15$ and that at $t = 0$: $x(0) = 15$, $y(0) = -60$.

(a) $S^2 = x^2 + y^2$. Hence

$$2S\frac{dS}{dt} = 2x\frac{dx}{dt} + 2y\frac{dy}{dt}.$$

The rate at which the ships are separating at time t is $\frac{dS}{dt} = \frac{x}{S}\frac{dx}{dt} + \frac{y}{S}\frac{dy}{dt}$.

$$\frac{dx}{dt} = -20 \Rightarrow x(t) = -20t + \text{const.},$$

$$x(0) = 15 \leftrightarrow \text{const.} = 15.$$

$$\frac{dy}{dt} = 15 \Rightarrow y(t) = 15t + \text{const.},$$

$$y(0) = -60 \Rightarrow \text{const.} = -60.$$

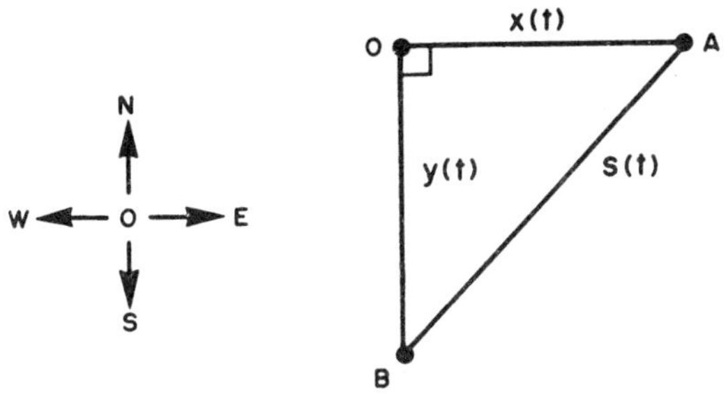

Figure II.22*a*

Hence

$$x(t) = -20t + 15,$$
$$y(t) = 15t - 60.$$

At $t = 1$: $x(1) = -5$, $y(1) = -45$; hence $S(1) = 5\sqrt{82}$. Thus at $t = 1$,

$$\frac{dS}{dt} = -\frac{1}{\sqrt{82}}(-20) + \frac{(-9)}{\sqrt{82}}(15)$$
$$= -\frac{115}{\sqrt{82}}.$$

Hence after one hour the ships are *approaching* each other at the rate of $\boxed{\frac{115}{\sqrt{82}} \text{ miles per hour}}$.

(b) The aim is to find the minimum value of $S(t)$ on its domain $0 \le t < \infty$.

$$\frac{dS}{dt} = 0 \leftrightarrow x\frac{dx}{dt} + y\frac{dy}{dt} = 0$$
$$\leftrightarrow (-20t + 15)(-20) + (15t - 60)(15) = 0$$
$$\leftrightarrow t = \frac{48}{25} \text{ hours.}$$

When $t = 0$, $\frac{dS}{dt} < 0$. As $t \to +\infty$, $\frac{dS}{dt} \to +\infty$. Hence $\boxed{t = \frac{48}{25} \text{ hours}}$ is the time when the ships are nearest one another (Fig. II.22*b*).

23. A ship sails east from a point O for 30 miles, and then turns north. If the ship sails at a constant speed of 10 miles per hour, how fast is its

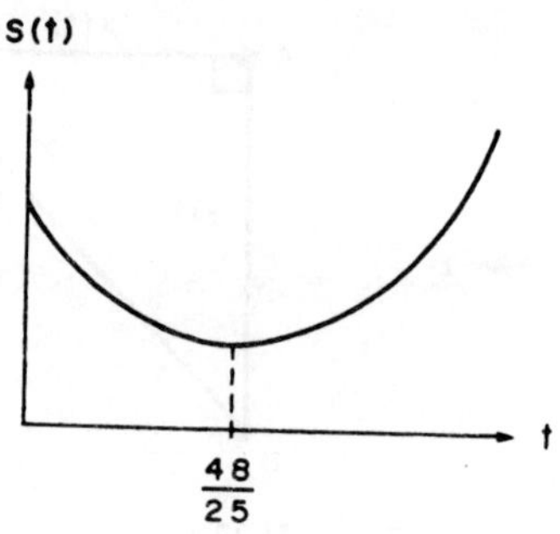

Figure II.22*b*

distance from the point O increasing

(a) 2 hours after leaving point O?

(b) 7 hours after leaving point O?

(a) Since the ship sails east from point O for 30 miles at a rate of 10 miles per hour, it sails east for the first three hours after leaving point O. Hence 2 hours after leaving point O the distance from point O continues to increase at a rate of **10 miles per hour**.

(b) For $t \geq 3$:

let $A \leftrightarrow$ position of ship,

let $S(t) \leftrightarrow$ distance from ship to point O,

let $y(t) \leftrightarrow$ distance ship has travelled north (Fig. II.23).

Aim: to find $\dfrac{dS}{dt}$ when $t = 7$ given that

$$\frac{dy}{dt} = \left\{\begin{matrix} 0, & 0 \leq t < 3 \\ 10, & 3 \leq t \leq 7 \end{matrix}\right\},$$

$$y(3) = 0.$$

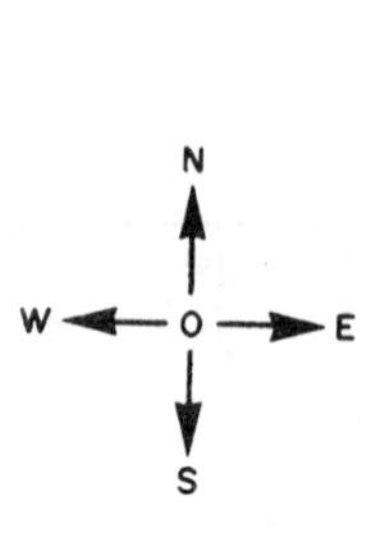

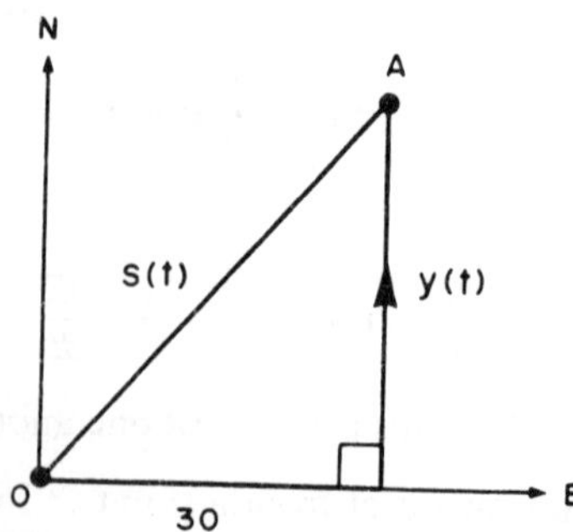

Figure II.23

Hence if $t \geq 3$, $y(t) = 10(t-3) \Rightarrow$

$$y(7) = 40.$$

$$S^2 = y^2 + (30)^2 \Rightarrow S(7) = 50.$$

Now

$$\frac{dS}{dt} = \frac{y}{S}\frac{dy}{dt}:$$

$$\text{at} \quad t=7, \ \frac{dS}{dt} = \frac{40}{50}(10) = \boxed{8 \text{ miles per hour}}.$$

24. At the moment a cyclist passes directly beneath a balloon, the balloon is 40 ft above the ground. The cyclist is travelling along a straight road at a constant speed of 20 ft/sec and the balloon is rising at a constant rate of 8 ft/sec. How fast is the distance between them changing one second after the balloon is directly above the cyclist?

Let $h(t) \leftrightarrow$ location of balloon B at time t,
$x(t) \leftrightarrow$ location of cyclist C at time t,
$S(t) \leftrightarrow$ distance between balloon and cyclist at time t (Fig. II.24).

At $t = 0$: $h(0) = 40$, $x(0) = 0$. For any t: $\frac{dx}{dt} = 20$, $\frac{dh}{dt} = 8$. Hence $x(t) = 20t$, $h(t) = 8t + 40$. Aim: to find $\frac{dS}{dt}$ when $t = 1$.

$$S^2 = x^2 + h^2$$

$$\Rightarrow \frac{dS}{dt} = \frac{x}{S}\frac{dx}{dt} + \frac{h}{S}\frac{dh}{dt}.$$

At $t = 1$: $x(1) = 20$, $h(1) = 48 \Rightarrow S(1) = 52$

$$\Rightarrow \frac{dS}{dt} = \left(\frac{20}{52}\right)(20) + \left(\frac{48}{52}\right)(8) = \boxed{\frac{196}{13} \text{ ft/sec}}.$$

25. A point P is moving along the part of the curve $x = y^2$ which is in the 1st quadrant in such a way that its x coordinate is increasing at 5 units/second. A rectangle with two sides along the coordinate axes has OP

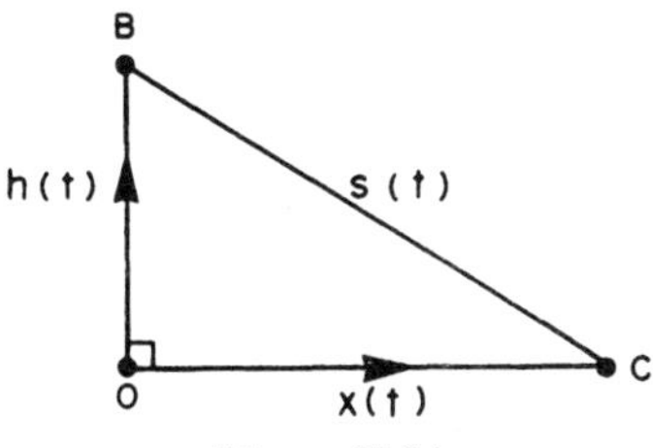

Figure II.24

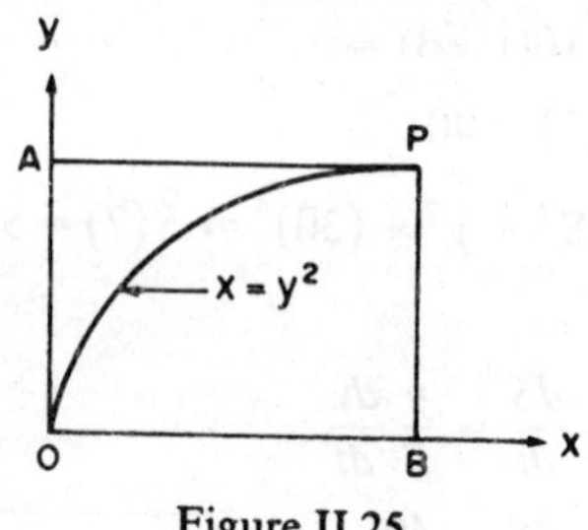

Figure II.25

as one diagonal and AB as the other (see Fig. II.25). At what rate is the area of the rectangle $OAPB$ changing when $x = 9$?

The rectangle determined by the point P at time t has horizontal dimension $x(t)$ and vertical dimension $y(t) = \sqrt{x}$. Thus the area of the rectangle at time t is

$$A(t) = xy = x^{3/2}.$$

One must find $\dfrac{dA}{dt}$ at $x = 9$. From the chain rule:

$$\frac{dA}{dt} = \frac{3}{2}x^{1/2}\frac{dx}{dt}.$$

One is given: $\dfrac{dx}{dt} = 5$. Hence when $x = 9$,

$$\frac{dA}{dt} = \frac{3}{2}(\sqrt{9})(5) = \boxed{\frac{45}{2}\,\frac{(\text{units})^2}{\text{second}}}.$$

26. Water is running into a conical reservoir 20 feet high with a base radius of 10 feet at the rate of 9 ft^3/min. How fast is the water level rising when it is 6 feet deep?

Let the radius of the circular surface of water be r when the depth of water is h (Fig. II.26). From similar triangles one obtains r in terms of h:

$$\frac{r}{h} = \frac{10}{20} \leftrightarrow r = \frac{1}{2}h.$$

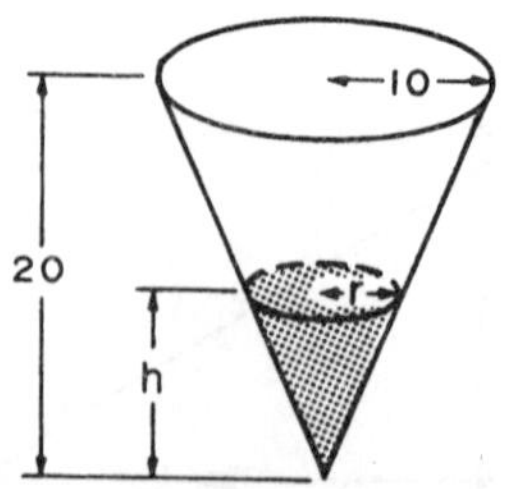

Figure II.26

The volume V of water at height h is

$$V = \frac{1}{3}\pi r^2 h = \frac{1}{3}\pi \frac{h^2}{4} h = \frac{\pi}{12} h^3.$$

Aim: to find $\dfrac{dh}{dt}$ when $h = 6$, given $\dfrac{dV}{dt} = 9$ ft³/min.

$$\frac{dV}{dt} = \frac{\pi}{4} h^2 \frac{dh}{dt} \leftrightarrow \frac{dh}{dt} = \frac{4}{\pi h^2} \frac{dV}{dt}.$$

Hence when $h = 6$ feet,

$$\frac{dh}{dt} = \left(\frac{4}{\pi (6 \text{ feet})^2}\right)(9 \text{ ft}^3/\text{min}) = \boxed{\frac{1}{\pi} \text{ ft/min}}.$$

27. The spring runoff from the melting of snow from a certain region is being accumulated in a reservoir. The reservoir is 450 metres long; and its cross-section is an isosceles triangle 40 metres wide at the top and $20\sqrt{3}$ metres high. Water is flowing into the reservoir at the rate of 200 cubic metres per hour and, simultaneously, water is leaking out from the reservoir at the rate of 50 cubic metres per hour. Find the rate at which the level of water is rising when the depth of water is $10\sqrt{3}$ metres.

Suppose the reservoir is filled to a depth h (Fig. II.27*a*). At that depth let the width of a cross-sectional slab of water be w at the surface (Fig. II.27*b*). One obtains w in terms of h from similar triangles:

$$\frac{w}{h} = \frac{40}{20\sqrt{3}} \leftrightarrow w = \frac{2}{\sqrt{3}} h.$$

Thus the area of a cross-sectional slab is

$$\frac{1}{2} wh = \frac{h^2}{\sqrt{3}}.$$

Hence the total volume at depth h is

$$V = (\text{cross-sectional area})\ (\text{length of reservoir})$$
$$= \left(\frac{h^2}{\sqrt{3}}\right)(450).$$

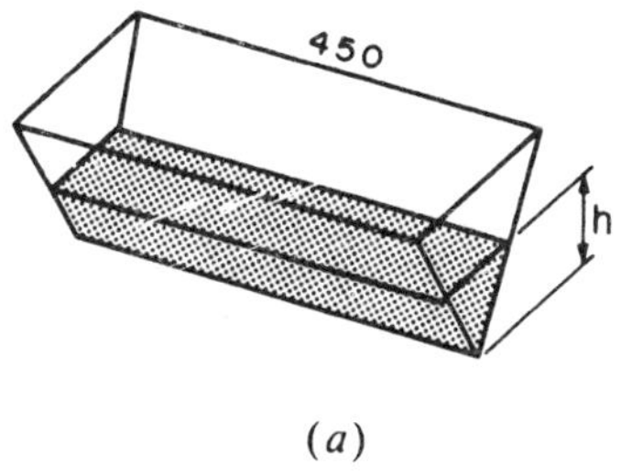

(*a*)

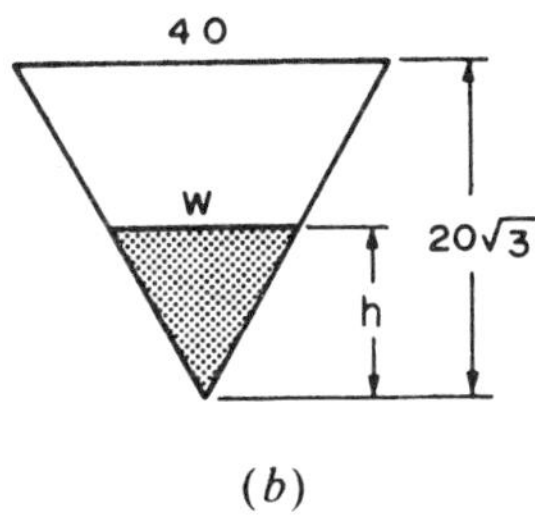

(*b*)

Figures II.27*a* and *b*.

One must find $\frac{dh}{dt}$ when $h = 10\sqrt{3}$ metres:

$$\frac{dV}{dt} = \frac{900h}{\sqrt{3}}\frac{dh}{dt} \leftrightarrow \frac{dh}{dt} = \frac{\sqrt{3}}{900h}\frac{dV}{dt}.$$

Observe that at any time, $\frac{dV}{dt} = (200-50)\ \text{m}^3/\text{hr} = 150\ \text{m}^3/\text{hr}$. Hence at the desired instant,

$$\frac{dh}{dt} = \frac{\sqrt{3}}{(900)(10\sqrt{3})}(150)\ \text{metres/hour} = \boxed{\frac{1}{60}\ \text{metres/hour}}.$$

(*)28. A swimming pool is 30 ft wide and 45 ft long. When full it would be 2 ft deep at the shallow end and 7 ft deep at the deep end. The bottom is an inclined plane.

Water is pumped into the (empty) pool at a rate of 27 cubic feet per minute. How long will it take for the depth at the deep end to reach 2 ft, and at what rate will the water level be rising then?

Consider a similar pool of more general dimensions: When full, its surface would have dimensions a by b; it would be c deep at the deep end, and $c - c^*$ deep at the shallow end. Therefore c^* is the critical depth at which the pool changes shape (Fig. II.28a). Suppose that at time t, the depth of water is h, and the surface width of the cross-sectional slab of water is w (Fig. II.28b). There are two cases to consider in the analysis of this problem:

Case I. $\boxed{h < c^*}$:

From similar triangles,

$$\frac{w}{h} = \frac{b}{c^*} \leftrightarrow w = \frac{b}{c^*}h,$$

and the area of a cross-section is $\frac{1}{2}wh = \frac{1}{2}\frac{b}{c^*}h^2$. Hence the total volume of

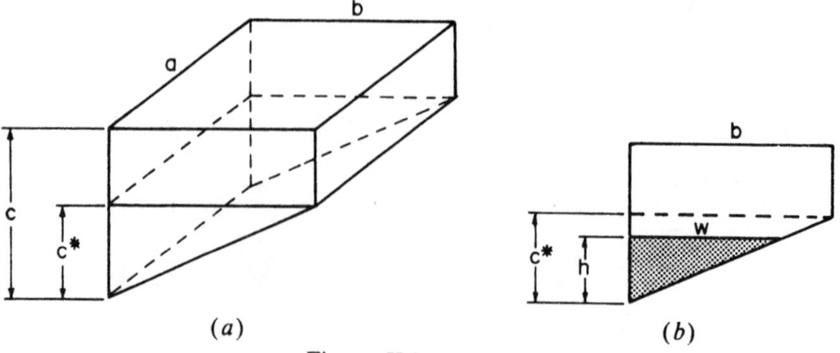

Figures II.28a and b.

water at depth h is

$$V = (\text{cross-sectional area})\,(\text{width of pool})$$
$$= \frac{ab}{2c^*}h^2.$$

Then $\dfrac{dV}{dt} = \dfrac{ab}{c^*}h\dfrac{dh}{dt} \leftrightarrow \dfrac{dh}{dt} = \dfrac{c^*}{ab}\dfrac{1}{h}\dfrac{dV}{dt}$.

Case II. $\boxed{h \geq c^*}$:

Here $w = b$.

When $h = c^*$, the volume is $\dfrac{ab}{2c^*}(c^*)^2 = \dfrac{1}{2}abc^*$. Furthermore, at this point one can ignore the fact that the pool previously had a different shape, as far as $\dfrac{dh}{dt}$ is concerned, since

$$V = \tfrac{1}{2}abc^* + (h - c^*)ab$$
$$\Rightarrow \frac{dV}{dt} = ab\frac{dh}{dt} \leftrightarrow \frac{dh}{dt} = \frac{1}{ab}\frac{dV}{dt}$$

which depends only on the dimensions relevant for $h \geq c^*$. Observe that for all h, $\dfrac{dh}{dt}$ is a continuous function of h.

For the given problem, $\dfrac{dV}{dt} = 27$, $c = 7$ and $c^* = 5$. Note that there is some question as to whether $a = 30$ and $b = 45$ or whether $a = 45$ and $b = 30$. However, the distinction is *not important* in any of the formulas for V, $\dfrac{dV}{dt}$, or $\dfrac{dh}{dt}$.

One need consider only *Case I* in answering the given questions since $h = 2 < c^*$. The time required for the depth to reach 2 feet is

$$t = \frac{\text{volume when } h = 2}{\text{rate of change of volume}} = \frac{V(h=2)}{dV/dt}$$

$$= \left.\frac{\dfrac{ab}{2c^*}h^2}{\dfrac{dV}{dt}}\right|_{h=2} = \frac{(30)(45)(2^2)}{(2)(5)(27)} = \boxed{20 \text{ minutes}}.$$

The rate of change of depth at this instant is

$$\frac{dh}{dt} = \frac{c^*}{ab}\frac{1}{h}\frac{dV}{dt} = \frac{5}{(30)(45)(2)}(27) = \boxed{\frac{1}{20}\text{ ft/min}}.$$

29. A pebble dropped into a still pool of water sends out concentric ripples. If the outer ripple spreads at a rate of 4 ft/sec, how fast is the area of this disturbance increasing after 8 seconds?

Suppose at time t the radius of the circle corresponding to the outer ripple is r. Then $r(t)=4t$, and the area of the circle at that time is

$$A=\pi r^2=16\pi t^2.$$

Then
$$\frac{dA}{dt}=32\pi t.$$

One wants the rate of change of the area at $t=8$. At this instant

$$\frac{dA}{dt}=(32\pi)(8)=\boxed{256\pi \text{ ft}^2/\text{sec}}.$$

30. A spherical iron ball 10 cm in radius is coated with a layer of ice of uniform thickness. If the ice melts at a rate of 50 cm³/min, how fast is the thickness of ice decreasing when the ice is 5 cm thick?

For the purposes of this problem, one might as well consider the ball to be completely made of ice, because the thickness of ice never approaches zero.

Therefore let the ball have radius r at time t; the volume of ice is

$$V=\tfrac{4}{3}\pi r^3$$

$$\Rightarrow \frac{dV}{dt}=4\pi r^2\frac{dr}{dt}.$$

One wants $\dfrac{dr}{dt}$ when $r=15$ cm, knowing that

$$\frac{dV}{dt}=-50 \text{ cm}^3/\text{min}.$$

$$\text{Hence } \frac{dr}{dt}=\frac{1}{4\pi r^2}\frac{dV}{dt}=\frac{1}{4\pi(15\text{ cm})^2}(-50)\text{ cm}^3/\text{min}$$

$$=\frac{-1}{18\pi}\text{ cm/min}.$$

At the given instant, the thickness of ice decreases at a rate of $\boxed{\dfrac{1}{18\pi}\text{ cm/min}}$.

31. A man is walking away from a street light at the rate of 5 ft/sec. If the man is 6 ft tall and the light is 15 ft high, how fast is the man's shadow lengthening when he is 10 ft from the base of the street light?

Let y be the length of the man's shadow at time t, x be the distance the man is from the base of the street light at time t (Fig. II.31). Aim: to find $\dfrac{dy}{dt}$ when $x=10$, given that $\dfrac{dx}{dt}=5$. The right-angled triangles in Fig. II.31 are similar, so

$$\frac{x+y}{15}=\frac{y}{6}.$$

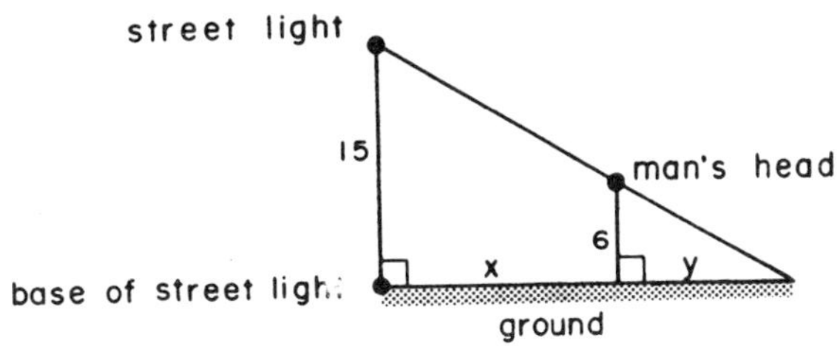

Figure II.31

Hence

$$y = \tfrac{2}{3}x$$

$$\Rightarrow \frac{dy}{dt} = \tfrac{2}{3}\frac{dx}{dt}.$$

Thus the man's shadow is lengthening at the rate of $\boxed{\tfrac{10}{3} \text{ ft/sec}}$ *no matter how far he is from the base of the street light.*

(*)**32.** A revolving beam three miles from a straight shore line makes 8 rpm. Find the velocity of the beam of light along the shore when it makes an angle of 45° with the shore line.

Let $L \leftrightarrow$ location of light source (revolving beam),
$B \leftrightarrow$ location of beam of light along the shore line at time t,
$x =$ position of B along the shore line at time t,
$\theta =$ angle of beam of light with the shore line at time t,
$\phi =$ angle of revolving beam with the perpendicular from L to the shore line at time t (Fig. II.32).

Aim: to find $\dfrac{dx}{dt}$ when $\theta = 45°$, given that $\dfrac{d\phi}{dt} = 8$ rpm. It is necessary to work with common angle measures:

$$1 \text{ rpm} = 2\pi \text{ radians/minute}$$

$$\Rightarrow \frac{d\phi}{dt} = 16\pi \text{ radians/minute;}$$

$$45° = \frac{\pi}{4} \text{ radians.}$$

$$x = 3\tan\phi \Rightarrow \frac{dx}{dt} = 3\sec^2\phi\frac{d\phi}{dt}.$$

$$\phi + \theta = \frac{\pi}{2} \Rightarrow \phi = \frac{\pi}{4} \text{ when } \theta = \frac{\pi}{4}.$$

Hence $\dfrac{dx}{dt} = 3(\sqrt{2})^2(16\pi) = \boxed{96\pi \text{ miles/minute}}$ is the velocity of the beam of light along the shore line when $\theta = 45°$.

33. A wine barrel has the shape of an ellipsoid of revolution, with the ends cut off. More specifically, it is formed geometrically by rotating the truncated half-ellipse (cf. Fig. II.33a) about the horizontal edge. Calculate the volume of the barrel.

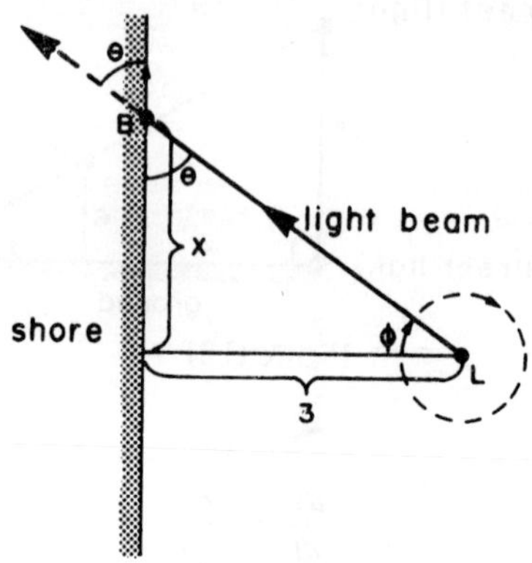

Figure II.32

Locate the ellipse in the xy-plane as shown in Fig. II.33b; then the horizontal semi-axis has length 4, and the vertical semi-axis has length 2. The standard form for the equation of the ellipse is

$$\frac{x^2}{4^2}+\frac{y^2}{2^2}=1 \leftrightarrow |y|=2\sqrt{1-\frac{x^2}{4^2}}\,.$$

If one revolves the graph of $y=2\sqrt{1-\frac{x^2}{4^2}}$, $-3<x<3$, about the x-axis, one obtains the desired wine barrel.

The cross-section of this barrel at a distance x along the x-axis is a circle of radius $|y|=2\sqrt{1-\frac{x^2}{4^2}}$. Therefore the cross-sectional area A at x is

$$A(x)=\pi|y|^2\equiv\pi y^2=2^2\pi\left[1-\frac{x^2}{4^2}\right].$$

Hence the desired volume is

$$V=\int_{-3}^{3}A(x)\,dx=\int_{-3}^{3}4\pi\left[1-\frac{x^2}{4^2}\right]dx=\boxed{\frac{39}{2}\pi}\,.$$

34. Using an integral, find the volume of a pyramid having a height of 5 inches and whose base is a square of side 4 inches.

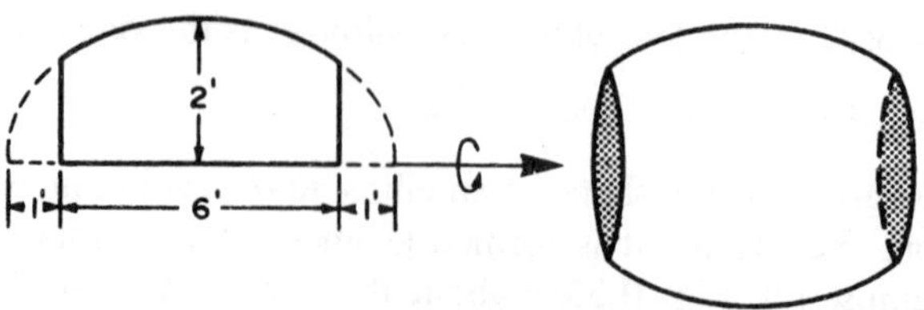

Figure II.33a

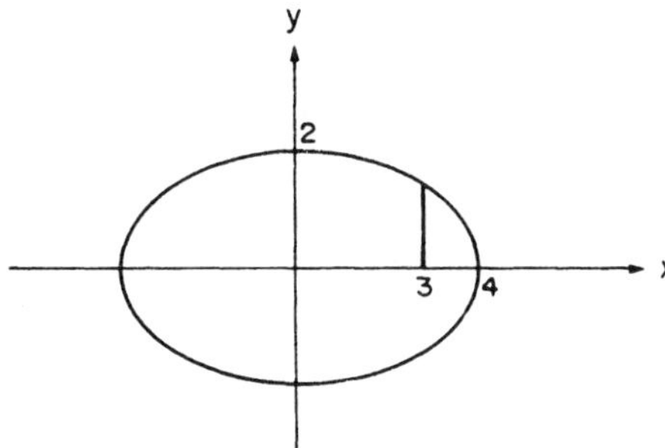

Figure II.33*b*

Locate the pyramid on the x-axis as shown in Fig. II.34*a*, i.e., let the x-axis correspond to the axis of symmetry. Consider the segment AB, extending from the apex A to the midpoint B of one of the sides of the base. Say the pyramid has height h and base side length a. Then the vertical distance from any point x on the x-axis $(0 \le x \le h)$ to the segment AB corresponds to half the side length of the *cross-sectional square* at x.

The side length $S(x)$ can be computed from similar triangles (Fig. II.34*b*):

$$\frac{S/2}{h-x} = \frac{a/2}{h} \leftrightarrow S(x) = a\left(1 - \frac{x}{h}\right).$$

Hence the cross-sectional area at x is

$$A(x) = S^2 = a^2\left(1 - \frac{x}{h}\right)^2$$

and the volume of the pyramid is

$$V = \int_0^h A(x)\,dx = \int_0^h a^2\left(1 - \frac{x}{h}\right)^2 dx = \frac{1}{3}a^2 h.$$

Here $a = 4$, $h = 5$; hence $V = \boxed{\tfrac{80}{3}\text{ in}^3}$.

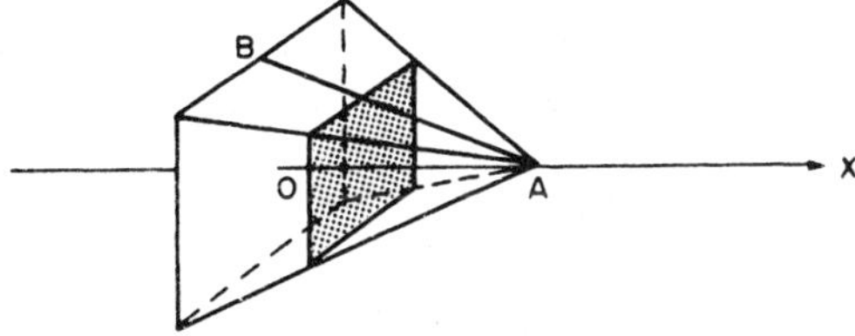

Figure II.34*a*

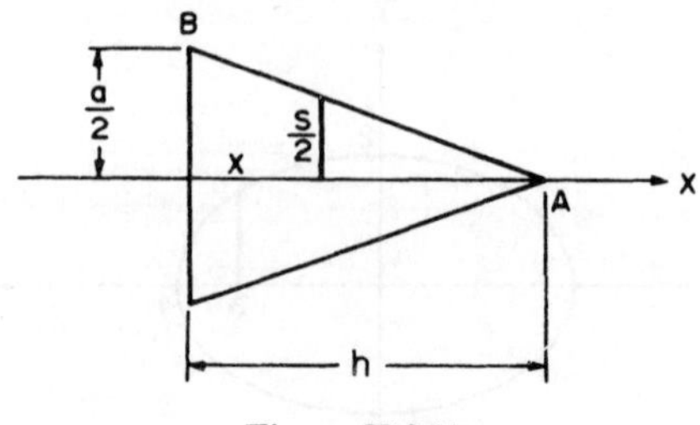

Figure II.34b

35. Find the volume of the solid defined as follows: the base is a quadrilateral with vertices at (3,0), (0,3), (−3,0) and (0, −3). Each cross-section perpendicular to the x-axis is a semi-circle.

Method 1

Since the base is symmetric with respect to the y-axis, one need only consider $x \geq 0$. At a point x along the x-axis, the cross-section perpendicular to the x-axis is a semi-circle, whose radius $r(x)$ is determined through similar triangles (Fig. II.35a):

$$\frac{r}{3-x} = \frac{3}{3} \leftrightarrow r(x) = 3 - x.$$

Thus the area of the cross-section at x is

$$A(x) = \frac{1}{2}\pi r^2 = \frac{\pi}{2}(3-x)^2$$

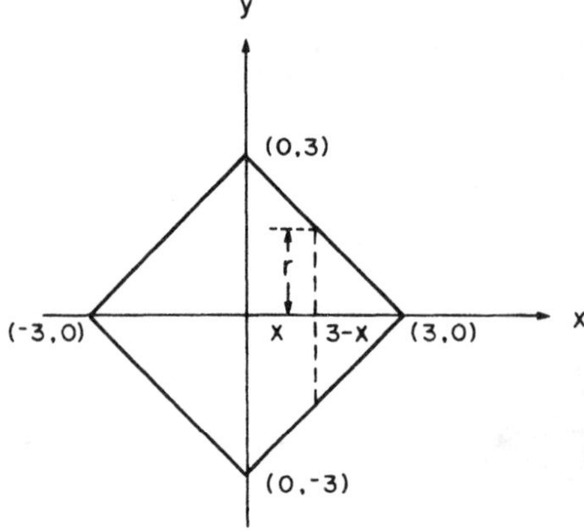

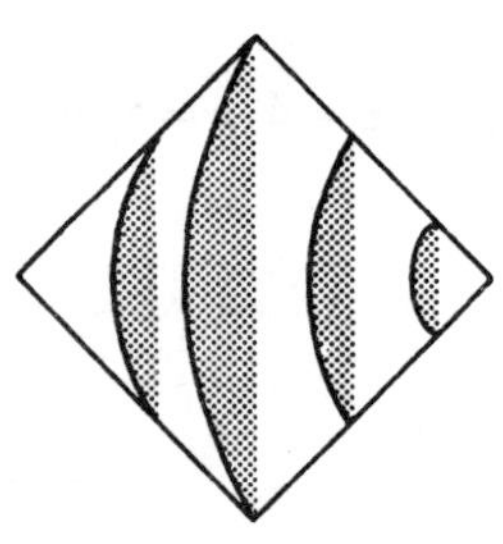

Figure II.35a

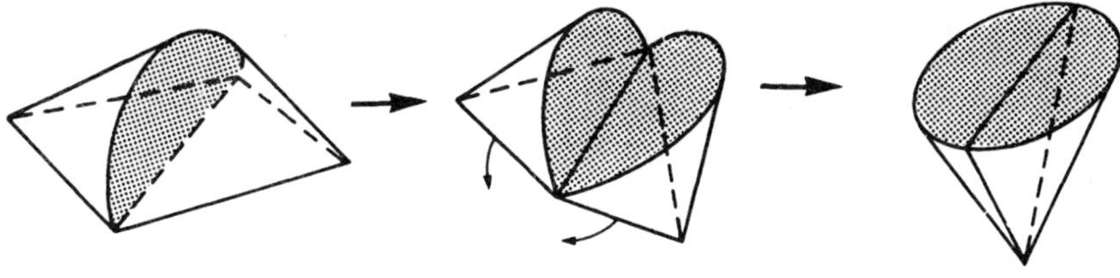

Figure II.35*b*

and the volume is twice that for $0 \le x \le 3$:

$$V = 2\int_0^3 A(x)\,dx = \pi\int_0^3 (3-x)^2\,dx = \boxed{9\pi}.$$

Method 2

The volume can be obtained much more easily. If one slices the solid in half with a plane perpendicular to the x-axis at $x = 0$, and folds opposite corners together, one obtains a cone with base radius 3 and height 3 (Fig. II.35*b*).

36. A solid has for its base the region of the xy-plane bounded by the curves $x = y$ and $x = y^2$. Find its volume if every section perpendicular to the x-axis is a semi-circle with diameter in the xy-plane.

The endpoints of the given region correspond to the points where the boundary curves intersect:

$$\left.\begin{array}{l} x = y \\ x = y^2 \end{array}\right\} \leftrightarrow y = y^2 \leftrightarrow y - y^2 = y(1-y) = 0 \leftrightarrow y = 0, 1.$$

When $y = 0, 1$, $x = 0, 1$, respectively; hence the given region exists only for $0 \le x \le 1$ and $0 \le y \le 1$. In this region the boundary curves can be written as $f_1(x) = x$ and $f_2(x) = \sqrt{x}$ (Figure II.36). The diameter D of the cross-section at x has a length equal to the distance $f_2(x) - f_1(x)$ between the

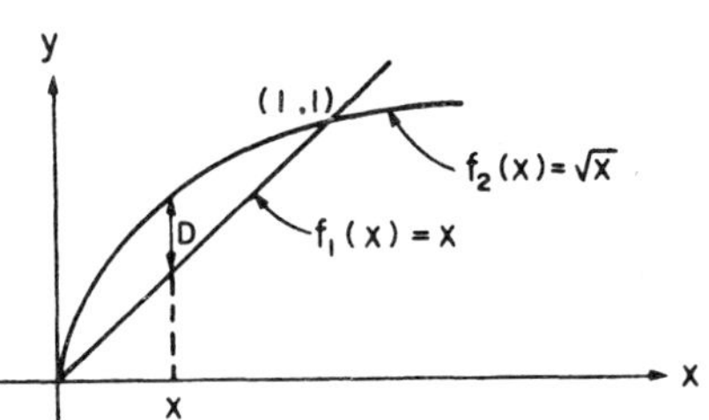

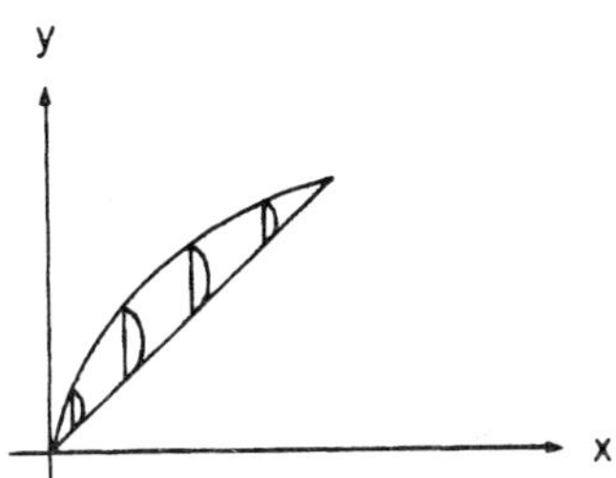

Figure II.36

boundary curves at x; hence the cross-sectional area at x is

$$A(x) = \frac{1}{2}\pi\left(\frac{D}{2}\right)^2 = \frac{\pi}{8}(\sqrt{x} - x)^2$$

and the volume of the solid is

$$V = \int_0^1 A(x)\,dx = \frac{\pi}{8}\int_0^1 (\sqrt{x} - x)^2\,dx = \boxed{\frac{\pi}{240}}.$$

37. Find the volume of the solid in Fig. II.37*a* whose base is the region bounded by the curves $y = x^2 - 1$ and $y = 0$ and whose cross-sections taken perpendicular to the x-axis are equilateral triangles.

Endpoints: $x^2 - 1 = 0 \leftrightarrow x = \pm 1$.

Hence the region is defined for $-1 \le x \le 1$. At x the distance between $f_1(x) = x^2 - 1$ and $f_2(x) = 0$ corresponds to the side length of the cross-sectional triangle. That is, the side length $L(x)$ at x is

$$L(x) = f_2(x) - f_1(x) = 1 - x^2.$$

Consider an equilateral triangle of side length L (Fig. II.37*b*). From trigonometry, the height h of the triangle is

$$h = \frac{L}{2}\tan 60° = \frac{\sqrt{3}}{2}L,$$

and hence the area of the triangle is

$$\frac{1}{2}Lh = \frac{\sqrt{3}}{4}L^2.$$

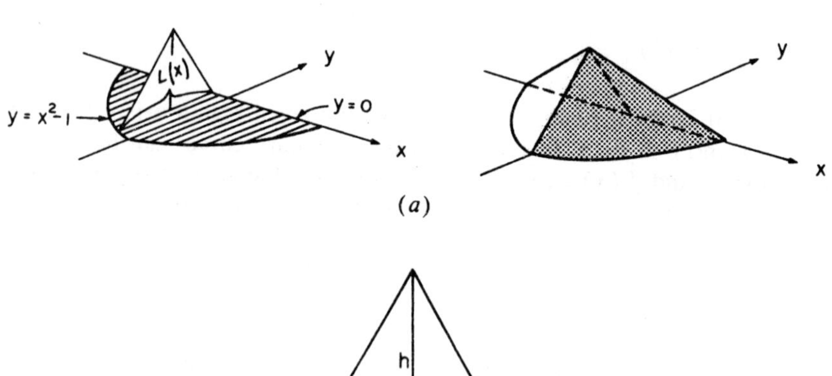

(*a*)

(*b*)

Figures II.37*a* and *b*.

Thus the area of the cross-section at x is

$$A(x) = \frac{\sqrt{3}}{4}L^2 = \frac{\sqrt{3}}{4}(1 - x^2)^2$$

and the volume of the solid is

$$V = \int_{-1}^{1} A(x)\,dx = \int_{-1}^{1} \frac{\sqrt{3}}{4}(1 - x^2)^2\,dx = \boxed{\frac{4\sqrt{3}}{15}}.$$

38. Compute the volume of the frustum of a cone as shown in Fig. II.38*a*.

Method 1

The frustum can be obtained by revolving the curve shown in Fig. II.38*b*, defined on the domain $0 \le x \le h$, about the x-axis. Using the two-point formula, the equation of the line is

$$y - R = \frac{(r - R)}{h}(x - 0)$$

$$\leftrightarrow y = \frac{r - R}{h}x + R.$$

At any point x along the x-axis, the cross-section is a circle of radius $y(x)$. Hence the cross-sectional area at x is

$$A(x) = \pi\left(\frac{r - R}{h}x + R\right)^2$$

and the volume of the solid is

$$V = \int_0^h \pi\left(\frac{r - R}{h}x + R\right)^2 dx.$$

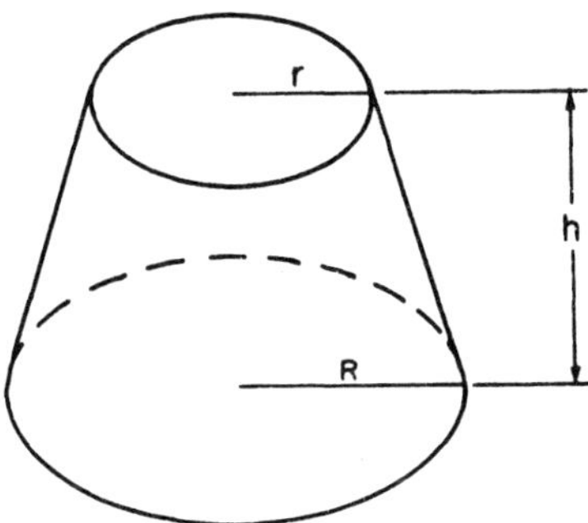

Figure II.38*a*

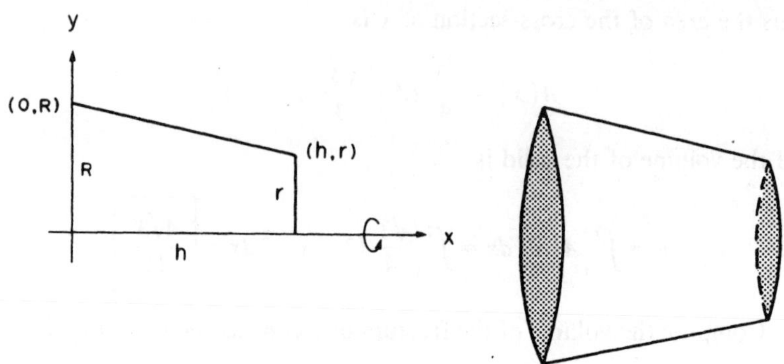

Figure II.38*b*

To evaluate this integral, use the substitution $u = \frac{r-R}{h}x + R$:

$$x = 0 \leftrightarrow u = R,$$
$$x = h \leftrightarrow u = r,$$
$$du = \left(\frac{r-R}{h}\right) dx,$$

$$\Rightarrow V = \frac{\pi h}{r-R}\int_R^r u^2\,du = \frac{\pi h}{3}\left(\frac{r^3 - R^3}{r-R}\right) = \boxed{\frac{\pi h}{3}(R^2 + Rr + r^2)}.$$

(Observe what happens when $r = 0$, or when $R = 0$.)

Method 2

This problem can be solved without calculus: consider the frustum as a difference of two cones (see Fig. II.38*c*) with dimension x to be determined. By similar triangles,

$$\frac{x}{r} = \frac{x+h}{R} \leftrightarrow x = \frac{rh}{R-r}.$$

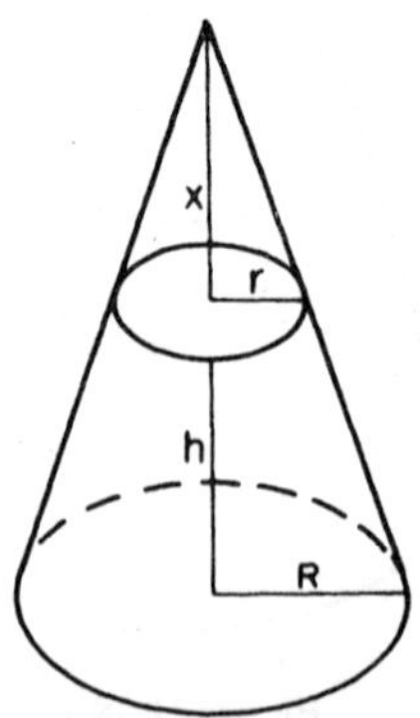

Figure II.38*c*

Hence the volume of the frustum is

$$\begin{aligned}V &= \tfrac{1}{3}\pi R^2(x+h) - \tfrac{1}{3}\pi r^2 x \\ &= \frac{\pi}{3}\left(R^2\left(\frac{rh}{R-r}\right) + R^2 h - \frac{r^3 h}{R-r}\right) \\ &= \frac{\pi h}{3}\left(R^2\left(\frac{r}{R-r}\right) + R^2 - \frac{r^3}{R-r}\right) \\ &= \frac{\pi h}{3}\left(\frac{R^3 - r^3}{R-r}\right) \\ &= \boxed{\frac{\pi h}{3}(R^2 + Rr + r^2)}.\end{aligned}$$

39. A semi-circular region of radius r is rotated about the line L which is the tangent line parallel to the diameter. Find the volume generated.

Locate the semi-circular region in the xy-plane as shown in Fig. II.39a. The equation of the semi-circle is

$$\begin{aligned} & x^2 + (y-r)^2 = r^2, \qquad y \le r \\ \leftrightarrow\ & y - r = \pm\sqrt{r^2 - x^2}, \qquad y \le r \\ \leftrightarrow\ & y = r - \sqrt{r^2 - x^2}. \end{aligned}$$

Hence the semi-circle is defined for $|x| \le r$.

The volume is obtained by revolving Fig. II.39a about the x-axis. A cut perpendicular to the x-axis reveals a *ring*, with outer radius r and inner radius $y(x)$. That is, the inner radius varies with x while the outer radius remains constant (cf. Fig. II.39b). Hence the cross-sectional area at x is

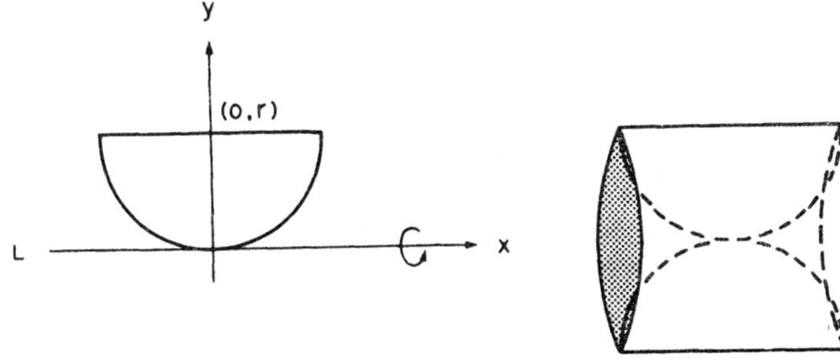

Figure II.39a

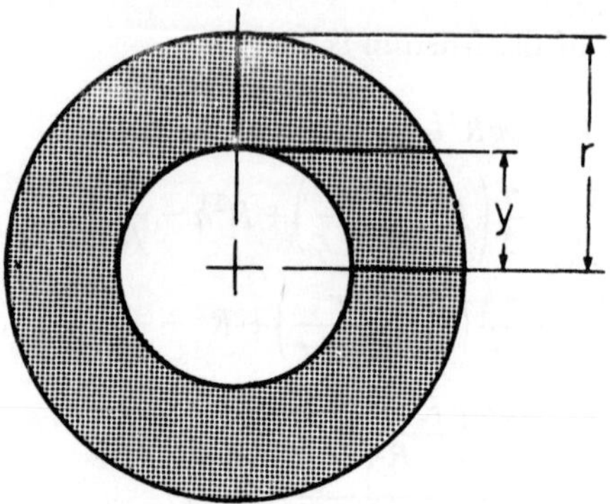

Figure II.39b

$$\begin{aligned} A(x) &= \pi r^2 - \pi y^2 \\ &= \pi(r-y)(r+y) \\ &= \pi\left[\sqrt{r^2-x^2}\right]\left[2r-\sqrt{r^2-x^2}\right] \\ &= \pi\left[2r\sqrt{r^2-x^2}-(r^2-x^2)\right] \end{aligned}$$

and the volume is

$$\begin{aligned} V &= \int_{-r}^{r} \pi\left[2r\sqrt{r^2-x^2}-(r^2-x^2)\right] dx \\ &= 2\pi\left[2r\int_0^r \sqrt{r^2-x^2}\,dx - \int_0^r (r^2-x^2)\,dx\right] \\ &= 2\pi\left[2r\left(\frac{\pi r^2}{4}\right)-\left(r^2x-\frac{x^3}{3}\right)\Bigg|_{x=0}^{x=r}\right] \\ &= 2\pi\left[\frac{\pi r^3}{2}-\frac{2r^3}{3}\right] \\ &= \boxed{\pi r^3\left(\pi-\tfrac{4}{3}\right)}. \end{aligned}$$

(Observe that $\int_0^r \sqrt{r^2-x^2}\,dx$ is the area of a quarter-circle with radius r.)

40. A hemispherical bowl of radius 12 inches is filled to a depth of 3 inches with water. Find the volume of water it contains.

Consider a circle of radius 12 centred at the origin. The desired volume can be obtained by revolving part of that circle ($9 \le x \le 12$) about the x-axis (Fig. II.40). The cross-section at x is a circle of radius $y(x)$; hence the area

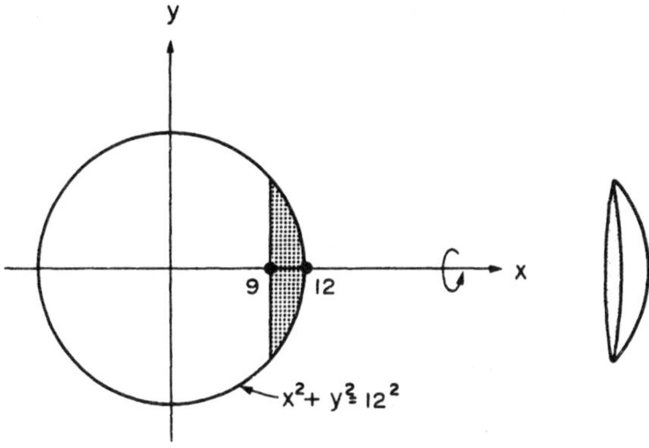

Figure II.40

at x is

$$A(x) = \pi y^2 = \pi\left((12)^2 - x^2\right)$$

and the volume is

$$V = \int_9^{12} A(x)\,dx = \pi\int_9^{12}\left((12)^2 - x^2\right)dx = \boxed{99\pi}.$$

Supplementary Problems

1. A poster of total area 500 sq. inches is to have a margin of 6 inches at the top and 4 inches at each side and the bottom. What dimensions yield the largest printed area?

2. An editor wants each page for his new mathematics book to contain 360 square centimetres of print. The margins at the top and bottom of a page must each be 5 cm, and at the sides must each be 2 cm. What is the smallest area of paper which can be used for a page?

3. An open box having a square base is to be constructed from 108 square centimetres of material. What should the dimensions of the box be to obtain the maximum volume?

4. Equal squares are cut off at each corner of a rectangular piece of cardboard 24 cm wide by 45 cm long and an open-topped box is formed by turning up the sides. Find the length x of the sides of the squares that must be cut off to produce a box of maximum volume.

5. A rectangular piece of cardboard, 8 cm by 15 cm, is to be made into an open box by cutting 4 equal squares out of the corners of the rectangle and folding up the sides.

(a) Translate the above words into a clear diagram. Label all knowns and unknowns.

(b) Find the dimensions of the square cut out if the volume of the box is to be maximized.

6. A fish tank is to be made from plate glass. The base is to be a square and the volume 4 ft^3. What is the minimum amount of glass which can be used to build such a tank?

7. A cylindrical tin can is to be made to contain a volume V of tomato juice. Find the dimensions so that the least amount of tin shall be used.

8. A manufacturer wishes to construct cylindrical metal cans with a fixed volume of 1000 cubic centimetres. There is no waste involved in cutting the material for the curved surfaces of the cans. However, in order to make the circular end pieces for the cans, the manufacturer cuts them from appropriately-sized squares, leaving four waste pieces for each end piece. Find the height and the radius of the can which uses the least amount of metal, including all waste materials.

9. A window is in the shape of a rectangle surmounted by a semi-circle. Find the dimensions of the window of maximum area whose perimeter is 12 feet.

10. A sports field consists of a rectangular region with semi-circular regions adjoined at two opposing sides. If the perimeter is to be 1000 feet, find the area of the largest possible field.

11. Find the maximum area of a rectangle enclosed in a semi-circle of radius 10, with one side of the rectangle along the diameter.

12. Find the area of the largest rectangle with sides parallel to the coordinate axes that can be enclosed in the semi-circular region bounded by $y = (1 - x^2)^{1/2}$ and $y = 0$.

13. A window has the shape of a rectangle surmounted by an equilateral triangle. If the perimeter of the window is L feet, find the dimensions which will allow the maximum amount of light to enter.

14. Let ABC be a triangle inscribed in a semi-circle as shown in Fig. S.14. Let the length of the side AC be 10 units. What is the maximum possible perimeter of the triangle ABC? Your answer must be justified using differential calculus.

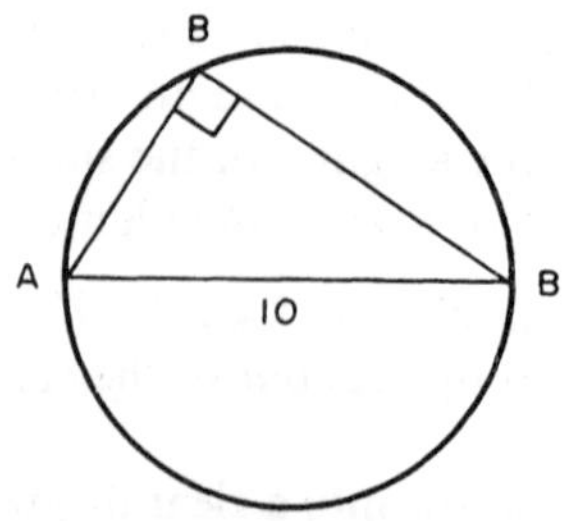

Figure S.14

15. An isosceles triangle has a base 10″ and an altitude 20″. Rectangles are inscribed with one side on the base of the triangle. What are the dimensions of the rectangle with the largest area?

16. Find the maximum volume of a cone if the sum of the base radius and slant height is equal to a constant a.

17. Find the rectangle with the largest possible area that can be inscribed in the ellipse $\frac{x^2}{16}+\frac{y^2}{9}=1$.

(*)18. A conical paper cup is to be made from some sector of a circle of given radius R, as shown in Fig. S.18. Find the volume of the largest such cup.

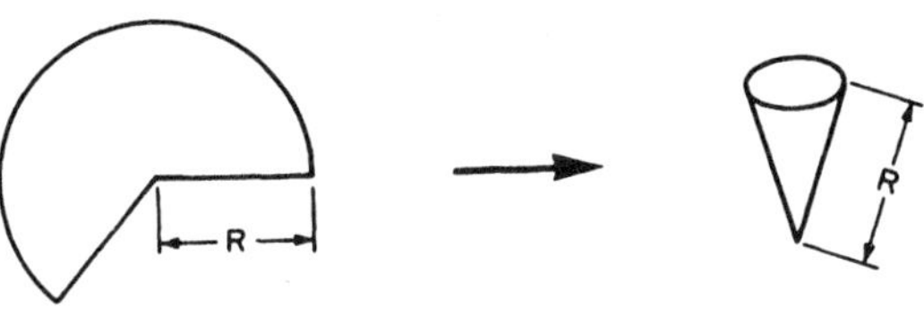

Figure S.18

19. A rectangle has the length of its diagonal fixed at L. What should be the lengths of its sides x, y in order to make the volume of a right circular cylinder obtained by rotating the rectangle about an edge of length y as large as possible?

(*)20. A crucifix has the shape of a symmetrical cross. Find the maximum area of a crucifix which can be cut from a circular disc of metal of radius 1 (Fig. S.20).

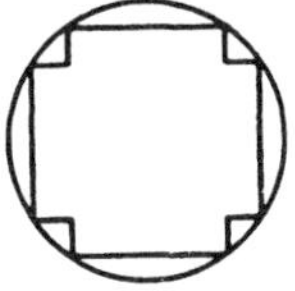

Figure S.20

(*)21. A silo of fixed capacity V is to be built in the form of a right circular cylinder surmounted by a hemisphere. The same building material is used for the floor, walls, and top. Using r and h as defined in Fig. S.21:

(a) Express the total surface area S as a function of r.

(b) Find the value of r that minimizes S.

(c) Prove that your answer for r yields the absolute minimum.

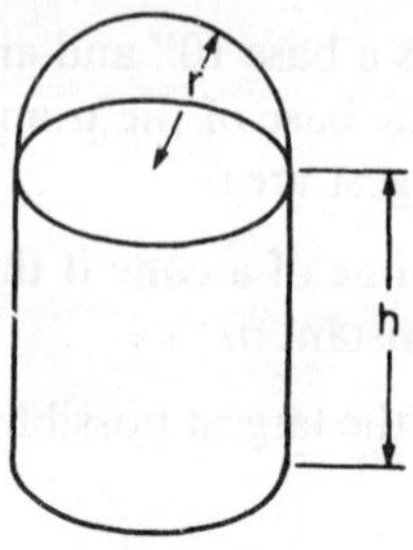

Figure S.21

22. The strength of a wooden beam of rectangular cross-section is proportional to the width of the beam and to the cube of its depth. Find the dimensions of the strongest beam which can be cut from a circular log of radius 0.5 metres.

23. A farmer wishes to build an E-shaped fence along a straight river bank so as to create two identical rectangular pastures. With 300 feet of fencing material, what is the largest area he can enclose?

24. It is desired to fence a rectangular area, and to divide it into three equal parts by two fences parallel to one of its sides (Fig. S.24). If the total area to be enclosed is 900 square feet, what should be the dimensions so as to make the cost of fencing a minimum?

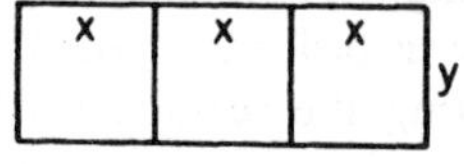

Figure S.24

25. A farmer has a long strip of land 100 metres wide along a straight river bank. He has 500 metres of fence with which to build two equal sized pens, one for horses and one for cows. The horses must not reach the river or they will swim away while the cows must be able to get down to the river to drink. Therefore the pens must be built to the design shown in Fig. S.25.

(a) How can the farmer use his fence to provide the most room for his animals on his land? (i.e. what are the outside dimensions of the structure enclosing the most area?)

(b) How much area can he enclose?

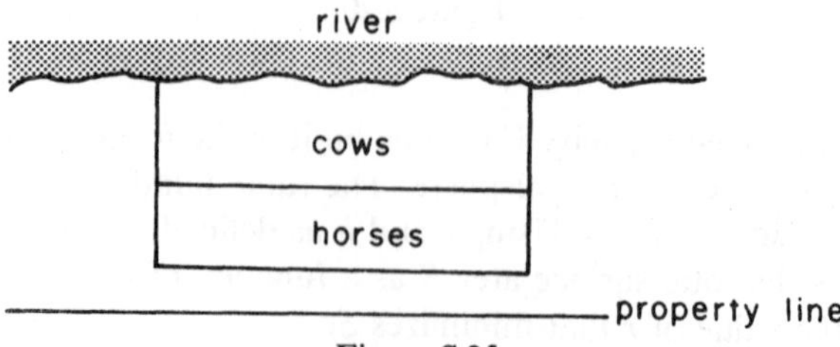

Figure S.25

26. Two poles 5 metres apart are 3 and 2 metres high respectively. A wire fastened to the top of each pole is held to the ground at a point between the poles and is tightened so that there is no sag (Fig. S.26). How far from the tallest pole will the wire touch the ground if the length of the wire is a minimum?

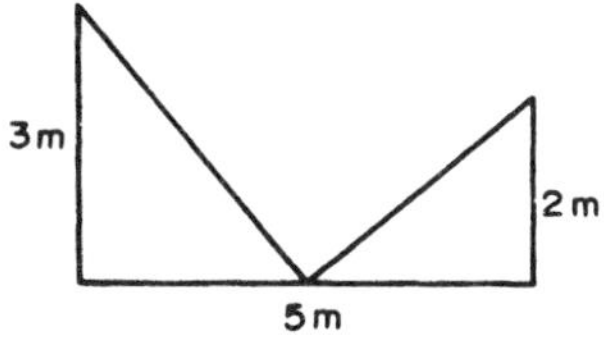

Figure S.26

27. Somebody wants to go from a point A to a point B in such a way that he first goes to a point P on the shoreline of a straight channel (presumably to fetch water) and then from P to B. Show that in order to walk the shortest distance the angle at P between the line AP and the shoreline is equal to the angle at P between the line BP and the shoreline.

28. A metal rain gutter is to have four-inch sides and a four-inch horizontal bottom, the sides making equal angles with the bottom. How wide should the opening across the top be for a maximum carrying capacity? Make *careful* arguments to show that your answer is in fact maximum.

29. A rectangular box has sides of length x, x, and y. The post office requires that the length y plus the girth (which equals $4x$) be 72 inches. What are the dimensions of the box of maximum volume that the post office will accept?

30. A window is to be constructed in the form of a rectangle surmounted by a semi-circle. The rectangle will be of clear glass and the semi-circle of coloured glass that transmits only half as much light per square foot. The total height of the window is to be 6 ft. Find the width of the window if the total quantity of light admitted is to be a maximum. How do you know that your answer yields a maximum and not a minimum quantity?

31. A clear rectangle of glass is put in a coloured semi-circular glass window. The radius of the window is 3 ft and the clear glass transmits twice as much light per square foot as the coloured glass. Find the dimensions of the rectangular glass so that the window transmits the most light.

32. A man on a boat 3 miles from the nearest point A on a straight shore wishes to reach, in minimum time, a point B on the shore 6 miles from A. How far from A should he land if he can row 4 miles per hour and walk 5 miles per hour?

33. A small island is three miles from the nearest point P on the straight shoreline of a large lake. If a man can row his boat 2.5 miles per hour and can walk 4 miles per hour, where should he land his boat in order to arrive in the shortest time at a town 12 miles down the shore from P?

34. A motorist finds herself in a desert 5 miles from a point A, which is the point on a long straight road nearest to her. She wishes to get to a point B on the road 10 miles from A. If she can travel at 15 miles per hour on the desert and 39 miles per hour on the road, find the point at which she must meet the road to get to B in the shortest possible time.

35. I am on a train travelling on a straight track at a speed of v ft/sec. Ahead, at a (perpendicular) distance D feet from the track is a maiden in distress. I wish to reach her as quickly as possible. If I can run at a speed of S feet/second, at what point should I leap off the train? You may assume $v > S$. A picture will be of assistance both in solving the problem, and in communicating your answer.

(*)36. At what point along the x-axis does the line segment joining $(0,1)$ to $(1,2)$ subtend the maximum angle? Verify that you have found the absolute maximum.

(*)37. Two posts are located on one side of a straight road at distances of 40 metres and 160 metres, respectively, from the road. The line joining the posts is perpendicular to the road. Find the position on the road for which the angle between the lines of sight to the two posts is a maximum.

(*)38. A tablet 7 feet high is placed on a wall with its base 9 feet above the level of an observer's eye. How far from the wall should an observer stand in order that the angle subtended at his (her) eye by the tablet be a maximum?

(*)39. A vacant plot of land is situated at the corner of two streets which intersect at right angles. A tree stands 64 feet from one street and 27 feet from the other. What is the shortest possible length of a straight line path cutting across the corner of the plot and passing by the tree (see Fig. S.39)?

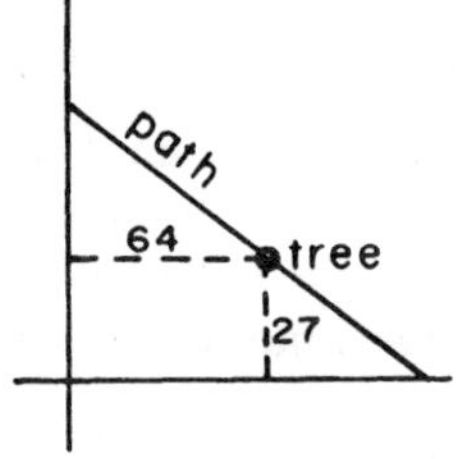

Figure S.39

(*)40. How would you fold a long rectangular piece of paper 8 inches wide to bring the upper left corner to the right-hand edge and to minimize the length of the fold—i.e., choose x to minimize y (Fig. S.40)?

Figure S.40

41. Find the coordinates of the point P on the line $y = 3x + 5$ which minimizes the distance $|\overline{OP}|$ from the origin to the point P and show that $\overrightarrow{OP}$ is perpendicular to the given line.

42. An isosceles triangle is drawn with a vertex at the origin, its base parallel to and above the x-axis, and the vertices of its base on the curve $12y = 36 - x^2$. Determine the area of the largest such triangle.

43. Find the minimum distance from the point (2,0) to the curve $y^2 = x^2 + 1$.

44. Show that if a rectangle has its base on the x-axis and two of its vertices on the curve $y = e^{-x^2}$, then the rectangle will have the largest possible area when the two vertices are at the points of inflection of the curve.

45. The area of a circle is increasing at the rate of 4 cm^2 each second. At what rate is the radius increasing at the exact moment when the area is 10 cm^2?

46. The area of a circle is increasing at the rate of 10 cm^2 per second. How fast is the radius changing when the area is $\frac{1}{\pi}$ cm^2?

47. What is the rate of change of the area of a circle with respect to its radius? Its diameter? Its circumference?

48. A hot-air balloon travelling horizontally at a constant rate of 100 ft per minute passes directly over an observer on the ground 300 ft below. How fast is the angle of elevation of the balloon changing when the balloon is 500 ft from the observer?

49. At a certain instant a balloon is released (from ground level) at a point 150 m away from an observer (at ground level). If the balloon rises vertically at a rate of 4 m/s, how rapidly will it be receding from the observer 50 seconds later?

50. A girl flies a kite at a height of 300 ft, the wind carrying the kite horizontally away from her at a rate of 25 ft/sec. How fast must she let out the string when the kite is 500 ft away from her?

51. A girl flying a kite lets out the string at the rate of 0.3 metres per second as the kite moves horizontally at an altitude of 40 metres. Find the rate at which the kite is moving when 50 metres of string have been let out.

52. Two mules start from point A at the same time. One travels west at 6 mph and the other travels north at 3 mph. How fast is the distance between them increasing 3 hours later?

53. A rope extends from a boat to a point on a wharf 25 ft above the water level. If the rope is pulled in at the constant rate of 3 ft per second, how fast is the angle between the rope and the water changing when 40 ft of rope are out?

54. Ship A is 30 nautical miles west of point P and is headed toward ship B at 11 knots (1 knot = 1 nautical mile per hour). Ship B is 20 nautical miles north of P and headed toward P at 16 knots. At this instant, how fast is the distance between the two ships decreasing?

55. At 8:00 a.m. a ship is sailing due north at the constant rate of 10 mph. A second ship, 22 miles due north of the first ship at 8:00 a.m., sails due east at 16 mph. What is the rate of change of the distance between the two ships at 9:00 a.m.?

56. A person is standing $\frac{1}{4}$ mile from a railroad track as a train goes by at 80 miles/hour. When the train is 2 miles past the point on the track nearest the person, at what speed is it moving away from him?

57. An automobile travelling at a rate of 60 ft/sec is approaching an intersection. When the automobile is 120 ft from the intersection, a truck travelling at the rate of 40 ft/sec crosses the intersection. The automobile and the truck are on roads that are at right angles to each other. How fast are the automobile and the truck separating 2 sec after the truck leaves the intersection?

58. Two cars, one going due east at 55 ft/sec and the other going due south at 40 ft/sec, are travelling toward the intersection of the two roads. At what rate are the two cars approaching each other when the first car is 400 ft and the second car is 300 ft from the intersection?

59. An airplane is flying east at 4 miles per minute. At the same altitude and 78 miles directly ahead of it is a second airplane flying north at 6 miles per minute.

(a) After 5 minutes is the distance between the airplanes increasing or decreasing and at what rate?

(b) Find when the two airplanes are closest together and their distance apart at that time.

60. A baseball diamond is a square 90 ft on each side (Fig. S.60). A player is running from first to second base at the rate of 30 ft/sec. At what rate is his distance from home plate changing when he is $\frac{2}{3}$ of the way to second base?

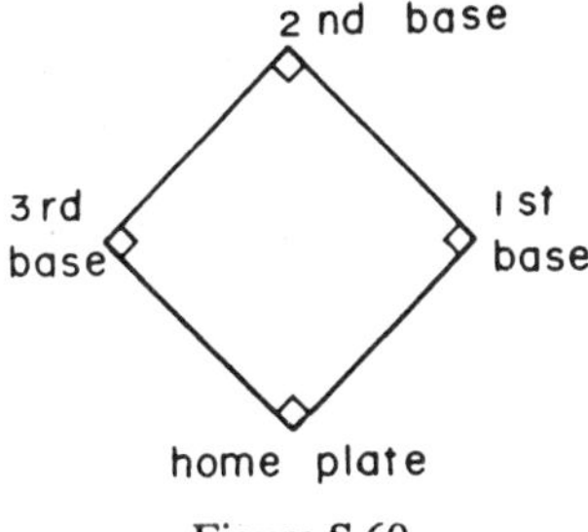

Figure S.60

61. The base of a right-angled triangle is 8 cm. If the height of the triangle is increasing at the rate of 5 cm/s, how fast is its hypotenuse changing when the height is 6 cm?

62. A point P is moving along the part of the curve $x = y^2$ which is in the first quadrant in such a way that its x-coordinate is increasing at 5 units/sec (Fig. S.62). At what rate is the distance P from the origin O changing when $x = 9$?

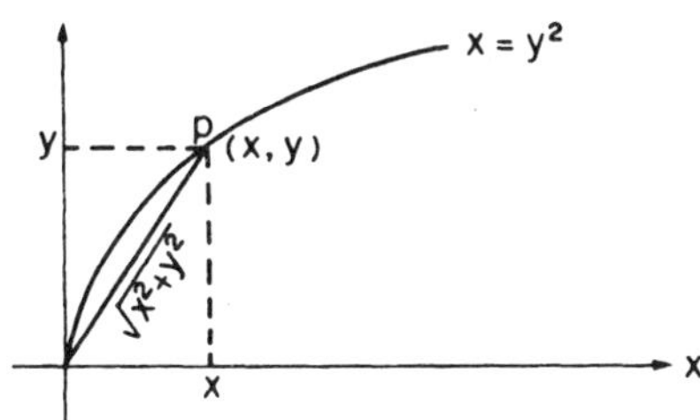

Figure S.62

63. A stone dropped into a pool of water creates a ripple which has the shape of an expanding circle. If the radius of this circle increases at the rate of 3 feet per second, find how fast the area of the circle is increasing

(a) at the instant when the radius is 5 feet;

(b) at t sec after the stone hits the water.

(*)64. Let A and B be two circles having the same centre and let the radius of A be the square of the radius of B. If the radius of B is increasing at 2 cm/s, how fast is the area between the two circles increasing when the radius of B is 10 cm?

65. Air is being pumped into a spherical balloon at the rate of 10 ft^3/min. At what rate is the balloon's radius increasing when the balloon has a volume of 36π ft^3?

66. Air is pumped into a spherical balloon so that its volume increases by 4 in^3/sec. How fast is the diameter expanding after 8 seconds?

67. The volume of a spherical soap bubble is increasing at the rate of 10 cubic centimetres per second. How fast is the surface area of the bubble increasing at the instant when its radius is 6 centimetres?

68. A pipe 10 feet long with an outer radius of 6 inches is covered with a shell of ice. (Exclude the ends of the pipe.) If the ice is melting at the rate of 50 in^3/min, at what rate is the thickness of the ice decreasing when it is 4 inches thick?

69. A spherical snowball is melting at the rate of $\frac{1}{2}$ in^3/hr. Assuming that the snowball maintains a spherical shape, how fast is the diameter decreasing when the snowball is 4 inches in diameter?

70. Water is pumped into a cylindrical tank of radius 10 feet at a rate of 50 ft^3 per minute. How fast does the water level in the tank rise?

(*)71. Water is pumped into a cylindrical tank of radius 10 feet and length L lying on its length at a rate of 50 ft^3/min. How fast does the water level in the tank rise?

72. Liquid is being poured into a parabolic bowl at a constant rate of 60π cm^3/s. The volume of the liquid in the bowl is given by (see Fig. S.72) $V = \frac{\pi x^4}{2}$ where the equation of the parabola is $y = x^2$. Find the rate of increase of the height of the liquid in the bowl (y in the diagram), when the height is 10 cm.

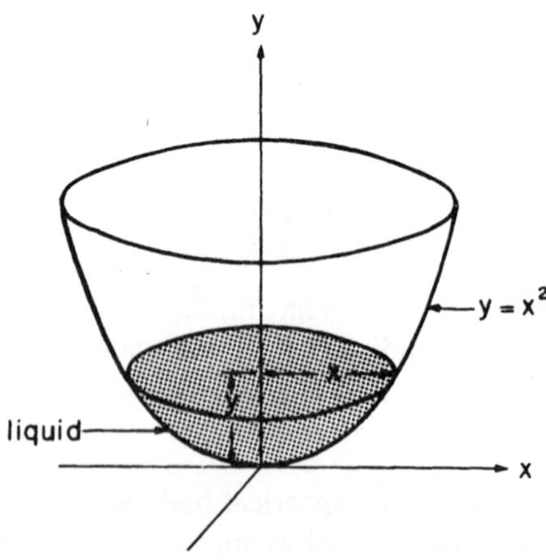

Figure S.72

73. A conical tank with vertical axis of symmetry open at the top and with vertex at the bottom has base radius 4 ft and altitude 10 ft. If water is flowing into the tank at the rate of 9 ft^3/min, how fast is the water level rising when the water is 5 ft deep?

74. A rectangular swimming pool is 40 ft long, 20 ft wide and 8 ft deep at the deep end and 3 ft deep at the shallow end. If the pool is filled by pumping water into it at the rate of 40 ft^3/min, how fast is the water level rising

(a) when it is 3 ft deep at the deep end,
(b) when it is 6 ft deep at the deep end?

75. In Prince George, sawdust is being dumped from a pipe at a rate of 100 ft^3/min. The sawdust forms a cone whose height is always equal to the diameter of its base. Find the rate at which the height of the cone increases when the height is 20 feet.

76. Sand is being poured onto the ground at the rate of 6 m^3/min. It forms a pile in the form of a cone whose diameter is always three times the altitude.

(a) Find the rate of increase of the radius of the base when the pile is 2 metres high.
(b) How rapidly is the area of the base increasing at the same instant?

77. A light is 20 ft from a wall and 10 ft above the centre of a path which is perpendicular to the wall. A man 6 ft tall is walking on the path towards the wall at the rate of 2 ft/sec. When he is 4 ft from the wall, how fast is the shadow of his head moving up the wall?

78. The shadow cast by a man standing 3 ft from a lamp post is 4 ft long. If the man is 6 ft tall and walks away from the lamp post at a speed of 400 ft/min, at what rate will his shadow be lengthening:

(a) a quarter of a minute later,
(b) when he is 20 ft from the lamp post?

79. A man is walking away from a wall at a rate of 1 metre per second. There is a window in the wall at a height of 40 metres above ground. When the man is 30 metres away from the building, at what rate is his distance from the window changing?

80. A light is on the ground 20 metres from a building. A man 2 metres tall walks from the light towards the building at 2 metres/second. How rapidly is his shadow on the building growing shorter when he is 8 metres from the building?

81. A 30-ft. ladder is leaning against a vertical wall. The foot of the ladder is being pulled away from the wall at the rate of 5 ft/sec. Find how fast the top of the ladder is sliding down the wall when the foot of the ladder is 18 feet from the wall.

82. A ladder 5 metres long leans against a vertical wall 4 metres high. The lower end of the ladder is being pulled horizontally away from the wall at a speed of 2 metres/sec. What is the rate of change of the angle θ between the ground and the ladder when $\theta = 45°$?

83. A ladder 9 metres long is placed vertically against a vertical wall which is 3 metres tall. The lower end of the ladder is then pulled horizontally away from the wall at a constant rate of $\frac{2}{3}$ metres/sec. If the top of the wall always remains in contact with the ladder, find the vertical component of the velocity of the top end of the ladder when the angle θ (see Fig. S.83) is $\pi/3$ (radians).

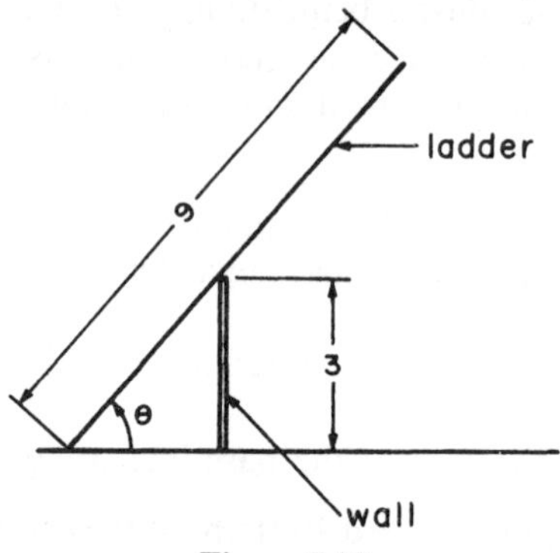

Figure S.83

84. An airplane is flying in a straight line at 500 ft/sec at an altitude of 4000 ft. A searchlight on the ground is trained on the plane. How fast is the searchlight revolving (in rad/sec) 4 seconds after the plane has passed directly over the light?

85. A helicopter leaves the ground at a point 800 ft from an observer and rises vertically at 25 ft/sec. Find the time rate of change of the observer's angle of elevation of the helicopter when the helicopter is 600 ft above the ground.

86. A light in a lighthouse one mile from a straight shoreline is rotating at 2 revolutions per minute. How fast is the beam of light moving along the shore when it passes the point a half mile from the point on the shore closest to the lighthouse?

87. A revolving light, 100 ft from a straight wall, makes one revolution every four seconds. Find the velocity of the light, along the wall, $\frac{1}{2}$ second after it passes the point on the wall nearest to the light.

88. The cross-sections of a certain solid by planes perpendicular to the x-axis are circles with diameters extending from the curve $y = x^2$ to the curve $y = 2 - x^2$. The solid lies between the points of intersection of the two curves. Find the volume of this solid.

89. The two faces of a lens are parabolic (i.e. they are formed by rotating a parabola about its axis of symmetry). The central cross-section has the dimensions shown in Fig. S.89. Determine the volume of glass required.

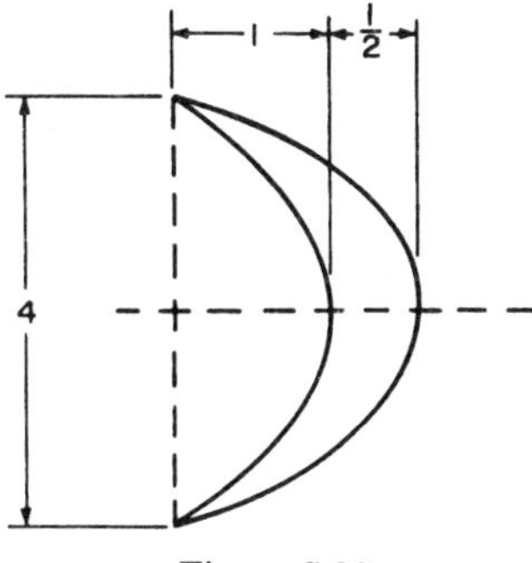

Figure S.89

90. The base of a solid lies in the xy-plane and is bounded by $y = \frac{x^2}{4}$, $x = 4$ and the x-axis. Find the volume if each cross-section perpendicular to the x-axis is a square.

91. Find, by integration, the volume of a pyramid with square base of side a, and altitude h.

92. Find, by integration, the volume of the pyramid in Fig. S.92, with the base a rectangle of dimensions a and b and with height h. Hint: Find the dimensions of a section parallel to the base, at a distance x from the apex, using similar triangles.

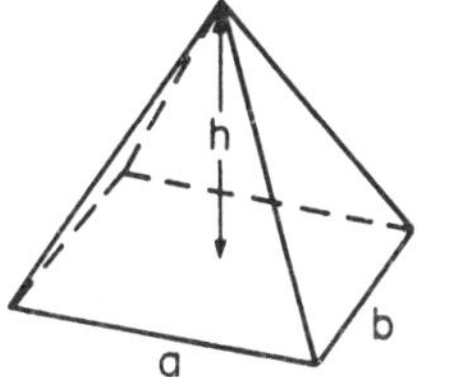

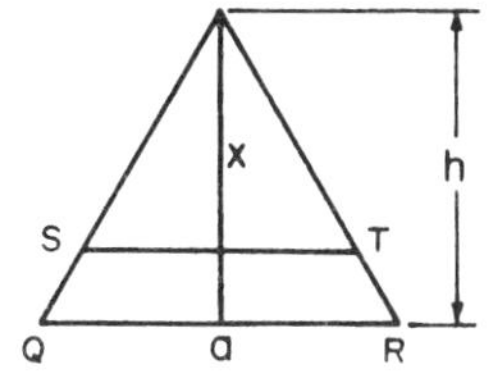

ST = ?

Figure S.92

93. The base of a solid lies in the xy-plane and is bounded by $xy = 1$, the y-axis, the line $y = 2$ and the line $y = 5$. Find the volume of the solid if every cross-section perpendicular to the y-axis is a square.

94. A three-foot loudspeaker horn has a square cross-section all along its length. If it is cut through its central axis by a plane parallel to one of the sides of its square mouth, the upper and lower curves are given by $y = \frac{x^2}{10}$ and $y = -\frac{x^2}{10}$, for x from 1 to 4 (Fig. S.94). What is the volume enclosed by the horn?

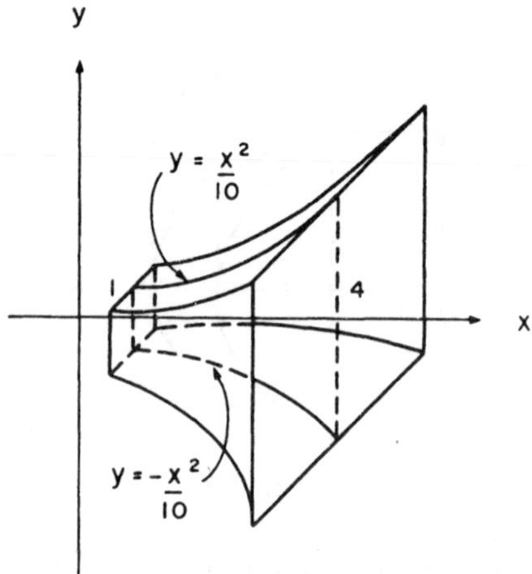

Figure S.94

95. Find the volume of the solid whose base is a circle of radius 1 and whose perpendicular cross-sections are squares.

96. A solid has as its base the triangle cut from the first quadrant by the line $3x+4y=12$. Every plane section of the solid perpendicular to the x-axis is a semi-circle. Find the volume of the solid.

97. The base of a certain solid is a semi-circle of radius R. Thus, one side of the base is a straight line segment L, of length $2R$. Cross-sections of the solid, which are perpendicular to line segment L, are semi-circles. Find the volume of this solid.

98. The base of a solid is the square with vertices at $(1,0)$, $(0,1)$, $(-1,0)$ and $(0,-1)$. Cross-sections of the solid perpendicular to the x-axis are triangles with altitude x^4-2x^2+1 (for $-1 \le x \le 1$). Compute the volume of this solid.

(*)99. The base of a solid is the disc $x^2+y^2 \le 1$. Cross-sections of the solid perpendicular to the x-axis are rectangles of height $|x|$. Find the volume of this solid.

100. A circular hole of radius a is drilled through a sphere of radius $2a$, the centre of the hole coinciding with the centre of the sphere. What is the volume of material removed to make the hole?

Chapter III

Physics and Engineering

In this chapter we consider elementary calculus problems related to the fields of physics and engineering. These problems involve applications of both differential and integral calculus. Almost all of them lead to working with functions which are no more complicated than polynomials. Thus few of the problems require more than a rudimentary knowledge of the techniques of calculus.

The difficulties which students encounter with a physical problem, as with other applied problems, are in its interpretation and translation to a mathematical form. A careful study of the various solved problems in this chapter should be a major help for resolving such difficulties. The broad areas covered by the problems of this chapter include:

1. *Position, Velocity, Acceleration, Distance*

In one dimension, at time t: let $x(t)$ be the *position* of an object, $v(t)$ its *velocity*, and $a(t)$ its *acceleration*. Then

$$v(t)=\frac{dx}{dt},$$

$$a(t)=\frac{d^2x}{dt^2}.$$

If the position of an object is given as a function of t, i.e., $f(t)$ is given where $x=f(t)$, then its velocity $v(t)=f'(t)$ and its acceleration $a(t)=f''(t)$.

Conversely:

(a) If $v(t)$ is given, then from integral calculus, $x(t)=\int v(t)\,dt+c$ where c is an arbitrary constant to be determined from an *initial condition* which specifies the position of the object at some particular (initial) time

$t = t_0$, i.e., $x(t_0)$ is a given datum. For example, say $v(t) = t^2$ and $x(0) = 3$. Then $x(t) = \int v(t)\,dt + c = \frac{t^3}{3} + c$; $x(0) = 3 \Rightarrow c = 3$; hence $x(t) = \frac{t^3}{3} + 3$.

(b) If $a(t)$ is given, then $v(t) = \int a(t)\,dt + c_1$, and $x(t) = \int v(t)\,dt + c_2$. Here two initial conditions at a particular time $t = t_0$ corresponding to the position $(x(t_0))$ and velocity $(v(t_0))$ of an object will determine the values of the arbitrary constants c_1 and c_2. For example say $a(t) = t$, $x(1) = 1$, $v(1) = 5$. Then $v(t) = \int t\,dt + c_1 = \frac{t^2}{2} + c_1$; $v(1) = 5 \Rightarrow c_1 = \frac{9}{2}$; hence $v(t) = \frac{t^2 + 9}{2}$. Then $x(t) = \int v(t)\,dt + c_2 = \frac{t^3}{6} + \frac{9}{2}t + c_2$; $x(1) = 1 \Rightarrow c_2 = -\frac{11}{3}$; hence $x(t) = \frac{t^3}{6} + \frac{9}{2}t - \frac{11}{3}$.

The *distance* S travelled by an object with position $x(t)$ at time t over the time interval $t_1 \le t \le t_2$ is

$$S = \int_{t_1}^{t_2} \left| \frac{dx}{dt} \right| dt = \int_{t_1}^{t_2} |v(t)|\,dt$$

where $|v(t)|$ is the *speed* of the object at time t.

In two dimensions the position of an object can be represented by a *position vector* $\underline{r}(t) = x(t)\mathbf{i} + y(t)\mathbf{j}$; its velocity by the *velocity vector* $\underline{v}(t) = \frac{d\underline{r}}{dt} = \dot{\underline{r}} = \frac{dx}{dt}\mathbf{i} + \frac{dy}{dt}\mathbf{j} = v_x\mathbf{i} + v_y\mathbf{j}$ ($v_x = \frac{dx}{dt}$ is the x-component or horizontal component of velocity; $v_y = \frac{dy}{dt}$ is the y-component or vertical component of velocity); and its acceleration by the *acceleration vector*

$$\underline{a}(t) = \frac{d\underline{v}}{dt} = \dot{\underline{v}} = \frac{d^2\underline{r}}{dt^2} = \ddot{\underline{r}} = \frac{d^2x}{dt^2}\mathbf{i} + \frac{d^2y}{dt^2}\mathbf{j} = a_x\mathbf{i} + a_y\mathbf{j}.$$

The *distance* S travelled by such an object over the time interval $t_1 \le t \le t_2$ is $S = \int_{t_1}^{t_2} \left| \frac{d\underline{r}}{dt} \right| dt = \int_{t_1}^{t_2} |\underline{v}(t)|\,dt = \int_{t_1}^{t_2} \sqrt{\left(\frac{dx}{dt}\right)^2 + \left(\frac{dy}{dt}\right)^2}\,dt$ where $|\underline{v}(t)| = \sqrt{\left(\frac{dx}{dt}\right)^2 + \left(\frac{dy}{dt}\right)^2}$ is the *speed* of the object at time t.

2. *Maxima and Minima*: See Chapter II
3. *Related Rates*: See Chapter II
4. *Centre of Mass* (*Centre of Gravity, Centroid*)

Consider a system of n particles located in the xy-plane at $(x_1, y_1), (x_2, y_2), \ldots, (x_n, y_n)$ (Fig. III.0a) with respective masses $m_1, m_2, \ldots, m_n$. The *centre of mass* (*centre of gravity*) of this system is

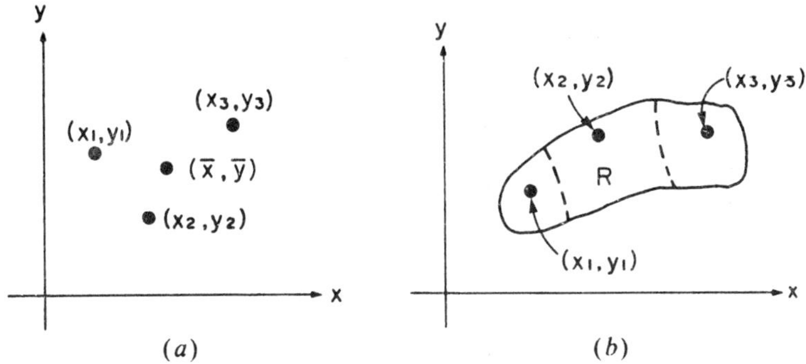

Figures III.0*a* and *b*.

located at $(\bar{x}, \bar{y})$ where

$$\bar{x} = \frac{\sum_{i=1}^{n} m_i x_i}{M}, \qquad \bar{y} = \frac{\sum_{i=1}^{n} m_i y_i}{M}, \quad \text{and}$$

$M = \sum_{i=1}^{n} m_i$ is the *total mass* of the system.

In the case of a continuous distribution of matter occupying region R of area A with mass density ρ (mass per unit area), one breaks up R into n subregions (see Fig. III.0*b*). Let $P_i = (x_i, y_i)$ be a point lying in the ith subregion of area ΔA_i, $i = 1, 2, \ldots, n$. Consider each subregion as a particle of mass $m_i = \rho(P_i)\,\Delta A_i$ located at $P_i = (x_i, y_i)$ where $\rho(P_i)$ is the mass density at P_i. The *centre of mass* (*centre of gravity*) of R is $(\bar{x}, \bar{y})$ where

$$\bar{x} = \frac{\lim\limits_{n \to \infty} \sum_{i=1}^{n} m_i x_i}{\lim\limits_{n \to \infty} \sum_{i=1}^{n} m_i} = \frac{\int_R \rho x \, dA}{M},$$

$$\bar{y} = \frac{\lim\limits_{n \to \infty} \sum_{i=1}^{n} m_i y_i}{\lim\limits_{n \to \infty} \sum_{i=1}^{n} m_i} = \frac{\int_R \rho y \, dA}{M},$$

dA is a differential element of area, and $M = \int_R \rho \, dA$ is the mass of R.

If the mass density ρ is constant over R, the centre of mass of R corresponds to its geometric centre or *centroid*

$$\bar{x} = \frac{\int_R x \, dA}{A}, \qquad \bar{y} = \frac{\int_R y \, dA}{A},$$

where A is the area of R. Two special types of regions are considered for constant mass density.

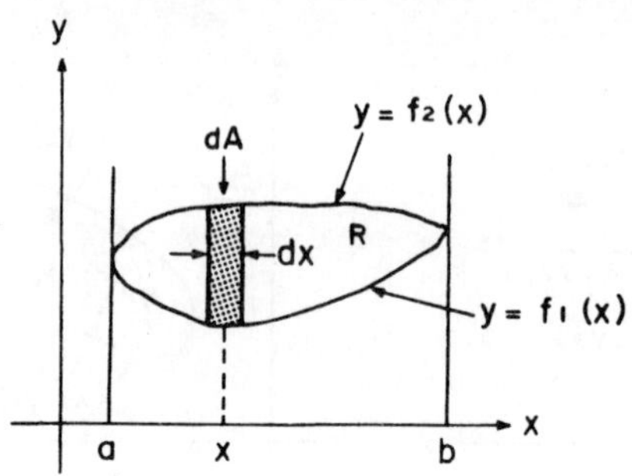

Figure III.0c. Region of Type (i).

Region of Type (*i*). Say R can be described by: $f_1(x) \le y \le f_2(x)$, $a \le x \le b$, (Fig. III.0c). Here consider each subregion as a rectangular strip parallel to the y-axis with $dA = [f_2(x) - f_1(x)]\,dx$; A = area of $R = \int_a^b [f_2(x) - f_1(x)]\,dx$;

$$\bar{x} = \frac{\int_a^b x[f_2(x) - f_1(x)]\,dx}{A};$$

$$\bar{y} = \frac{\frac{1}{2}\int_a^b [f_2(x) + f_1(x)][f_2(x) - f_1(x)]\,dx}{A}$$

$$= \frac{\int_a^b \{[f_2(x)]^2 - [f_1(x)]^2\}\,dx}{2A}.$$

Region of Type (*ii*). Say R can be described by: $g_1(y) \le x \le g_2(y)$, $c \le y \le d$. Then (Fig. III.0d) consider each subregion as a rectangular strip parallel to the x-axis with $dA = [g_2(y) - g_1(y)]\,dy$; A = area of $R = \int_c^d [g_2(y) - g_1(y)]\,dy$;

$$\bar{y} = \frac{\int_c^d y[g_2(y) - g_1(y)]\,dy}{A}; \qquad \bar{x} = \frac{\int_c^d \{[g_2(y)]^2 - [g_1(y)]^2\}\,dy}{2A}.$$

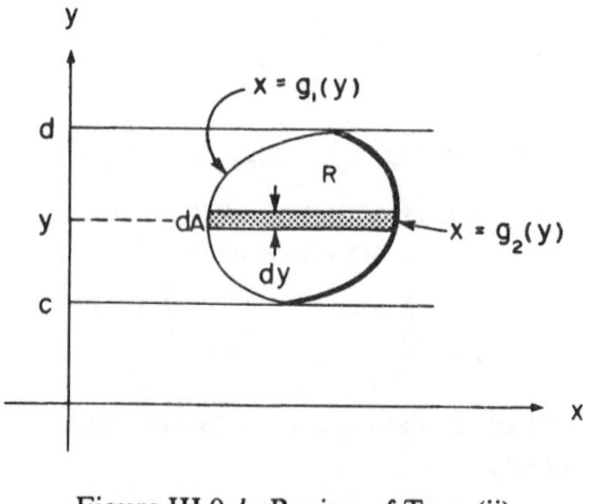

Figure III.0d. Region of Type (ii).

If R can be described as a union of a finite number of regions of Type (i), i.e., $R = R_1 \cup R_2 \cup \cdots \cup R_n$ where each R_i is a region of Type (i), then

$$\bar{x} = \frac{\int_R x\,dA}{A} = \frac{\sum_{i=1}^{n} \int_{R_i} x\,dA}{A}.$$

Similarly if R can be composed of a finite number of Type (ii) regions the above analysis leads to an algorithm to compute $\bar{y}$.

The First Theorem of Pappus. Consider a uniform distribution of matter occupying a region R of the xy-plane and bounded by a curve C (Fig. III.0e).

Assume C does not cross the x-axis. Let $(\bar{x}, \bar{y})$ be the centroid of R, and A the area of R. Consider the solid of revolution formed by rotating R about the x-axis. Let V be the volume of this solid. Then the First Theorem of Pappus states that $V = 2\pi\bar{y}A$. This first theorem of Pappus is proved in Solved Problem 16.

Physical applications. (Assume ρ is constant; otherwise replace centroid by centre of mass.)

(a) The potential energy of a body is equal to that of a single particle located at the centroid of the body with its mass equal to the total mass of the body.

(b) The linear momentum of a body is equal to that of a single particle located at the centroid of the body whose mass is equal to the total mass of the body and whose velocity is that of its centroid.

(c) See sections 6 and 7 on fluid pressure and work.

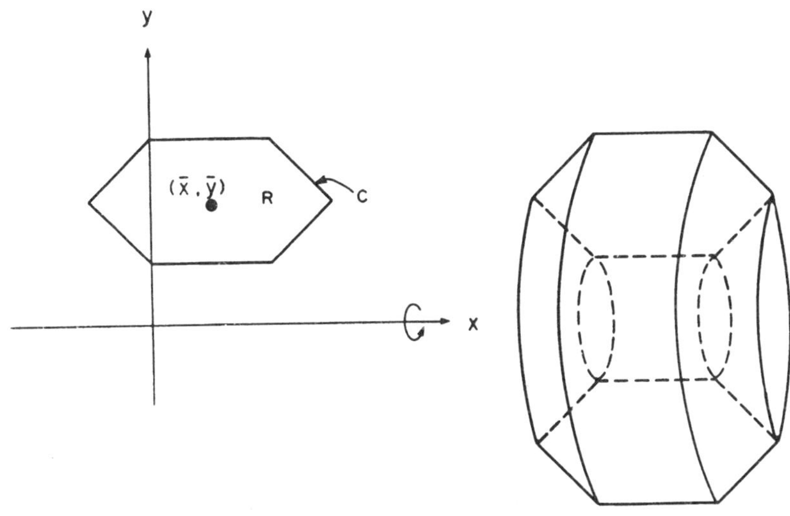

Figure III.0e

5. *Moment of Inertia*

Consider a body of constant mass density ρ (mass per unit area) occupying region R (as in Fig. III.0b). Its *moment of inertia* about the y-axis is

$$I_y = \rho \int_R x^2 \, dA.$$

If R is a region of Type (i) then

$$I_y = \rho \int_a^b x^2 [f_2(x) - f_1(x)] \, dx.$$

If the region R does not cross the y-axis and is rotating about the y-axis with angular velocity ω, then the kinetic energy of this rotating body is $\frac{1}{2} I_y \omega^2$.

6. *Fluid Pressure*

Consider a flat plate of area A submerged horizontally (Fig. III.0f) in a fluid of weight-density w ($w \simeq 62.5$ lb/ft^3 for water). Then the *force* F acting on this plate is $F = whA$ where h is the depth of the plate in the fluid. The *pressure* P (force per unit area) on the plate is $P = wh$. By Pascal's principle the pressure at a depth h of any fluid is the same in all directions. Hence for a flat plate of area A submerged *vertically* in a fluid (Fig. III.0g) the total force of fluid pressing on either side of the plate is $F = w\bar{h}A$ where $\bar{h}$ is the depth of the *centroid* of the submerged plate.

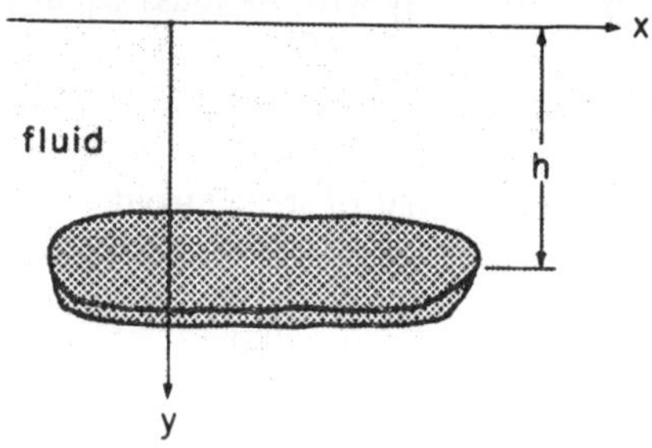

Figure III.0f. Flat plate submerged horizontally in a fluid.

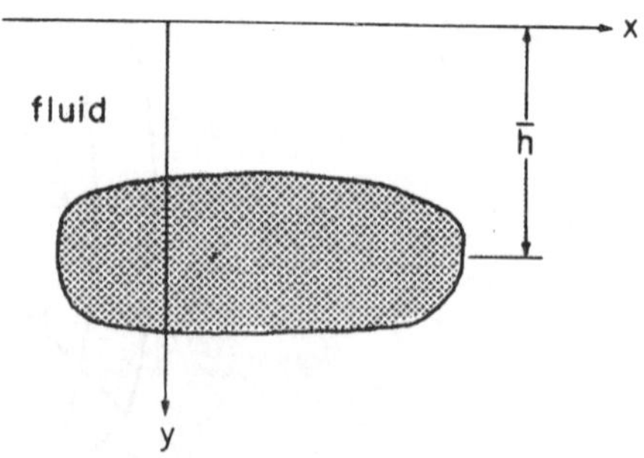

Figure III.0g. Flat plate submerged vertically in a fluid.

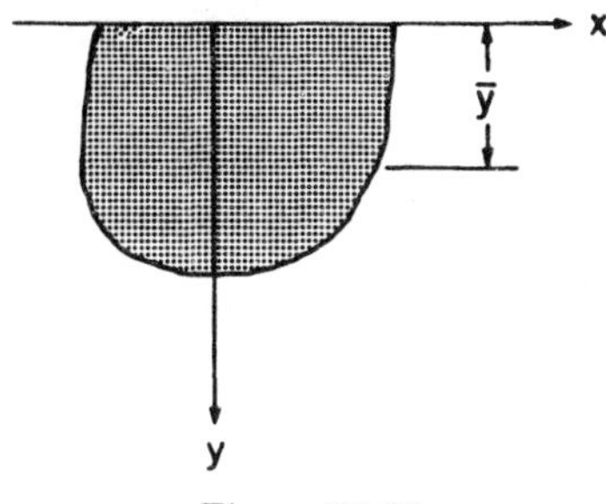

Figure III.0h

7. *Work*

   If a constant force F acts on a particle moving in a straight line, then the *work* done W in giving the particle a displacement x is $W = Fx$. If F is variable, i.e., $F = F(x)$, $a \le x \le b$, then the work done W in moving a particle from a to b is

   $$W = \int_a^b F(x)\, dx.$$

   In the case of pumping liquid out of a container (Fig. III.0h), the work done W in pumping all of the liquid out of a filled container is

   $$W = wV\bar{y}$$

   where w is the weight density of the liquid, V is the volume of the container, and $\bar{y}$ is the depth of the centroid of the space enclosed by the container.
8. *Hooke's Law*: See Solved Problem 29
9. *Newton's Law of Cooling*: See Solved Problem 30

Solved Problems

1. Assume that the brakes of an automobile produce a constant deceleration of k ft/sec^2, i.e. $\dfrac{d^2x}{dt^2} = -k$.

(a) By solving this differential equation, determine what value k must have in order to bring an automobile travelling 88 ft/sec to rest in a distance of 100 ft from the point where the brakes are applied.

(b) With this same value of k, how far would a car travelling at 44 ft/sec travel before being brought to a stop by applying the brakes?

(a) Here $x(t)$ is the position of the automobile at time t, $v(t) = \dfrac{dx}{dt}$ is its velocity. $\dfrac{d^2x}{dt^2} = \dfrac{dv}{dt} = \dfrac{dv}{dx}\dfrac{dx}{dt}$, by the chain rule. Equally well, one may think of $v = v(x)$. In this case the given differential equation simplifies to $v\frac{dv}{dx} = -k$.

For the given problem when $x = 0$, $v = 88$; when $x = 100$, $v = 0$. Separating variables, the simplified differential equation becomes the differential relation $v\,dv = -k\,dx$.

The given data imply that $\int_{88}^{0} v\,dv = \int_{0}^{100} -k\,dx$.

This leads to $k = \boxed{38.72 \text{ ft/sec}^2}$.

(b) Here when $v = 44$, $x = 0$. The aim is to find $x = X$ when $v = 0$. Applying this data to the differential relation, one has

$$\int_{44}^{0} v\,dv = \int_{0}^{X} -k\,dx.$$

This leads to $X = \boxed{25 \text{ ft}}$.

2. If the brakes of a vehicle can produce a constant deceleration of 20 ft/sec^2, what is the maximum allowable velocity if the vehicle is to be stopped in 60 ft or less after the brakes are applied?

Let v be the velocity of the vehicle and let a be its acceleration. $a = \frac{dv}{dt} = -20$. Let X be the stopping distance and let V be the velocity at the instant the brakes are applied; $X \leq 60$.

$v\frac{dv}{dx} = -20 \leftrightarrow v\,dv = -20\,dx$. From the given data, $v = V$ when $x = 0$; $v = 0$ when $x = X$. Hence $\int_{V}^{0} v\,dv = \int_{0}^{X} -20\;dx$. This leads to

$X = \frac{1}{40}V^2$. Then $X \leq 60 \Rightarrow V^2 \leq (40)(60)$
$\Rightarrow V \leq 20\sqrt{6}$. Hence the maximum allowable velocity
is $\boxed{20\sqrt{6} \text{ ft/sec}}$.

(*)3. A particle moves in a straight line with acceleration $a = (2t - 3)$ ft/sec^2. If $x = 0$ and $v = -4$ when $t = 0$, find

(a) when the particle reverses direction;

(b) the total distance the particle moves in the first 5 seconds.

The particle's acceleration is $a(t) = 2t - 3$, with initial position $x(0) = 0$ and initial velocity $v(0) = -4$. Hence its velocity is $v(t) = \int_0^t a(t')\,dt' - 4$ $= \int_0^t (2t' - 3)\,dt' - 4 = t^2 - 3t - 4$.

(a) The particle reverses direction when $v(t) = \frac{dx}{dt} = 0 \leftrightarrow t^2 - 3t - 4 = (t - 4)(t + 1) = 0 \leftrightarrow t = 4$, $t = -1$. Since $t \geq 0$, the particle reverses direction after $\boxed{\text{4 seconds}}$.

(b) The total distance S travelled by the particle in the time interval $0 \le t \le 5$ is $\int_0^5 |v(t)|\, dt$.

$$\text{For } 0 \le t \le 4: \quad v(t) \le 0.$$
$$\text{For } 4 \le t \le 5: \quad v(t) \ge 0.$$

Hence
$$S = \int_0^4 -v(t)\, dt + \int_4^5 v(t)\, dt$$
$$= \int_0^4 -(t^2 - 3t - 4)\, dt + \int_4^5 (t^2 - 3t - 4)\, dt$$
$$= \boxed{21\tfrac{1}{2}\text{ ft}}.$$

(*)**4.** A sprinter who runs the 100 metre dash in 10.2 seconds accelerates at a constant rate for the first 25 metres and then continues at a constant speed for the rest of the race. Find his acceleration.

Let T be the time period over which the sprinter accelerates at a constant rate a. Let $a(t)$ be the acceleration of the runner and $v(t)$ his speed at time t. Then

$$a(t) = \begin{cases} a, & 0 \le t \le T \\ 0, & T < t \le t^* \end{cases}$$

where $t^* = 10.2$;

$$v(t) = \int_0^t a(t')\, dt' = \begin{cases} \int_0^t a\, dt' = at, & 0 \le t \le T \\ \int_0^T a\, dt' = aT, & T \le t \le t^* \end{cases}.$$

Let $x(t)$ be the distance travelled by the runner at time t.

Then
$$x(t) = \int_0^t v(t')\, dt'$$
$$= \begin{cases} \int_0^t at'\, dt' = \dfrac{at^2}{2}, \; 0 \le t \le T \\ a\dfrac{T^2}{2} + \int_T^t aT\, dt' = aT\left[t - \dfrac{T}{2}\right], \; T \le t \le t^* \end{cases}.$$

For the given data,
$$\left.\begin{aligned} \frac{aT^2}{2} &= 25 \\ aT\left[t^* - \frac{T}{2}\right] &= 100 \end{aligned}\right\} \qquad (1)$$

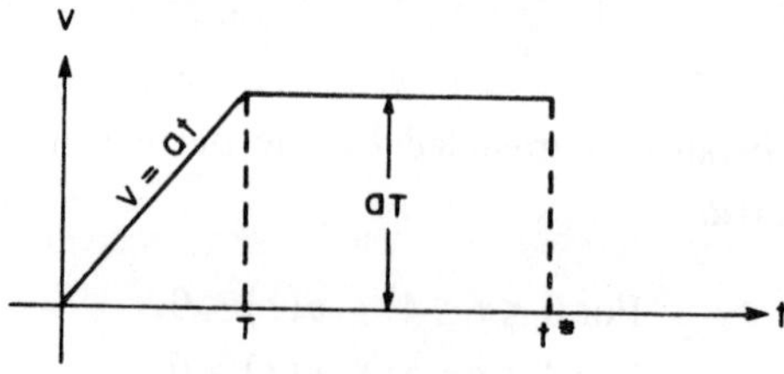

Figure III.4

corresponding to two equations in the two unknowns a and T. Solving these equations, one finds that $a \simeq \boxed{3.00\ \text{m/sec}^2}$.

Alternatively one can derive the system of equations (1) using geometry (Fig. III.4). The area under the $v(t)$ curve corresponds to the covered distance. Hence the area of the triangle leads to the first equation of (1) and the area of the rectangle plus the area of the triangle corresponds to the second equation of (1).

5. A projectile is shot upward from the edge of a cliff 2400 feet high with an initial velocity of 400 ft/sec. How high does it rise, and how many seconds elapse before it hits the ground? (Acceleration of gravity = 32 ft/sec^2)

Let $x(t)$ be the height of the projectile above the ground at time t. Then its acceleration $a(t) = \dfrac{d^2x}{dt^2} = g = -32$. One is given the data $x(0) = 2400$, $v(0) = \dfrac{dx}{dt}\ (t = 0) = 400$.

$$a(t) = -32 \Rightarrow v(t) = \int_0^t -32\, dt' + 400 = -32t + 400.$$

$$\text{Hence } x(t) = \int_0^t v(t')\, dt' + 2400 = -16t^2 + 400t + 2400$$

$$= -16[t^2 - 25t - 150] = -16[t+5][t-30]$$

$$= -16\left\{\left(t - \tfrac{25}{2}\right)^2 - \tfrac{1225}{4}\right\}.$$

Clearly $t = \frac{25}{2}$ corresponds to the highest point for the projectile (Fig. III.5).

$$x\left(\tfrac{25}{2}\right) = (16)\left(\tfrac{1225}{4}\right) = \boxed{4900\ \text{ft}}.$$

The projectile hits the ground at the time $t = T$ when $x = 0$; $T = \boxed{30\ \text{seconds}}$.

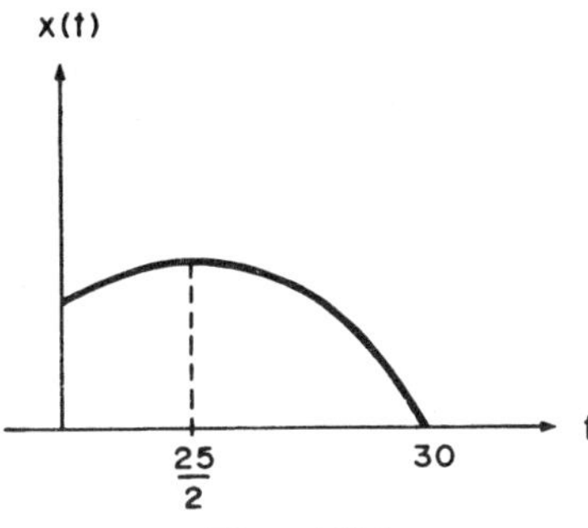

Figure III.5

6. Ace Los Angeles Dodgers' pitcher Sandy Koufax could launch a baseball horizontally 110′ from a point 8′ above the ground (Fig. III.6). How far could he throw a baseball from a point on a bridge 100′ above the water? (Neglect air resistance.)

In this problem there are two components to the motion: The acceleration vector

$$\underline{a}(t) = a_x\mathbf{i} + a_y\mathbf{j} \text{ where } a_x = 0,\ a_y = -g, \text{ i.e., } \frac{d^2x}{dt^2} = 0,\ \frac{d^2y}{dt^2} = -g = -32$$

where g is the acceleration due to gravity; $x(0) = 0$, $y(0) = 8$. Let T be the time of flight of the given Koufax pitch. Then $x(T) = 110$, $y(T) = 0$. Moreover $v_x(0) = v_0$, $v_y(0) = 0$. First we should determine v_0 = the speed of a Koufax pitch.

$$\frac{d^2x}{dt^2} = 0 \Rightarrow x = v_0t + c_1; \qquad x(0) = 0 \Rightarrow c_1 = 0.$$

$$\frac{d^2y}{dt^2} = -g \Rightarrow y = -\frac{gt^2}{2} + c_2t + c_3;$$

$$v_y(0) = 0 \Rightarrow c_2 = 0;\ y(0) = 8 \Rightarrow c_3 = 8.$$

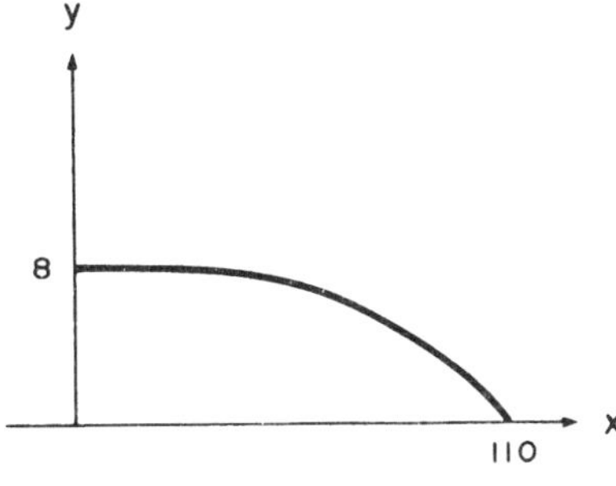

Figure III.6. Trajectory of a Koufax pitch.

Hence the trajectory satisfies the parametric equations

$$x = v_0 t, \qquad y = \frac{-gt^2}{2} + 8.$$

$$x(T) = 110, \quad y(T) = 0 \Rightarrow v_0 T = 110, \quad -\frac{gT^2}{2} + 8 = 0.$$

Hence $v_0 = 110\sqrt{2}$ ft/sec.

In the case of Koufax throwing a baseball from a point on a bridge 100′ above the water, the equations become

$$x = v_0 t \quad \text{where } v_0 = 110\sqrt{2},$$

and
$$y = -\frac{gt^2}{2} + 100.$$

$y = 0$ corresponds to the time $t = t^*$ when the ball hits the water. Thus $t^* = \sqrt{\frac{200}{g}}$. Hence the baseball travels a horizontal distance of $110\sqrt{\frac{400}{g}} = \boxed{275\sqrt{2}\ \text{ft}}$.

Alternatively one can solve this problem by noting that the trajectory lies on a fixed parabola with the motion beginning from the vertex of the parabola in each case.

7. A gun inclined 30° to the horizontal is fired from ground level. Find parametric equations of the path of the bullet if the initial speed of the bullet is 1000 m/s.

As in the previous problem the equations of motion are

$$\frac{d^2x}{dt^2} = 0, \qquad \frac{d^2y}{dt^2} = -g = -9.8 \text{ m/s}^2.$$

v_0 = initial speed of bullet = 1000,

v_1 = initial horizontal component of velocity of the bullet is $v_x(0) = v_0 \cos\theta = 500\sqrt{3}$,

v_2 = initial vertical component of velocity of the bullet is $v_y(0) = v_0 \sin\theta = 500$,

$x(0) = y(0) = 0$ (Fig. III.7).

Hence $x(t) = v_1 t$, $y(t) = -\frac{1}{2}gt^2 + v_2 t$. Thus parametric equations for the

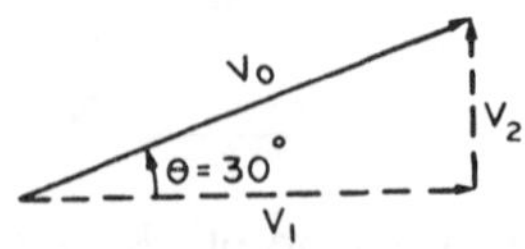

Figure III.7

bullet's path are

$$\begin{cases} x = \boxed{500\sqrt{3}\,t}, \\ y = \boxed{-4.9t^2 + 500t}. \end{cases}$$

Here t is measured in seconds, x and y in metres.

8. A particle is travelling along the branch of the hyperbola $x^2 - y^2 = 1$ containing the point (1,0) (Fig. III.8). It moves so that its y-component of velocity is always equal to 2 metres per second. Find the speed and the acceleration of the particle at the point $(2, \sqrt{3})$.

$x = x(t)$, $y = y(t)$ represents the particle's motion where

$$x^2 - y^2 = 1. \tag{1}$$

The differentiation of equation (1) with respect to t leads to $x\dfrac{dx}{dt} = y\dfrac{dy}{dt}$. (2) At any time t, $\dfrac{dy}{dt} = 2$. Thus $\dfrac{dx}{dt} = \dfrac{2y}{x}$. At the point $(2, \sqrt{3})$, $\dfrac{dx}{dt} = \sqrt{3}$.

Hence at $(2, \sqrt{3})$, the speed of the particle is $\sqrt{\left(\dfrac{dx}{dt}\right)^2 + \left(\dfrac{dy}{dt}\right)^2} = \boxed{\sqrt{7}\text{ m/s}}$.

The differentiation of equation (2) with respect to t $\left(\text{Note that } \dfrac{d^2y}{dt^2} = 0.\right)$ leads to $\left(\dfrac{dx}{dt}\right)^2 + x\dfrac{d^2x}{dt^2} = \left(\dfrac{dy}{dt}\right)^2$. Hence at $(2, \sqrt{3})$, $\dfrac{d^2x}{dt^2} = \frac{1}{2}$ and the acceleration of the particle is $\boxed{(\frac{1}{2}\text{ m/s}^2)\mathbf{i}}$.

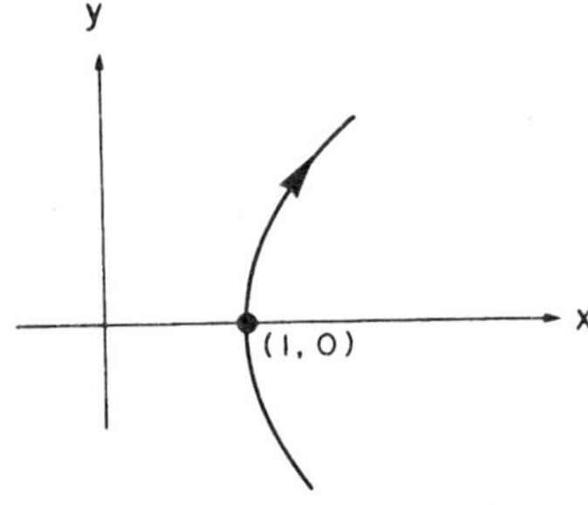

Figure III.8. Trajectory of particle in motion. Arrow indicates direction in which particle moves.

(*)9. A point moves along the curve $y = x^2$ in such a way that at time t $(0 < t < \frac{\pi}{2})$ the line joining the point to the origin makes an angle t with the positive x-axis. Find the position, velocity and acceleration of the point at time t.

First it should be noted that this problem does not make sense dimensionally unless one thinks of a nondimensional time.

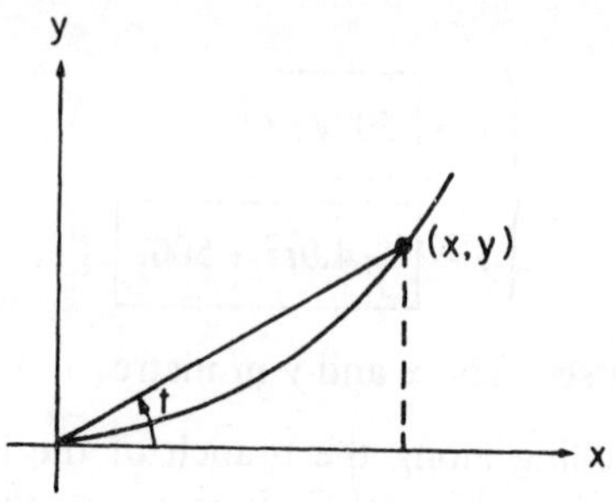

Figure III.9

Consider a point (x, y) along the curve (Fig. III.9). Then $\tan t = \frac{y}{x} = \frac{x^2}{x} = x$. This leads to the parametric equations $x = \tan t$, $y = \tan^2 t$, and hence the position vector of the point at time t is $\underline{r}(t) = \boxed{\tan t\mathbf{i} + \tan^2 t\mathbf{j}}$. The velocity components are

$$v_x = \frac{dx}{dt} = \sec^2 t, \qquad v_y = \frac{dy}{dt} = 2\tan t \sec^2 t.$$

(Alternatively, differentiating $y = x^2$ leads to $\frac{dy}{dt} = 2x\frac{dx}{dt}$.) Hence the velocity of the point at time t is

$\underline{v}(t) = \boxed{\sec^2 t[\mathbf{i} + 2\tan t\mathbf{j}]}$. The acceleration components are

$$a_x = \frac{dv_x}{dt} = 2\tan t \sec^2 t,$$

$$a_y = \frac{d^2y}{dt^2} = 2\left(\frac{dx}{dt}\right)^2 + 2x\frac{d^2x}{dt^2} = 2\sec^4 t + 4\tan^2 t \sec^2 t.$$

Hence the acceleration of the point at time t is

$$\underline{a}(t) = \boxed{2\sec^2 t\left[\tan t\mathbf{i} + (\sec^2 t + 2\tan^2 t)\mathbf{j}\right]}.$$

10. The strength of a beam of rectangular cross-section is proportional to its width and the square of its depth. What are the dimensions of the strongest beam that can be cut from a log of radius r?

Let x be the width of the beam and y its depth. Then (Fig. III.10)

$$x^2 + y^2 = 4r^2, \tag{1}$$

and the strength of the beam is $S = kxy^2$ where k is a constant of proportionality. The aim is to maximize S where y is treated as an implicit function of x.

$$\frac{dS}{dx} = k\left[y^2 + 2xy\frac{dy}{dx}\right].$$

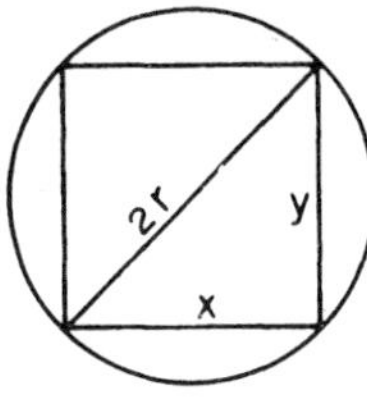

Figure III.10

Differentiation of equation (1) with respect to x leads to $\frac{dy}{dx} = -\frac{x}{y}$. Hence $\frac{dS}{dx} = k[y^2 - 2x^2]$.

$$\frac{dS}{dx} = 0 \leftrightarrow y^2 = 2x^2 \leftrightarrow 3x^2 = 4r^2.$$

Thus the strongest beam has width $\boxed{\frac{2}{\sqrt{3}}r}$ and depth $\boxed{\frac{2\sqrt{2}}{\sqrt{3}}r}$.

(For a proof that $x = \frac{2}{\sqrt{3}}r$ does indeed correspond to the maximum value of S see Solved Problem 5 of Chapter II, Geometry.)

11. A rain gutter is constructed from a sheet of metal 15 inches wide. One third of the metal's width on each side is to be bent up through an angle θ to form the gutter's sides (see Fig. III.11a). What should the value of θ be in order that the gutter have maximum cross-sectional area?

Consider Fig. III.11b in order to determine the cross-sectional area $A(\theta)$ of the gutter where $a = 5$ inches.

$A(\theta) = a^2 \sin\theta\cos\theta + a^2\sin\theta$, defined on the domain D: $0 \le \theta \le \frac{\pi}{2}$. The aim is to maximize $A(\theta)$ over D: $\frac{dA}{d\theta} = a^2[\cos^2\theta - \sin^2\theta + \cos\theta] = a^2[2\cos^2\theta + \cos\theta - 1] = a^2[2\cos\theta - 1][\cos\theta + 1]$; $\frac{dA}{d\theta} = 0 \leftrightarrow \cos\theta = \frac{1}{2}$ or $\cos\theta = -1$. The only critical point in D corresponds to $\cos\theta = \frac{1}{2} \leftrightarrow \theta = \frac{\pi}{3}$.

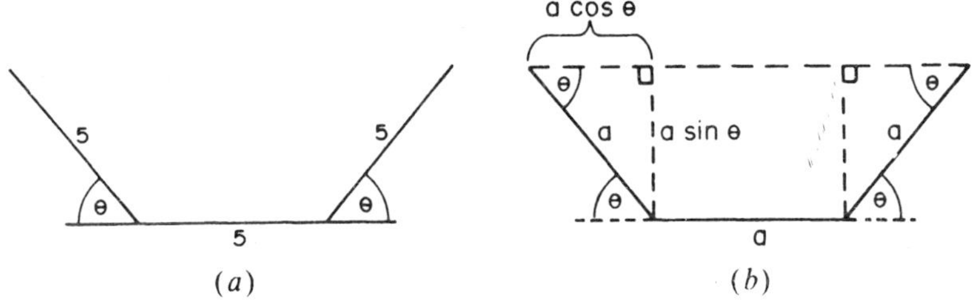

Figures III.11a and b.

Since at $\theta = 0$, $\dfrac{dA}{d\theta} = 2a^2 > 0$; and at $\theta = \dfrac{\pi}{2}$, $\dfrac{dA}{d\theta} = -a^2 < 0$, it follows that $\theta = \boxed{\dfrac{\pi}{3}}$ gives the gutter having the maximum cross-sectional area.

12. A tropical storm is approaching a straight coastline at a speed of 9 kilometres per hour in a direction perpendicular to the shore. If a meteorologist wishes to stay exactly 50 kilometres from the storm, how fast must he drive along the coast road when the storm is 40 kilometres from the coast?

At time t, let $x(t)$ be the distance from the storm to the coastline, let $y(t)$ be the position of the meteorologist along the coast road, and let $L = 50$ kilometres be the distance between the storm and the meteorologist (Fig. III.12). The aim is to find $\dfrac{dy}{dt}$ when $x = 40$, given that at any time t, $\dfrac{dx}{dt} = -9$. From Fig. III.12, $x^2 + y^2 = L^2$. Differentiating this expression with respect to t, we find $\dfrac{dy}{dt} = -\dfrac{x}{y}\dfrac{dx}{dt}$. When $x = 40$, $y = 30$ and hence $\dfrac{dy}{dt} = \left(\dfrac{4}{3}\right)(9) = \boxed{\text{12 kilometres per hour}}$.

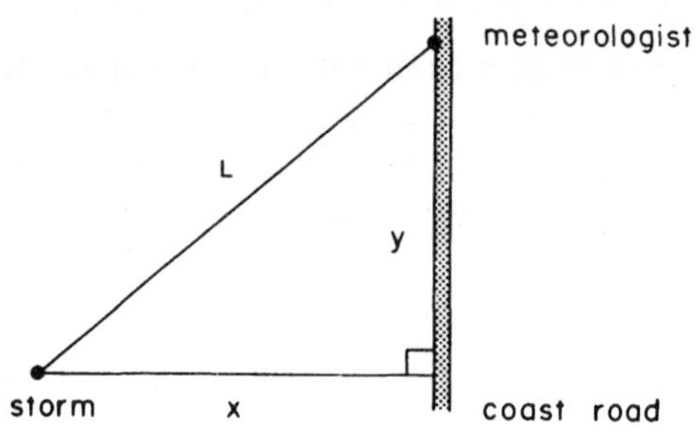

Figure III.12

(*)13. A radar antenna is mounted 500 metres horizontally away from, and is aimed at, a rocket sitting on a launching pad. The rocket blasts off at time $t = 0$ and thereafter climbs vertically with a constant acceleration of 10 m/s^2. If the antenna remains aimed directly at the rocket, how fast must it be rotating upward 10 sec after blast-off?

Let $x(t)$ be the height of the rocket at time t and let $\theta(t)$ be the angle that the antenna has turned at time t (Fig. III.13). We are given

$$\frac{d^2x}{dt^2} = 10;\ x(0) = 0,\ \frac{dx}{dt}(t = 0) = 0,\ b = 500.$$

The aim is to find $\dfrac{d\theta}{dt}$ when $t = 10$.

Note that $\tan\theta = \dfrac{x}{b}$. Differentiation of this equation gives $\sec^2\theta \dfrac{d\theta}{dt}$

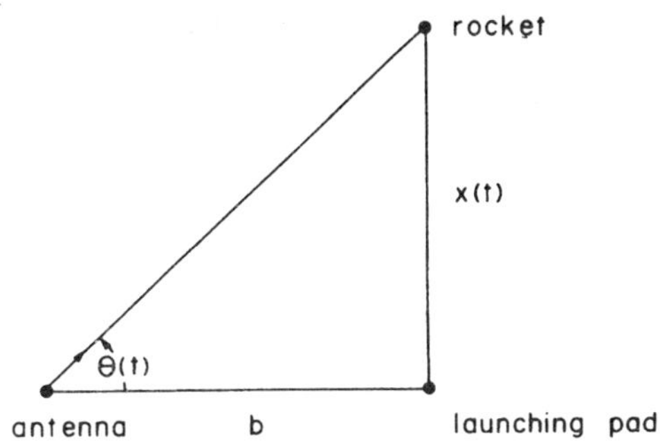

Figure III.13

$= \frac{1}{b}\frac{dx}{dt}$. Hence $\frac{d\theta}{dt} = \left(\frac{\cos^2\theta}{b}\right)\frac{dx}{dt} = \left(\frac{b}{x^2+b^2}\right)\frac{dx}{dt}$. Now $\frac{d^2x}{dt^2} = 10 \Rightarrow \frac{dx}{dt}$ $= 10t + c_1$. The initial data $\Rightarrow c_1 = 0$. Thus $x = \int \frac{dx}{dt}dt = 5t^2 + c_2$. The initial data $\Rightarrow c_2 = 0$. Hence $x(t) = 5t^2$, $\frac{dx}{dt} = 10t$. When $t = 10$, $x = 500$ and $\frac{dx}{dt} = 100$. Hence 10 sec after blast-off, the antenna is rotating upward at a rate of $\boxed{\frac{1}{10}\text{ radians/second}}$.

(*)14. Fig. III.14 shows the familiar mechanism of a reciprocating engine. Suppose that the flywheel has radius 2m and rotates with a constant angular velocity of 5 revolutions/sec in the clockwise direction. How fast is the piston moving to the right when the flywheel has turned through the angle $\alpha = \pi/2$?

In Fig. III.14 the piston is moving to the right at a speed $-\frac{dx}{dt}$, where $x(t)$ is the distance from the end of the piston to the centre of the flywheel.

From the law of cosines.

$$6^2 = 2^2 + x^2 - 4x\cos\alpha.$$

Differentiation of this equation with respect to t leads to

$$\frac{dx}{dt} = \frac{2x\sin\alpha\frac{d\alpha}{dt}}{2\cos\alpha - x}.$$

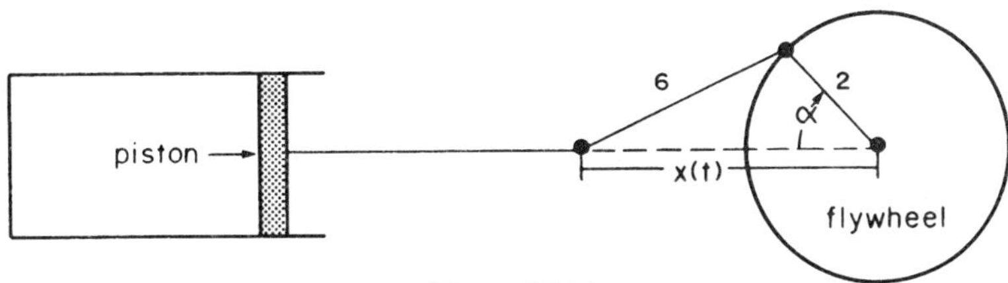

Figure III.14

At all times, $\frac{d\alpha}{dt} = 5$ revolutions/sec $= 10\pi$ radians/sec. Hence when $\alpha = \frac{\pi}{2}$, $\frac{dx}{dt} = -20\pi$. Thus at this angle the piston moves to the right at a speed of $\boxed{20\pi \text{ m/s}}$.

15. Find the centroid of the region

$$R = \left\{(x, y)\middle|0 \le x \le \frac{5}{2}, \frac{x}{2} \le y \le 3x - x^2\right\}.$$

The region R (Fig. III.15) lies between the curves $y = f_1(x) = \frac{x}{2}$ and $y = f_2(x) = 3x - x^2$. At $x = 0$, $f_1(0) = f_2(0) = 0$; at $x = \frac{5}{2}$, $f_1(\frac{5}{2}) = f_2(\frac{5}{2})$. R is a single region of Type (i).

Hence

$$\bar{x} = \frac{\int_0^{5/2} x[f_2(x) - f_1(x)]\,dx}{A}, \text{ and}$$

$$\bar{y} = \frac{\int_0^{5/2} \frac{[f_2(x) + f_1(x)]}{2}[f_2(x) - f_1(x)]\,dx}{A},$$

where $$A = \int_0^{5/2} [f_2(x) - f_1(x)]\,dx \text{ is the area of } R.$$

$$A = \int_0^{5/2} \left(\frac{5}{2}x - x^2\right) dx = \left(\frac{5}{4}x^2 - \frac{x^3}{3}\right)\Bigg|_0^{5/2} = \frac{125}{48};$$

$$A\bar{x} = \int_0^{5/2} x[f_2(x) - f_1(x)]\,dx = \int_0^{5/2} (\tfrac{5}{2}x^2 - x^3)\,dx = \tfrac{625}{192};$$

$$A\bar{y} = \int_0^{5/2} \frac{[f_2(x) + f_1(x)][f_2(x) - f_1(x)]}{2}\,dx$$

$$= \tfrac{1}{2}\int_0^{5/2} (\tfrac{7}{2}x - x^2)(\tfrac{5}{2}x - x^2)\,dx = \tfrac{625}{192}.$$

Hence $\bar{y} = \bar{x} = \boxed{\tfrac{5}{4}}$.

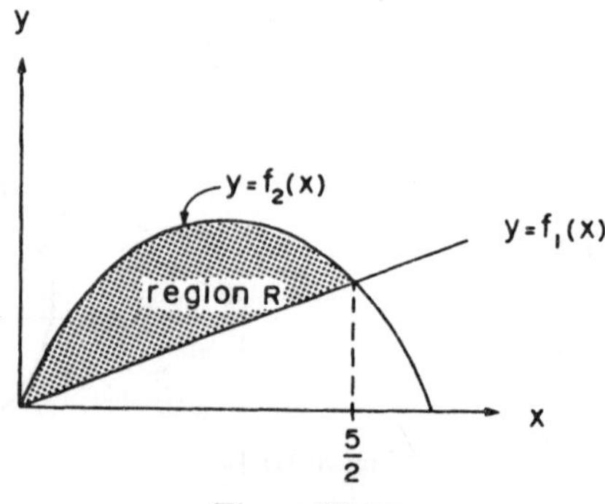

Figure III.15

(*)**16.** The shape of Fig. III.16a is to be revolved about the x-axis to generate a volume. A 1700 year old theorem says that the volume can be calculated by multiplying the shape's area by the distance travelled by the shape's centroid.

(a) Prove this "theorem of Pappus", i.e. Volume = (distance travelled by centroid)×(area of region): $V = (2\pi\bar{y})(A)$.

(b) Use part (a) to determine the centroid of a semi-circle.

(a) If the shaded element of area in Fig. III.16a is revolved about the x-axis, then the resulting differential solid of revolution has volume $dV = 2\pi yl\,dy$, where $l = l(y)$, $c \leq y \leq d$. (Here the element of area $dA = l(y)\,dy$.) Hence $V = 2\pi \int_c^d yl\,dy$ is the volume of the generated solid of revolution.

On the other hand

$$\bar{y} = \frac{\int_c^d yl\,dy}{\int_c^d l\,dy} = \frac{V}{2\pi A},$$

and thus $V = 2\pi\bar{y}A$.

(b) For the given semi-circle (Fig. III.16b), $\bar{x} = 0$, by symmetry. If this semi-circle is revolved about the x-axis, the resulting sphere has volume V

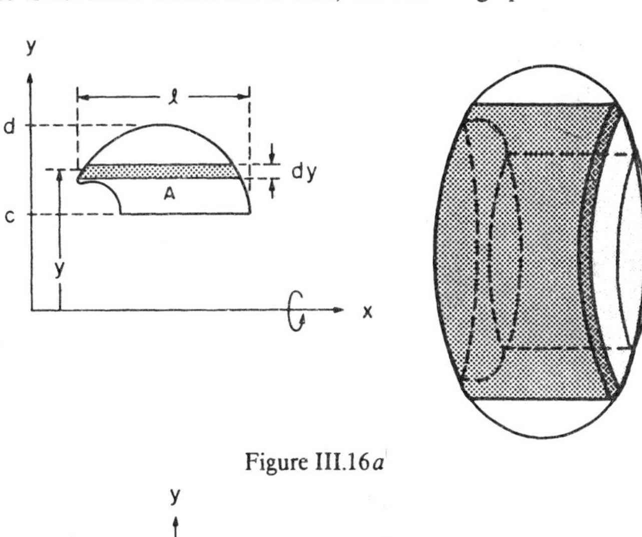

Figure III.16a

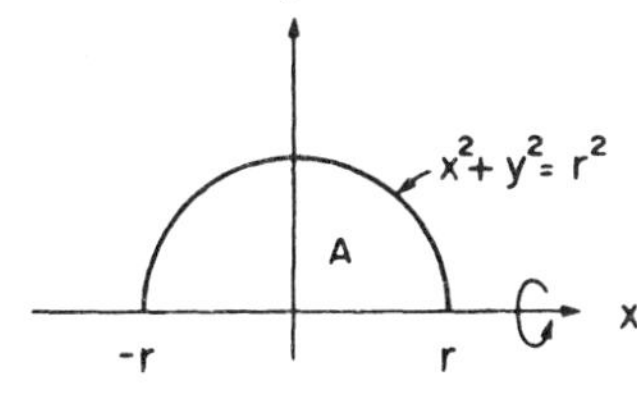

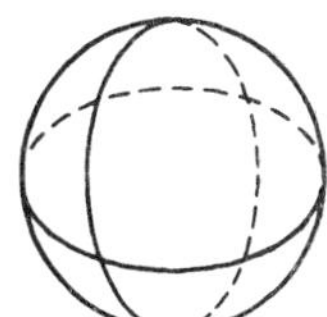

Figure III.16b

$=\frac{4}{3}\pi r^3$. $A=$ area of semi-circle $=\dfrac{\pi r^2}{2}$. Hence for a semi-circle, $\bar{y}=\dfrac{V}{2\pi A}$ $=\boxed{\dfrac{4r}{3\pi}}$.

17. Find the centroids for the shaded regions shown in Figures III.17*a* and III.17*b*.

(a) For the shaded region of Fig. III.17*a*, one can use the first theorem of Pappus to find $\bar{y}$. Consider the solid of revolution (two cones) formed by rotating the region $R=R_1\cup R_2$ about the x-axis. Then the volume of this solid is $V=(2)(\frac{1}{3}\pi)=2\pi\bar{y}A$, where A is the area of R; $A=1$. Hence $\bar{y}=\boxed{\frac{1}{3}}$.

$$\bar{x}=\int_{R_1} x\,dA+\int_{R_2} x\,dA;$$

$$R_1=\{(x,y)\,|\,-1\le x\le 0,\,0\le y\le x+1\},$$

$$R_2=\{(x,y)\,|\,0\le x\le 1,\,0\le y\le x\}.$$

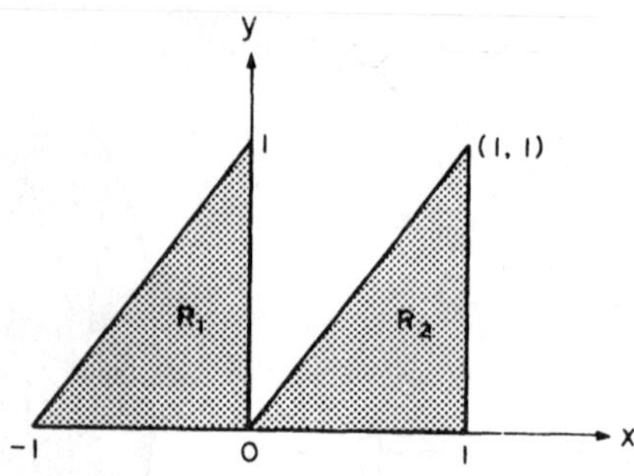

Figure III.17*a*

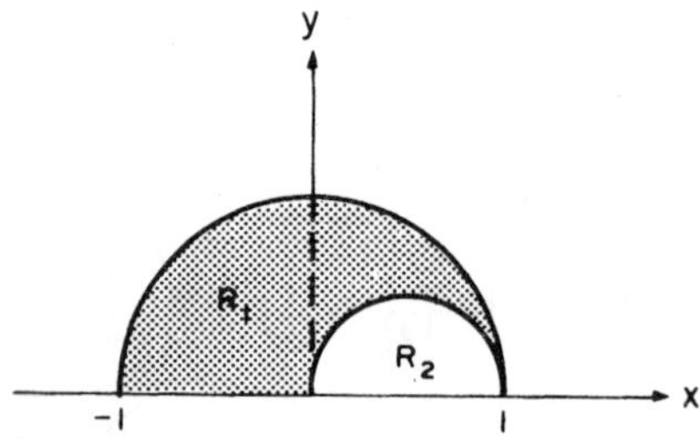

Figure III.17*b*

Hence $\quad \bar{x} = \int_{-1}^{0} x(x+1)\,dx + \int_{0}^{1} x^2\,dx = \int_{-1}^{1} x^2\,dx + \int_{-1}^{0} x\,dx$

$$= \tfrac{2}{3} - \tfrac{1}{2} = \boxed{\tfrac{1}{6}}.$$

(b) For the shaded region of Fig. III.17*b*, the large semi-circle occupying the region $R_1 \cup R_2$, has centroid $(0, \frac{4}{3\pi})$ and area $\frac{\pi}{2}$; the small semi-circle occupying region R_2, has centroid $(\frac{1}{2}, \frac{2}{3\pi})$ and area $\frac{\pi}{8}$. Now consider region R_1:

$$A = \text{area of } R_1 = \tfrac{\pi}{2} - \tfrac{\pi}{8} = \tfrac{3\pi}{8};$$

$$\bar{x} = \frac{\int_{R_1} x\,dA}{A} = \frac{\int_{R_1 \cup R_2} x\,dA - \int_{R_2} x\,dA}{A}$$

$$= -\frac{\int_{R_2} x\,dA}{A} = \frac{-(\bar{x}\text{ for } R_2)(\text{area of } R_2)}{A} = \boxed{-\tfrac{1}{6}};$$

$$\bar{y} = \frac{\int_{R_1 \cup R_2} y\,dA - \int_{R_2} y\,dA}{A} = \frac{(\bar{y}\text{ for } R_1 \cup R_2)(\text{area of } R_1 \cup R_2)}{A}$$

$$-\frac{(\bar{y}\text{ for } R_2)(\text{area of } R_2)}{A} = \boxed{\tfrac{14}{9\pi}}.$$

Alternatively $\bar{y}$ can be found using the first theorem of Pappus:

Revolve the region R_1 about the x-axis. The volume V of the resulting solid of revolution is the difference between the volume of a sphere of radius 1 (obtained by revolving region $R_1 \cup R_2$ about the x-axis) and the volume of a sphere of radius $\frac{1}{2}$ (obtained by revolving region R_2 about the x-axis).

Hence $V = \frac{4\pi}{3}(1-(\frac{1}{2})^3) = \frac{7\pi}{6}$. Thus $\bar{y} = \dfrac{V}{2\pi A} = \boxed{\tfrac{14}{9\pi}}$.

18. Find the moment of inertia about the x-axis of a thin plate of constant density ρ if its boundary is defined by the curves

$$y = \sqrt{x+1}, \qquad y = 0, \qquad x = 2.$$

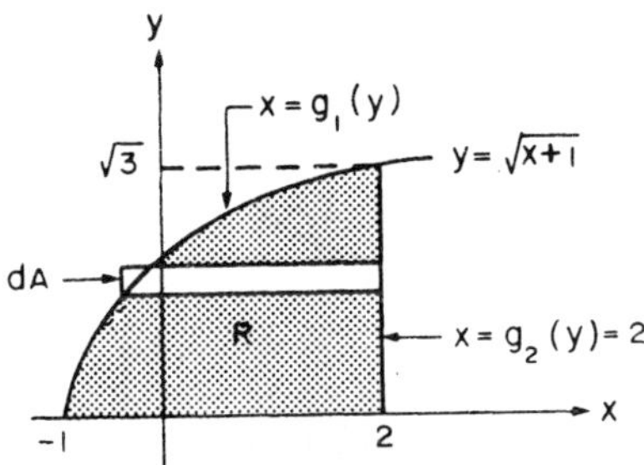

Figure III.18

The moment of inertia is $I_x = \rho\int_R y^2\,dA$. Treat R as a region of Type (ii) (Fig. III.18).

$$y = \sqrt{x+1} \Rightarrow y^2 = x+1 \leftrightarrow x = y^2 - 1 = g_1(y);$$

the differential element of area, $dA = [g_2(y) - g_1(y)]\,dy$; hence

$$I_x = \rho\int_0^{\sqrt{3}} y^2[3 - y^2]\,dy = \rho\int_0^{\sqrt{3}}[3y^2 - y^4]\,dy = \rho\left[y^3 - \frac{y^5}{5}\right]\Bigg|_0^{\sqrt{3}} = \boxed{\frac{6}{5}\sqrt{3}\,\rho}.$$

19. A semi-circular plate of radius 3 feet is suspended vertically in a liquid of density 30 lb per cubic foot. Find the total force on one side if the diameter is at the surface.

Let F be the total force acting on one side of the plate (Fig. III.19). Then $F = w\bar{h}A$ where $\bar{h}$ is the depth of the centroid of the submerged plate, the density of the liquid is $w = 30$, and A is the area of the submerged plate.

Here $$\bar{h} = \left(\frac{4}{3\pi}\right)(3) = \frac{4}{\pi}, \qquad A = \frac{9\pi}{2}.$$

Hence $$F = (30)\left(\frac{4}{\pi}\right)\left(\frac{9\pi}{2}\right) = \boxed{540 \text{ lb}}.$$

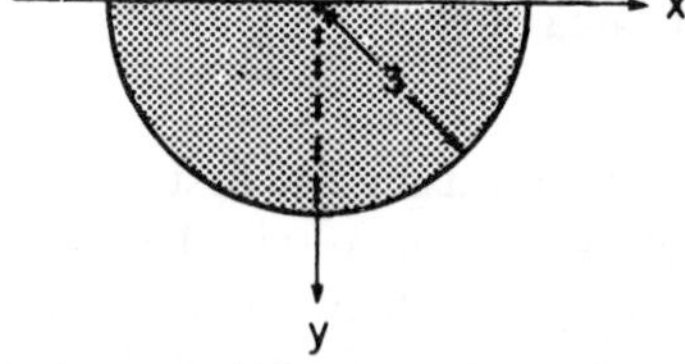

Figure III.19

20. A dam has a parabolic shape with its axis vertical. If the dam is 45 feet across at the water level and is 60 feet deep at its centre find the force that water exerts on the dam.

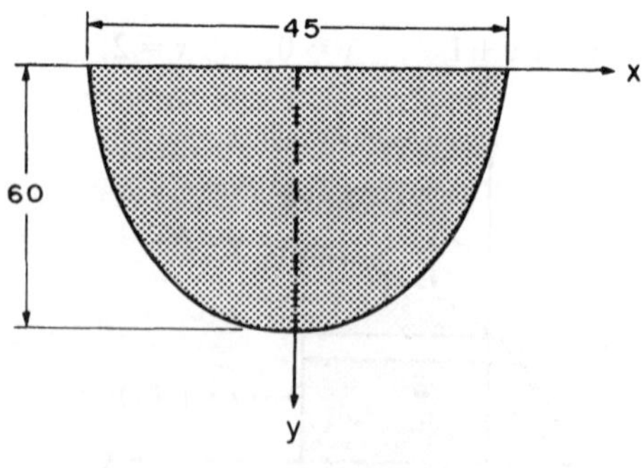

Figure III.20

Let F be the force that water exerts on the dam (Fig. III.20); $F = w\bar{h}A$ where $w = 62.5\ \text{lb/ft}^3$ is the density of water, $\bar{h}$ is the depth of the centroid of the dam, and A is the cross-sectional area of the dam.

First one must find the curve describing the perimeter of the dam: $y - 60 = ax^2$, where $y = 0$ when $x = \boxed{b = \frac{45}{2}}$. Hence $-60 = ab^2$ $\Rightarrow \boxed{a = \dfrac{-60}{b^2}}$. Then

$$\bar{h}A = \frac{1}{2}\int_{-b}^{b}[60 + ax^2]^2\,dx = \int_0^b[60 + ax^2]^2\,dx$$

$$= \int_0^b[3600 + 120ax^2 + a^2x^4]\,dx$$

$$= 3600b + 40ab^3 + \frac{a^2b^5}{5} = 1920b.$$

Hence $F = (62.5)(1920)(\frac{45}{2}) = \boxed{2.7 \times 10^6\ \text{lb}}$.

(*)21. Show that if a dam in the form of a vertical rectangle is divided in half by its diagonal, the force on one half of the dam is twice the force on the other half.

Consider the rectangle broken up, as shown in Figure III.21, into regions R_1 and R_2. $y = \dfrac{b}{a}x$ describes the diagonal. Since R_1 and R_2 have equal areas, to show the desired result one must prove that the depth of the centroid of R_2 is twice that of R_1.

Let H be the depth of the centroid of R_2, and h the depth of the centroid of R_1.

$$\frac{H}{h} = \frac{\int_{R_2} y\,dA}{\int_{R_1} y\,dA} = \frac{\int_0^a \frac{1}{2}\left[b + \frac{b}{a}x\right]\left[b - \frac{b}{a}x\right]dx}{\int_0^a \frac{1}{2}\left[\frac{b}{a}x\right]\left[\frac{b}{a}x\right]dx} = \frac{\frac{1}{2}b^2\int_0^a\left[1 - \left(\frac{x}{a}\right)^2\right]dx}{\frac{1}{2}b^2\int_0^a\left(\frac{x}{a}\right)^2 dx}.$$

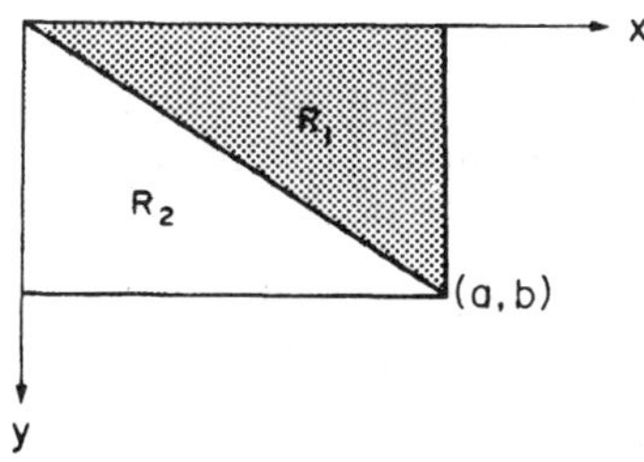

Figure III.21

Now make the substitution $u = \dfrac{x}{a}$ in each integrand. Hence

$$\frac{H}{h} = \frac{\displaystyle\int_0^1 (1-u^2)\,du}{\displaystyle\int_0^1 u^2\,du} = \frac{1-\frac{1}{3}}{\frac{1}{3}} = \boxed{2}.$$

22. A cable 100 metres long and weighing 3 Newtons/metre hangs over a cliff. Find the work done in raising the cable to the top of the cliff.

Consider a small segment of the cable of thickness dx (Fig. III.22) and weight dF, located a distance x from the top of the cliff; $dF = 3\,dx$. The work done in moving this segment to the top of the cliff is $dW = x dF = 3x\,dx$. Hence the total work done in raising the cable to the top of the cliff is

$$W = \int_0^{100} 3x\,dx = \boxed{15000 \text{ Joules}}.$$

Alternatively, the centre of mass is moved 50 metres. The weight of the cable is 300 Newtons. Hence the work done is $(50)(300) = \boxed{15000 \text{ Joules}}$.

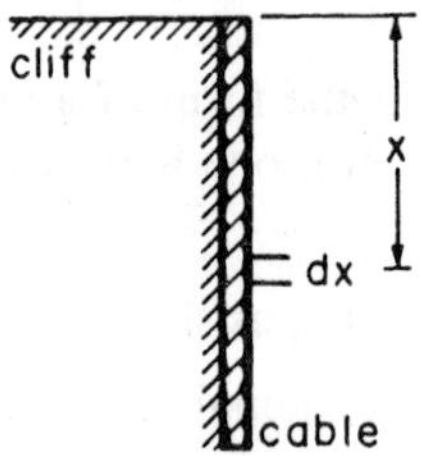

Figure III.22

23. A uniform rope 20 metres long, having a total mass of 5 kilograms, is stretched tightly on a floor. If one assumes that there is no friction between the rope and floor, determine the amount of work done (against gravity) to lift one end of the rope vertically to a height 50 metres above the floor.

The work done W is the same as that done in lifting the centre of mass (centroid) a distance of 40 metres (Fig. III.23). Hence

$$W = F\bar{h} = mg\bar{h} = (5\text{ kg})\left(9.8\frac{\text{N}}{\text{kg}}\right)(40\text{ m}) = \boxed{1960\text{ J}}.$$

24. A swimming pool has the shape of a rectangular parallelepiped which is 20 feet wide, 30 feet long and 6 feet deep. Originally it is full to a depth of 5 feet. How much work is required to empty the pool by pumping all the water out over the top rim of the pool?

The weight of water in the pool is $(62.5)(20)(30)(5) = 187{,}500$ lbs. The centroid of the water is lifted $3\frac{1}{2}$ feet (Fig. III.24). Hence the work done in emptying the pool is $(187{,}500)(\frac{7}{2}) = \boxed{656{,}250 \text{ foot-pounds}}$.

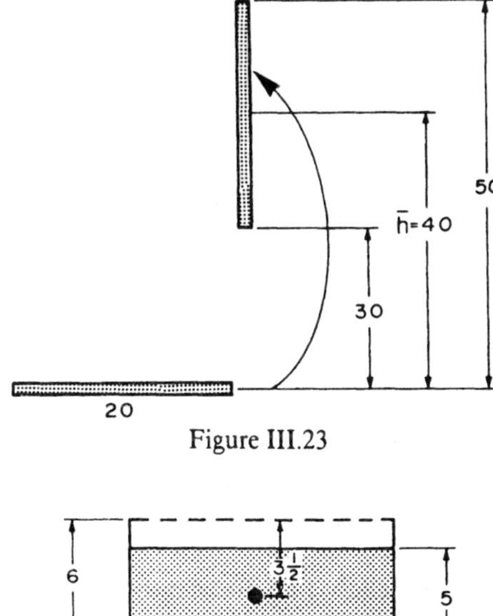

Figure III.23

Figure III.24

(*)**25.** A spherical tank of radius 3 feet is half full of water. Find the work required to pump all of this water to a point 7 feet above the top of the tank.

The distance from the centre of the tank to a point 7 feet above the top of the tank is $h = 10$ feet, the radius of the spherical tank is $r = 3$ feet (Fig. III.25a). Now consider a cylindrical slice of water located at a depth of $-y$, $-r \le y \le 0$. This slice is lifted a height $h - y$ (Fig. III.25b) and its weight is wdV, where dV is the volume of the slice and $w = 62.5$ lb/ft^3 is the density of water.

$dV = \pi x^2\,dy$ where $|x|$ is the radius of this slice. But $x^2 + y^2 = r^2$. Hence $dV = \pi(r^2 - y^2)\,dy$. Let dW be the work done in lifting this slice of water 7 feet above the top of the tank:

$$dW = (wdV)(h - y) = \pi w(r^2 - y^2)(h - y)\,dy.$$

Hence the work done W in pumping all the water 7 feet above the top of the tank is

$$W = \pi w \int_{-r}^{0} (r^2 - y^2)(h - y)\,dy = \pi w r^3\left[\frac{r}{4} + \frac{2}{3}h\right]$$

$$= \boxed{\frac{100{,}125}{8}\pi \text{ foot-pounds}}.$$

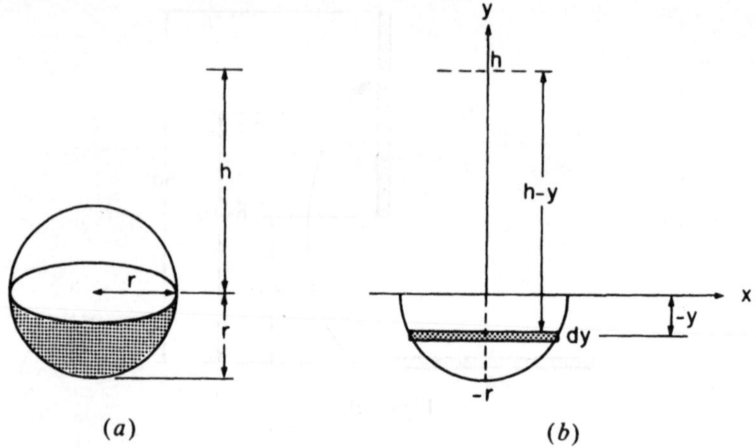

Figures III.25*a* and *b*.

Alternatively, one can show that the depth of the centroid of a hemisphere of radius r is $\frac{3}{8}r$. The weight of water in the tank is $\frac{2}{3}\pi r^3 w$. Hence the work done in pumping the water to the desired height is $\frac{2}{3}\pi r^3 w\left[\frac{3}{8}r+h\right]=\pi w r^3\left[\frac{r}{4}+\frac{2}{3}h\right]$.

(*)26. A tank containing 1000 lb of water is being lifted to the top of a building 80 ft high by a cable 80 ft long and weighing 160 lb. How much work is done in raising this tank of water 50 ft if the water is leaking out of the tank at a rate of one pound for each foot the tank is raised?

The cable has density $\frac{160}{80}=2$ lb/ft. When the tank is x ft from the top of the building (Fig. III.26) the weight of the rest of the cable is $2x$ lb, and the tank contains $(920+x)$ lb of water. Hence the force $F(x)$ required to lift the tank of water (assuming the tank is weightless) to a distance x ft from the top is $F(x)=3x+920$. Thus the work done in raising the tank of water

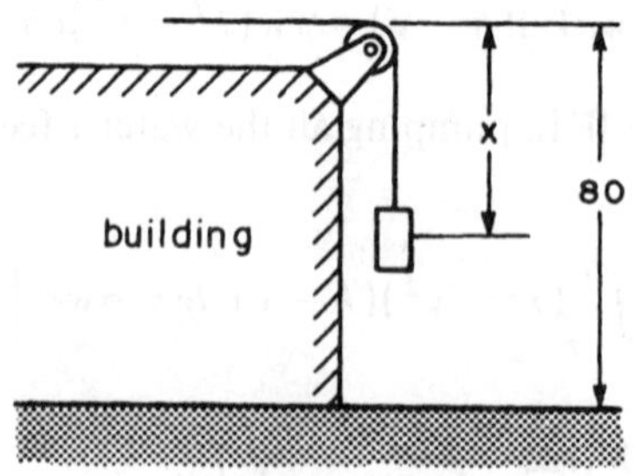

Figure III.26

50 ft is

$$W = \int_{30}^{80} F(x)\, dx = \int_{30}^{80} (3x + 920)\, dx$$

$$= \boxed{54{,}250 \text{ ft-lb}}.$$

(*)**27.** A hydroelectric storage dam (Fig. III.27*a*) has the form of a large trough with a rectangular base $CDHG$ ($CD = 100$ metres, $DH = 400$ metres), two rectangular walls $BCGF$ and $ADHE$ equally inclined to the vertical, and vertical trapezoidal ends $ABCD$, $EFGH$. When the dam is full, the water surface is a rectangle $ABFE$ with $AB = 250$ metres and $AE = 400$ metres, and the depth of water is 100 metres. Pipes lead from the base of the dam through a pump/generator P to a large lake L, 300 metres below the base of the dam. Find the work W that must be done to fill the dam (starting empty) by pumping water from the lake.

As a first step find the height $\bar{h}$ of the centroid of a cross-section (trapezoid) of the given trough where $\boxed{a = 100}$, $\boxed{b = 250}$, $\boxed{c = 100}$ (Fig. III.27*b*).

The area of this trapezoid is $\boxed{A = \left(\frac{b+c}{2}\right)a}$, l = length of trough = 400.

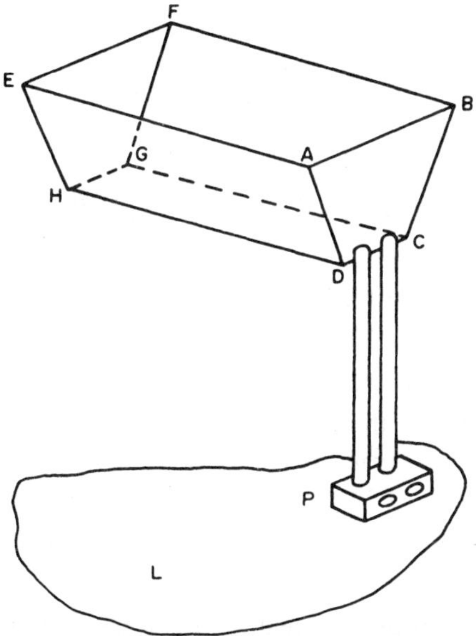

Figure III.27*a*

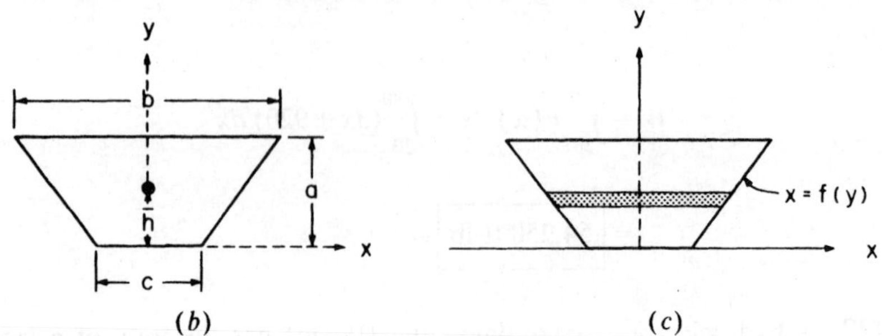

Figures III.27*b* and *c*.

$w = 1$ tonne/m³ is the density of water. Then the dam can contain $T = \left(\frac{b+c}{2}\right) alw = 7 \times 10^6$ tonnes of water, and the work done to fill the dam by pumping water from the lake is $W = T(\bar{h} + H)$ metre-tonnes, where $H = 300$ metres is the distance from the base of the dam to the lake.

Calculation of $\bar{h}$ *(Fig. III.27c)*:

$$\bar{h} = \frac{2\int_0^a yf(y)\,dy}{A},$$

where $x = f(y)$ satisfies the equation $y - a = -\frac{a(2x-b)}{c-b}$ $\leftrightarrow x = \left(\frac{b-c}{2a}\right)y + \frac{c}{2} = f(y)$. Hence

$$\bar{h} = \frac{\frac{b-c}{a}\int_0^a y^2\,dy + c\int_0^a y\,dy}{A}$$

$$= \frac{\frac{(b-c)a^2}{3} + \frac{ca^2}{2}}{A} = \left(\frac{2b+c}{b+c}\right)\frac{a}{3}.$$

Thus $W = \boxed{2.5 \times 10^9 \text{ metre-tonnes}}$.

(*)28. Two positive charges q_1 and q_2 are placed at positions $x = 5$ and $x = -2$ on the x-axis. A third positive charge q_3 is moved along the x-axis from $x = 1$ to $x = -1$ (Fig. III.28). Find the total work done by the electrostatic forces of q_1 and q_2 on q_3. (The force of repulsion between two positive charges q and Q has magnitude $\frac{kqQ}{r^2}$ where r is the distance between q and Q, and k is a positive constant.)

Let x be the position of the charge q_3, $-1 \le x \le 1$. The charge q_2 pushes the charge q_3 to the left whereas the charge q_1 pushes the charge q_3 to the right.

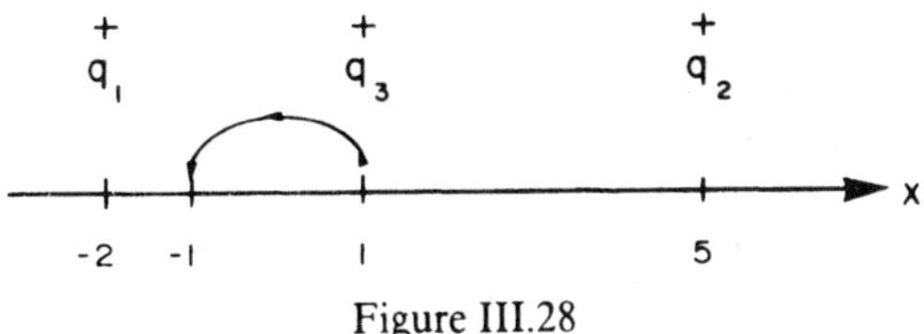

Figure III.28

The distance from q_3 to q_2 is $|x-5|$ and the distance from q_3 to q_1 is $|x+2|$.

The net force acting on q_3, when it is located at position x, is $F(x)=\frac{-kq_2q_3}{(x-5)^2}+\frac{kq_1q_3}{(x+2)^2}$. Hence the total work done by the electrostatic forces when q_3 is moved from $x=1$ to $x=-1$ is

$$\int_1^{-1} F(x)\,dx = kq_3\left[\int_1^{-1}\left(\frac{q_1}{(x+2)^2}-\frac{q_2}{(x-5)^2}\right)dx\right]$$

$$= kq_3\left[\frac{-q_1}{x+2}+\frac{q_2}{x-5}\right]\Bigg|_{x=1}^{x=-1} = \boxed{\frac{kq_2q_3}{12}-\frac{2kq_1q_3}{3}}.$$

29. Hooke's law states that if a spring is stretched S inches beyond its natural length, it is pulled back by a force equal to kS where k is a constant depending on the spring. How much work is done to stretch a spring 5 inches beyond its natural length if it takes 10 pounds of force to stretch it 1 inch?

The force $F(x)$ necessary to stretch a spring x inches beyond its natural length is kx. Hence the work done W to stretch a spring X inches beyond its natural length is

$$W=\int_0^X F(x)\,dx = k\int_0^X x\,dx = \frac{kX^2}{2}.$$

One is given that $k=10$, $X=5$. Hence $W=\boxed{125 \text{ inch-pounds}}$.

30. Suppose that the outside temperature is $-10°C$ and your heating system breaks down. A quick check verifies that the inside temperature falls from 20°C to 18°C in 14 minutes. How long will it be before there is any danger of the pipes freezing (i.e. how long before the inside temperature falls to 0°C)? (Note: If $T(t)$ is the inside temperature in °C at time t, then $y(t)=T(t)-(-10)$ is the temperature difference between inside and outside. Newton's law of cooling states that the rate of change of $y(t)$ is proportional to $y(t)$.)

By Newton's law of cooling, $\frac{dy}{dt} = ky$, where k is a proportionality constant. One is given that $T(0) = 20$, $T(14) = 18$. Hence $y(0) = 30$, $y(14) = 28$. The aim is to find the time $t = \tau$ such that $T(\tau) = 0$ corresponding to $y(\tau) = 10$.

$$\frac{dy}{dt} = ky \leftrightarrow \frac{dy}{y} = k\,dt.$$

Hence $\log y = kt + \text{const.} \leftrightarrow y = Ce^{kt}$ for some constant C.

$$y(0) = 30 \leftrightarrow 30 = C;$$

$$y(14) = 28 \leftrightarrow 28 = 30e^{14k} \Rightarrow 14k = \log \tfrac{28}{30} \Rightarrow k \simeq -0.005;$$

τ satisfies the equation $10 = 30e^{k\tau}$.

Hence $\log\frac{1}{3} = k\tau$ and thus

$$\tau \simeq \frac{-1.1}{-0.005} = \boxed{220 \text{ min}}.$$

(*)31. A water pipe of diameter 10 cm has its flow controlled by a valve consisting of a circular disk, of the same diameter as the pipe, which is moved back and forth across the pipe (Fig. III.31a). When the centre of the valve disk is 5 cm from the centre of the pipe, what percentage of the maximum water flow occurs?

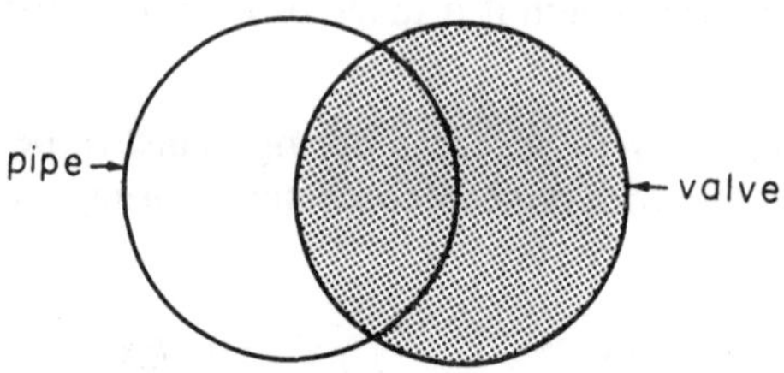

Figure III.31a

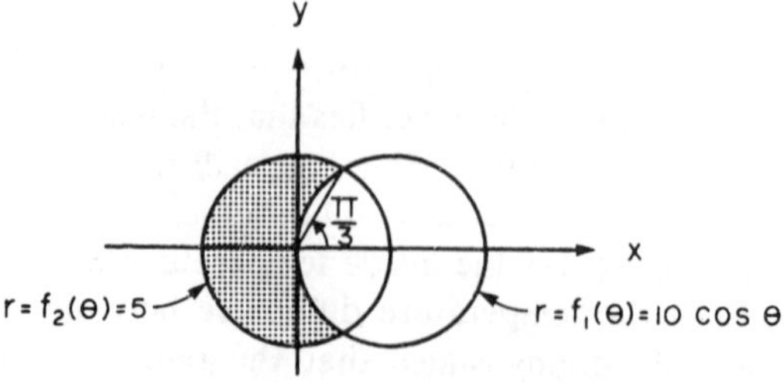

Figure III.31b

(Hint: Use polar coordinates, the equation of the pipe being $r = 5$ cm, that of the valve disk being $r = 10\cos\theta$ cm, and assume that the water flow is directly proportional to the area of the opening.)

Let A be the area of the shaded region (Fig. III.31b). Then the percentage of the maximum water flow occurring is given by $P = \dfrac{A}{\pi(5)^2} \times 100\%$. The circles intersect when $5 = 10\cos\theta \leftrightarrow \cos\theta = \frac{1}{2} \leftrightarrow \theta = \dfrac{\pi}{3}, \dfrac{5\pi}{3}$. Hence

$$A = 2\left[\int_{\frac{\pi}{3}}^{\pi} \frac{1}{2}[f_2(\theta)]^2\, d\theta - \int_{\frac{\pi}{3}}^{\frac{\pi}{2}} \frac{1}{2}[f_1(\theta)]^2\, d\theta\right]$$

$$= \int_{\frac{\pi}{3}}^{\pi} 25\, d\theta - \int_{\frac{\pi}{3}}^{\frac{\pi}{2}} 100\cos^2\theta\, d\theta$$

$$= \frac{50\pi}{3} - 50\int_{\frac{\pi}{3}}^{\frac{\pi}{2}} [1 + \cos 2\theta]\, d\theta$$

$$= \frac{25\pi}{3} - 50\int_{\frac{\pi}{3}}^{\frac{\pi}{2}} \cos 2\theta\, d\theta = \frac{25\pi}{3} - [25\sin 2\theta]\Big|_{\theta = \frac{\pi}{3}}^{\theta = \frac{\pi}{2}}$$

$$= \frac{25\pi}{3} + \frac{25\sqrt{3}}{2}.$$

Hence

$$P = 100\left[\frac{1}{3} + \frac{\sqrt{3}}{2\pi}\right]\% \simeq \boxed{61\%}.$$

32. The frequency of a simple pendulum is given by $f(L) = \dfrac{1}{2\pi}\sqrt{\dfrac{g}{L}}$, where L is the length of the pendulum and g is a constant (the local acceleration due to gravity). Estimate Δf (the change in frequency) if L changes to $(L + \Delta L)$. Show that the relative change in the frequency, $\dfrac{\Delta f}{f}$, is given approximately by $-\dfrac{1}{2}\dfrac{\Delta L}{L}$.

$$f(L) = \frac{g^{1/2}}{2\pi} L^{-1/2}. \text{ Hence } \frac{df}{dL} = -\frac{g^{1/2}}{4\pi} L^{-3/2} = \frac{-f}{2L},$$

$$\text{i.e., } df = -\frac{f}{2L}\, dL.$$

Hence if L changes to $L + \Delta L (\Delta L \ll L)$, then

$$\Delta f \simeq -\frac{f}{2}\frac{\Delta L}{L} \leftrightarrow \frac{\Delta f}{f} \simeq -\frac{\Delta L}{2L}.$$

(*)**33.** Design a reflector which has the property of focussing parallel rays in a given plane onto a single point. Assume the laws of geometrical optics (angle of incidence = angle of reflection) and set up an appropriate differential equation. "Guess" an appropriate solution (of course you must check your "guess"). Let $(p, 0)$ be the focal point F of the reflector.

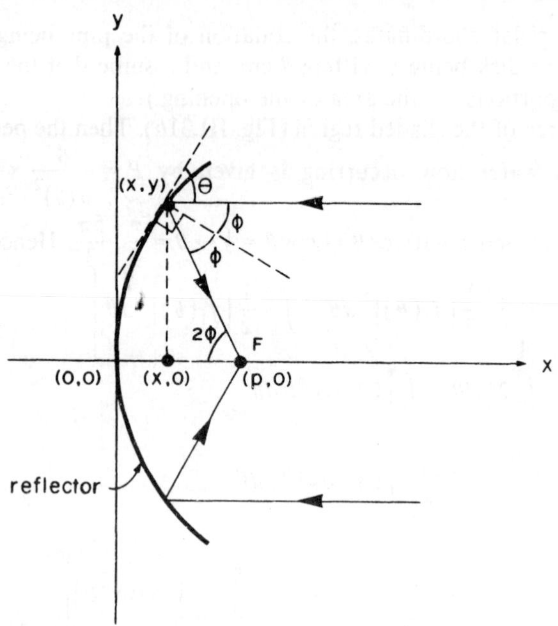

Figure III.33

Let $x = f(y)$ be the curve describing the shape of the reflector. Consider a representative incident ray with angle of incidence ϕ. Let θ be the angle between the incident ray and its contact with the reflector (Fig. III.33). Let $y' = \dfrac{dy}{dx}$ be the slope of the curve. Then

$$\tan\theta = y', \qquad \phi + \theta = \frac{\pi}{2}, \qquad \tan 2\phi = \frac{y}{p - x};$$

$$\tan\phi = \tan\left(\frac{\pi}{2} - \theta\right)$$

$$= \cot\theta = \frac{1}{\tan\theta} = \frac{1}{y'};$$

$$\tan 2\phi = \frac{2\tan\phi}{1 - \tan^2\phi} = \frac{2}{y'\left[1 - \dfrac{1}{(y')^2}\right]} = \frac{2y'}{(y')^2 - 1}.$$

Hence $x = f(y)$ satisfies the differential equation

$$\frac{2y'}{(y')^2 - 1} = \frac{y}{p - x}, \text{ where } y(0) = 0.$$

One might "guess" a solution of the form $y^2 = \alpha x$. Then $\boxed{2yy' = \alpha}$ and

$$\frac{2y'}{(y')^2 - 1} = \frac{2y^2y'}{y^2\left[(y')^2 - 1\right]} = \frac{2yy'y}{(yy')^2 - y^2} = \frac{\alpha y}{\left(\dfrac{\alpha}{2}\right)^2 - \alpha x} = \frac{y}{\dfrac{\alpha}{4} - x}.$$

Hence $p = \frac{\alpha}{4}$. Thus the shape of the mirror is given by the parabolic equation $\boxed{x = \frac{y^2}{4p}}$.

(*)34. A uniform cable hangs across two smooth pegs at the same height, the ends hanging down vertically. If the free ends are each 12 ft long and the tangent to the catenary at each peg makes an angle of 60° with the horizontal, find

(a) the total length of the cable,

(b) the sag in the cable.

Let $y = f(x)$ be the (unknown) curve describing the cable (Fig. III.34a). At point x along the cable, let $T(x)$ be the tension force which acts in the tangential direction and let $\theta(x)$ be the angle which the tangent line to the cable makes with the x-axis; $\tan\theta = \frac{dy}{dx}$. Let w be the weight per unit length of cable; $w = \text{const.}$

Now consider the balance of forces acting on a section of cable $(x, x + \Delta x)$ in Fig. III.34b. Since the cable is in static equilibrium, the horizontal force balance is

$$T(x+\Delta x)\cos\theta(x+\Delta x) = T(x)\cos\theta(x) \tag{1}$$

and the vertical force balance is

$$T(x+\Delta x)\sin\theta(x+\Delta x) = T(x)\sin\theta(x) + w\sqrt{(\Delta x)^2 + (\Delta y)^2} \tag{2}$$

where $\sqrt{(\Delta x)^2 + (\Delta y)^2}$ is approximately the length of a section of cable.

Dividing equations (1) and (2) by Δx and letting $\Delta x \to 0$ leads to the equations

$$T\cos\theta = H = \text{const.} \tag{1'}$$

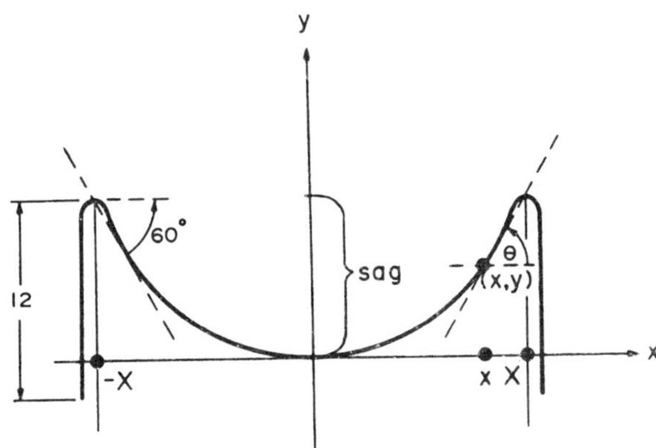

Figure III.34a

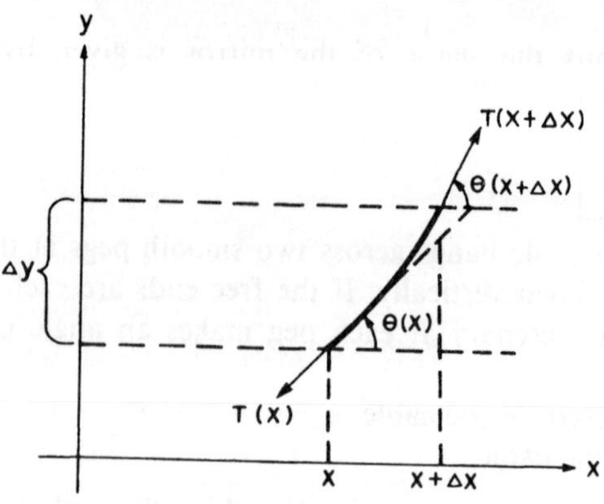

Figure III.34b

and

$$\frac{d}{dx}[T\sin\theta] = w\sqrt{1+\left(\frac{dy}{dx}\right)^2}. \tag{2'}$$

Substituting equation (1′) into equation (2′) one finds that

$$\frac{H}{w}\frac{d}{dx}(\tan\theta) = \sqrt{1+\left(\frac{dy}{dx}\right)^2}.$$

The cable $y = f(x)$ occupies the region $-X \le x \le X$, where $f(0) = 0$, $f(-X) = f(X)$, $f'(-X) = -f'(X)$, $\theta(X) = \frac{\pi}{3}$, $\theta(-X) = -\frac{\pi}{3}$. From the given data, $H = 12w\cos 60° = 6w$. Hence $y = f(x)$ satisfies the differential equation

$$6\frac{d}{dx}(\tan\theta) = \sqrt{1+\left(\frac{dy}{dx}\right)^2}$$

$$\leftrightarrow 6\frac{d}{dx}(\tan\theta) = \sqrt{1+\tan^2\theta}. \tag{3}$$

(a)

The total length of the cable is $L = \int_{-X}^{X}\sqrt{1+\left(\frac{dy}{dx}\right)^2}\,dx + 24$

$$= 6\int_{-X}^{X}\left[\frac{d}{dx}(\tan\theta)\right]dx + 24$$

$$= 6\tan\theta(x)\Big|_{x=-X}^{x=X} + 24$$

$$= 12\tan\frac{\pi}{3} + 24 = \boxed{(12\sqrt{3}+24)\text{ feet}}.$$

(b) Let $u = \tan\theta = \dfrac{dy}{dx}$. The aim is to find the sag $y(X) - y(0) = y(X)$.

With $u = \tan\theta$, equation (3) reduces to $\dfrac{du}{\sqrt{1+u^2}} = \dfrac{dx}{6}$.

Hence $u = \sinh\left(\dfrac{x-C}{6}\right)$ where C is some constant.

$$u(-X) = -u(X) \Rightarrow C = 0. \text{ Hence } \frac{dy}{dx} = u = \sinh\frac{x}{6}.$$

Integration of this equation gives $y = 6\cosh\dfrac{x}{6} + \text{const.}$ The given data yield const. $= -6$. Hence $y = f(x) = 6\cosh\dfrac{x}{6} - 6$. Note that

$$\sqrt{1+\left(\frac{dy}{dx}\right)^2} = \cosh\frac{x}{6},$$

and after integrating this equation from $x = -X$ to $x = X$, one gets

$$12\sqrt{3} = \int_{-X}^{X} \cosh\frac{x}{6}\,dx = 12\sinh\frac{X}{6},$$

$$\text{i.e., } \boxed{\sinh\frac{X}{6} = \sqrt{3}}.$$

Hence the sag in the cable is

$$y(X) = 6\cosh\frac{X}{6} - 6 = 6\left[\sqrt{1+\sinh^2\frac{X}{6}} - 1\right]$$

$$= \boxed{6 \text{ feet}}.$$

(*)35. What should the speed limit be for cars on the Lions Gate Bridge in Vancouver, British Columbia, during rush hour traffic, in order to maximize the flow of traffic?

This is of course an "open-ended" problem. The answer depends on one's assumptions. Here we make some simple-minded assumptions:

(a) The source of cars to the bridge is excellent.
(b) All cars move at the same speed v and have the same length l.
(c) The separation distance between each car is the same.

Let q be the traffic flow rate, i.e., the flux of cars, corresponding to the number of cars per hour passing a given point or the *capacity* of the bridge; let ρ be the density of cars, i.e., the number of cars per mile of road; let l^* be the effective space occupied by each car. l^* equals l plus the spacing between cars. Then $\rho = \dfrac{1}{l^*}$ and $q = \rho v$.

For a given bridge one can determine ρ by aerial photographs and q by traffic counts. The problem now is to determine the density $\rho = \rho(v)$ and find the optimal speed, $v = v_{\text{opt}}$, maximizing $q(v)$. $[q_{\text{opt}} = q(v_{\text{opt}})]$

Model 1. This model is based on radio and television advertisements in British Columbia which recommended that drivers should space out one car length for every 10 mph of speed, i.e. $l^* = l\left[1 + \frac{v}{10}\right]$ corresponding to $\rho = \frac{1}{l\left[1 + \frac{v}{10}\right]}$. (The time interval $\frac{l^*}{v}$ is called the *headway*.)

Hence $q(v) = \dfrac{v}{l\left[1 + \frac{v}{10}\right]}$.

Here $q(v)$ is a monotonically increasing function of v (Fig. III.35a), $\lim_{v \to +\infty} q(v) = \frac{10}{l}$.

This implies that there should be no speed limit!

For a typical Vancouver car such as a 1983 Ford Fairmont, $l \simeq \frac{1}{325}$ miles. Hence the optimal flow rate is $q_{\text{opt}} = \lim_{v \to +\infty} q(v) = 3250$ cars/hour. Note

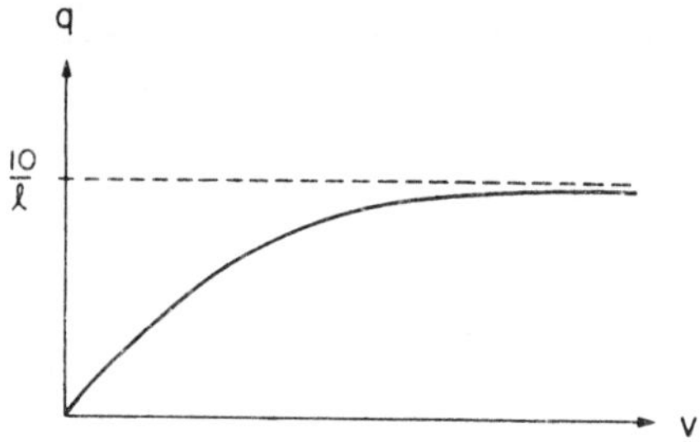

Figure III.35a. Flow vs. speed for Model 1.

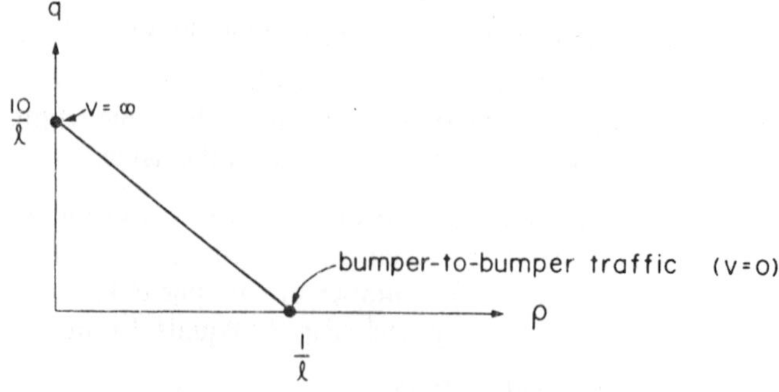

Figure III.35b. Flow vs. density for Model 1.

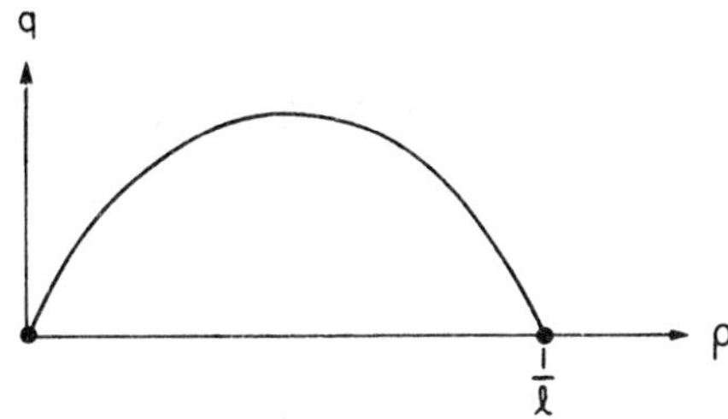

Figure III.35*c*. Typically observed flow vs. density curve.

that when $v = 15$ mph, $q = 1950$ cars/hour; when $v = 30$ mph, $q \simeq 2450$ cars/hour.

Since for a given situation one is able to determine q and ρ from simple measurements, a traffic engineer is interested in the flow vs. density curve (Fig. III.35*b*). For *Model 1*, $q = q(\rho) = \dfrac{10}{l} - 10\rho$.

Experimentally it appears that Fig. III.35*c* shows the shape of a typical curve representing flow vs. density on a throughway.

It is observed that the optimal flow rate during rush hours for each lane of traffic on the Lions Gate Bridge is about 1600 to 1800 cars/hour.

Model 2. This model could be described as "tailgating at low speeds" (see Figs. III.35*d* and III.35*e*).

Say the distance between cars is $l^* = l\left[1 + \left(\dfrac{v}{\alpha}\right)^2\right]$ where $\alpha \gg 10$ mph.

Then $q = \dfrac{v}{l\left[1 + \left(\dfrac{v}{\alpha}\right)^2\right]} = \alpha\sqrt{\dfrac{\rho}{l}}\sqrt{1 - \rho l}$.

$$\frac{dq}{d\rho} = 0 \leftrightarrow \rho = \frac{1}{2l}; \qquad \frac{dq}{dv} = 0 \leftrightarrow v = \alpha.$$

If $l = \dfrac{1}{325}$ and the optimal value of q is 1700 then the value of the parameter $\alpha \simeq 11$ mph, using $q_{\text{opt}} = \dfrac{\alpha}{2l}$. This violates the original assertion that $\alpha \gg 10$ mph!

(*d*)

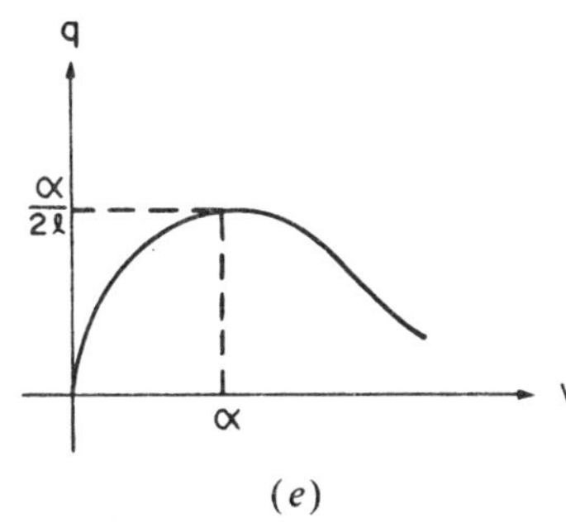

Figure III.35*d*. Flow vs. density for Model 2. *e*. Flow vs. speed for Model 2.

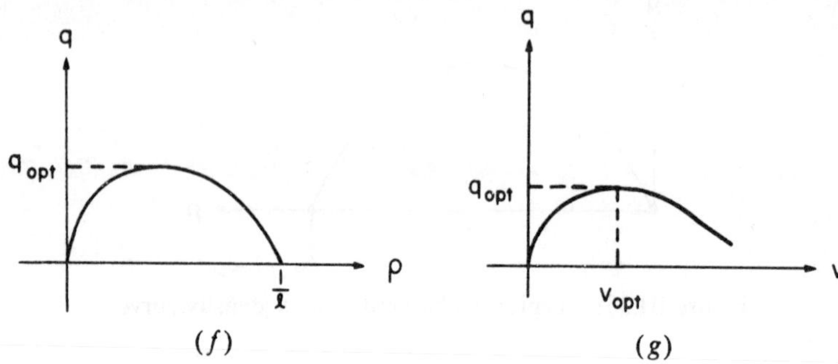

Figure III.35f. Flow vs. density for Model 3. g. Flow vs. speed for Model 3.

Model 3. Say each driver is very cautious and treats the car ahead of him as if it were stationary to determine his spacing. Then $l^* = l + v\tau + \frac{v^2}{A}$ where $l^* - l$ is the *braking distance*, A is a deceleration constant and τ is a driver's reaction time. From data in the British Columbia motor vehicle manual, on stopping distances at different speeds, one can determine that $\tau \simeq 0.7$ seconds and $A \simeq 88{,}000$ miles/(hr)^2.

Using differential calculus, one can easily show that the maximum (optimum) flow rate (Fig. III.35f) is given by

$$q = q_{opt} = \frac{1}{\tau + \sqrt{\frac{4l}{A}}} \simeq 1800 \text{ cars/hr if } l = \frac{1}{325} \text{ miles,}$$

and the optimal speed is (Fig. III.35g)

$$v = v_{opt} = \sqrt{lA} \simeq 16 \text{ mph.}$$

In each model it is interesting to note the dependence of the answer on the length l of a motor vehicle. (During rush hours, San Francisco has experimented with separate lanes for vehicles according to length.)

In general one might consider a model where $l^* = l + \alpha v + \beta v^2$ with the values of the parameters α and β determined experimentally for a particular throughway (freeway). Note that for such a model, the value of v_{opt} is *independent* of the value of α.

Supplementary Problems

1. Consider the legendary race of the hare and the tortoise. Suppose from the start of the race the hare is travelling at the speed of $30\sqrt{t}$ metres per hour while the tortoise is moving at the speed of $20t$ metres per hour. How far has each travelled after 1 hour, after 4 hours and after 9 hours?

(*)2. A rocket car accelerates from 0 km per hour to 480 km per hour in a test run of one kilometre. If the acceleration is not allowed to increase during the run, what is the longest time that the run can take?

3. In the year 2001 A.D. a spaceship is coming in for a *horizontal* landing on the moon at 2000 metres per second. (Aside: 2000 m/s is half way between the lunar circular velocity of 1680 m/s and the lunar escape velocity of 2376 m/s.)

The spaceship is to be arrested by an electromagnetic landing track so that during touchdown its velocity v will obey the law $v = 2000 - 20t$ where t is in seconds. (Note: When $t = 100$ sec the spaceship will have come to rest.)

(a) Using calculus and $v = 2000 - 20t$, compute the deceleration of the spaceship in m/s^2.

(b) Use calculus and $v = 2000 - 20t$ to compute the distance covered between touchdown and final stop.

4. A particle moves along the x-axis starting from $x = 12$ with a velocity of -4 units per second. Its acceleration depends on the time t and is given by $a(t) = 12t - 10$.

(a) What are the particle's velocity and position after 10 seconds?

(b) At what times $t \geq 0$ is the particle moving to the left? to the right? Note: Negative velocities are to the left, positive to the right.

5. A ball is thrown vertically upward from the ground with an initial speed of 64 feet/second.

(a) When does the ball reach its maximum height?

(b) How high does the ball rise?

(c) How many seconds does it take for the ball to reach the ground?

(d) What is the speed of the ball on impact with the ground?

(*)6. A rocket of mass 1 kg is launched vertically upward from the ground with zero initial velocity. The rocket engine supplies a constant upward force of F kg m/s^2 for T sec. After T sec the rocket is subject only to the force of gravity.

(a) Find the position and velocity of the rocket when the rocket engine stops.

(b) Find the time at which the rocket returns to the ground. (Neglect the air resistance and assume that the gravitational acceleration g is constant.)

(*)7. A home-made rocket has a total initial mass of $M = 100$ lb including 50 lb of fuel. It burns its fuel at the rate of $b = 10$ lb/sec and the fuel is ejected with a velocity of $u = 400$ ft/sec relative to the rocket. The vertical motion of the rocket is governed by Newton's law $\frac{d}{dt}(mV) = mg$ where $g = 32$ ft/sec^2. This yields the differential equation

$$M\left(1 - \frac{bt}{M}\right)\frac{dV}{dt} = ub - M\left(1 - \frac{bt}{M}\right)g \quad \text{with } V(0) = 0.$$

Find the velocity $V(t)$ for $0 \leq t \leq 5$. Explain why in the design of such a rocket one must have $ub - Mg \geq 0$ in order for the rocket to be airborne.

8. A shell leaves the mouth of a cannon at 4800 ft/sec. The shell was fired vertically upwards by mistake. If the acceleration due to gravity is 32 ft/sec^2 at all heights and air resistance is negligible, how high will the shell go before it begins to fall back to earth? How long does the incompetent gun crew have to make a getaway, assuming the shell takes as long to fall back to earth as it took to reach its maximum height?

9. A rocket far out in space (so that gravity may be neglected) is falling towards the earth with a velocity inversely proportional to $\sqrt{x}$, where x is its distance from the centre of the earth. Show that the acceleration is inversely proportional to x^2.

10. The acceleration of a falling body on the moon is one-sixth that of a body on the earth. Suppose that, from the top of a spaceship, 576 feet high, that is resting on the moon, a rock is thrown upward with a velocity of 80 feet per second. Obtain an equation which describes the position of the rock from the moment it is thrown and then at all subsequent times. What is the rock's maximum height? At what time does it hit the surface of the moon?

(*)11. A marble rolls along a straight line in such a manner that its velocity is directly proportional to the distance it has *yet* to roll. If the total distance it rolls is 5 feet, and after 1 second it has rolled 2 feet, express the distance it has rolled as a function of time.

12. The cross-section of a trough has the shape of an inverted isosceles triangle. If the lengths of the equal sides are 15 in, find the size of the vertex angle that will give maximum capacity for the trough. Show you have a maximum.

13. A particle moves on the x-axis such that its position at time t is $x = 1 - \tan t,\ 0 \leq t < \frac{\pi}{2}$. Another particle moves on the y-axis with its position given by $y = \sec t$ at the same time t. Find the minimum distance that the particles will be apart.

14. A missile is faulty and is destroyed by the range officer 3 seconds after launching. Telemetric data show that the missile height h, in metres, above the launching pad at time t was given by

$$h = 20t^3 - 90t^2 + 120t \text{ metres.}$$

(a) What was the greatest height it reached?

(b) What would it have been if the range officer had been nimbler and had destroyed the missile 2 seconds after launching?

15. We are given heat sources at points A and B, 8 units apart, with the ratio of the strengths of the sources being 27/8. If the heat received at a point is inversely proportional to the square of the distance from the heat source and directly proportional to the strength of the source, at what point on the line segment joining A and B will the heat received be a minimum?

(*)**16.** Highways (1) and (2) run along the x- and y-axes respectively, starting from points A and B. The interior of the ellipse $\frac{x^2}{a^2}+\frac{y^2}{b^2}=1$, with $a=OA$, $b=OB$, has been declared a national park in which no roads are allowed (Fig. S.16). Prove that any straight road linking up the two highways must exceed $a+b$ in length.

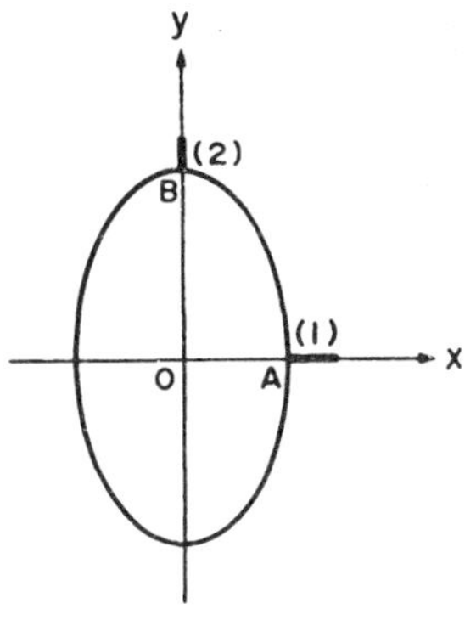

Figure S.16

17. A rocket is moving along a path with equation $y=\sqrt{x^2-4}$, $x\geq 4$, in a certain coordinate system. y represents the height in km, and x represents the horizontal distance from a fixed reference point, in km. If x is increasing at the rate of 5 km per second, how fast is y changing?

18. A cylinder of radius 10 cm contains gas at a constant temperature which is being compressed by a piston. When the piston is 20 cm from the end of the cylinder the pressure $p=100$ kilopascals. The piston moves at 3 cm/s. If pV is constant, where V is the volume of gas, what is the rate of change of pressure when the piston is 10 cm from the end of the cylinder?

(*)**19.** In the Second World War a German bomber had a cruising speed of 300 miles per hour. A searchlight in London catches a German bomber in its beam. The bomber is travelling at a constant height in a straight line which passes directly over the searchlight. If the searchlight must be turned at 75 radians per hour to keep the bomber in the beam when the angle between the ground and the beam is 45°, what is the height of the bomber?

(*)**20.** Suppose that at night an RCMP patrol boat is sailing due south and its crew notices a motor boat sailing due north and wants to board it to search it. The RCMP boat is travelling at 15 feet per second and the other boat at 10 feet per second. When first observed, the other boat is at a distance 200 feet in a direction 30° from the path of the RCMP boat.

(a) What will be the distance between the boats at closest approach if neither changes course? When will that occur?

(b) How fast must the RCMP patrol boat's searchlight be turned to keep the other boat in view when it is first seen?

21. Figure S.21 shows a rod OB rotating counterclockwise in the xy-plane about O at 2 revolutions per second. Attached to OB is a rod AB. A is confined to slide horizontally along the x-axis only.

(a) If OB has length l metres, and AB has a length greater than $2l$ metres, find the x-component of the velocity of A as a function of x and θ.

(b) At what position(s) of B is the velocity of A equal to zero?

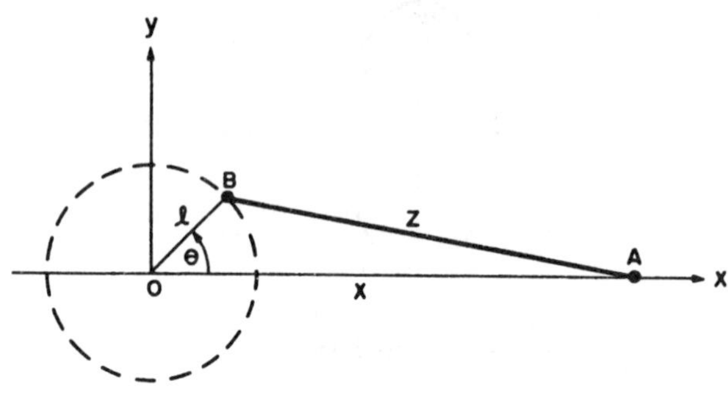

Figure S.21

22. A crankshaft OA of length 20 cm is connected by a connecting rod AP to a piston P which slides on the line OP. The crankshaft is turning around O at n rpm. What is the speed of the piston at a moment when $OP = 80$ cm and $< AOP = 45°$?

(*)23. You are the commander of a tank proceeding down the y-axis toward the origin at a rate proportional to your distance from the origin. At time $t = 0$ you are 4 km from the origin; 10 minutes later you are 2 km from the origin. Along the curve $y = \frac{1}{x}$ (all distances in km) there is a high wall. You know that the enemy is waiting somewhere along the positive x-axis with his gun aimed tangentially to the wall. For your safety's sake, how fast must the gun turret on your tank be capable of turning?

24. Determine the centroid of the region in the first quadrant bounded by the coordinate axes and the graph of $y = 4 - x^2$.

25. Let R be the plane region bounded by the curve $y = x^{1/3}$, the x-axis, and the line $x = 1$. Find the centroid of the solid of revolution generated by revolving R about the line $x = -1$.

26. Find the position of the centroid of the first quadrant of the circle $x^2 + y^2 = a^2$.

27. Determine the centroid of the region indicated in Figure S.27.

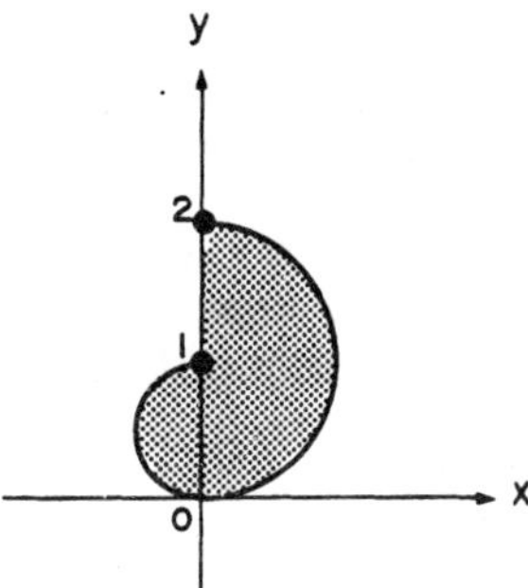

Figure S.27. Two semi-circles.

28. Set up, *but do not evaluate*, integrals for the coordinates of the centre of mass of the plane region bounded by

$$y = \cos^{-1} x, \qquad y = 0, \qquad x = -\frac{1}{2}, \qquad x = \frac{\sqrt{3}}{2},$$

assuming the density ρ is constant.

29. Find the centre of gravity of the region shown in Fig. S.29 consisting of a square of side length $2a$ topped by a semi-circle of radius a.

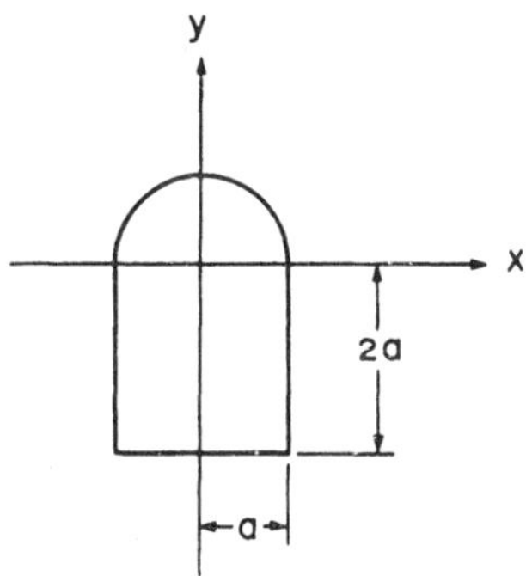

Figure S.29

30. The centre of gravity of a beer mug is strictly above the bottom of the mug, when the mug is empty. As beer is poured in the mug, the centre of gravity will initially descend, but will eventually rise. Show that when the centre of gravity is at its lowest point, it is exactly on the surface of the beer.

31. Find the moment of inertia about the x-axis of a thin plate of constant density ρ (mass per unit area) if its edges are defined by the curves

$$x = 0, \qquad y = 1, \qquad y = 3, \qquad x = \frac{1}{y\sqrt{y^2 + 7}}.$$

32. A vertical wall of a swimming pool is 10 ft wide and 8 ft high. Find the force of water on the wall when the pool is half full.

33. A one-foot cube is suspended in a lake, with the top of the cube 100 feet below the surface of the water. What is the total force on all sides of the cube, due to liquid pressure, if the top surface of the cube is horizontal?

34. An oil tank is in the shape of a right circular cylinder of diameter 4 ft with axis horizontal. Find the total force due to fluid pressure on one end of the tank when the tank is half full of oil weighing 40 pounds per cubic foot.

35. A boat is anchored so that the anchor is 100 feet directly below the drum about which the anchor chain is wound. The anchor weighs 3000 pounds and the chain weighs 20 pounds per foot. How much work is done in bringing up the anchor?

36. Find the work done in pumping all the water out of a conical reservoir of radius 10 m at the top, altitude 8 m, to a height of 6 m above the top of the reservoir.

37. A cylindrical tank 5 m high and 3 m in radius, is filled with oil weighing 50 Newtons/m^3. Find the work done in pumping all the oil out over the inner rim of the tank.

(*)38. A rectangular swimming pool full of water is 25 metres long and 10 metres wide at the top. It is 3 metres deep for its first 10 metres and then the depth decreases linearly to 1 metre at the shallow end. How much work is done in emptying the pool by pumping the water over the edge?

(*)39. A hemispherical reservoir is 20 metres in diameter and contains water with a depth of 5 metres. Find the work done in pumping the water out of the reservoir with a pump 2 metres above the top of the reservoir.

40. A dugout in the shape of a prism is shown in Fig. S.40 full of water. Find the work done in emptying the dugout if the water is pumped to a canal which is 8 metres above the top of the dugout.

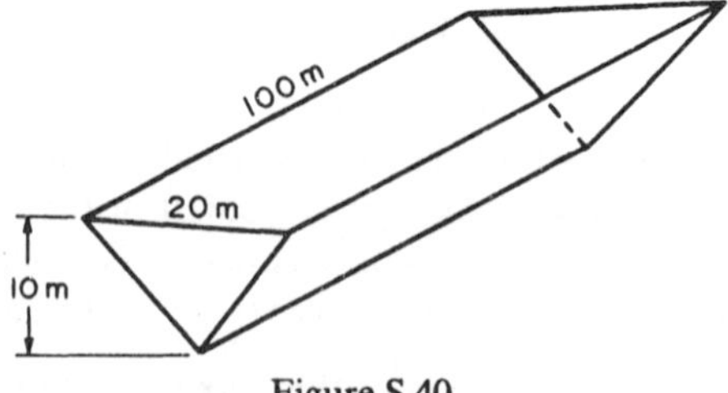

Figure S.40

41. A bucket is raised a distance of 20 feet from the bottom of a well with water leaking out of the bucket at a uniform rate. Find the work done if the bucket originally contains 30 pounds of water and if one-half of the water has leaked out by the time the bucket has been raised 20 feet.

42. A rope 10 feet long is hanging over the edge of a cliff. The (linear) density of the rope varies according to the formula $\rho(x) = e^{-x^2}$ lb/ft, where x is the vertical distance from the upper end of the rope (see Fig. S.42). What is the work required to raise the rope to the top of the cliff?

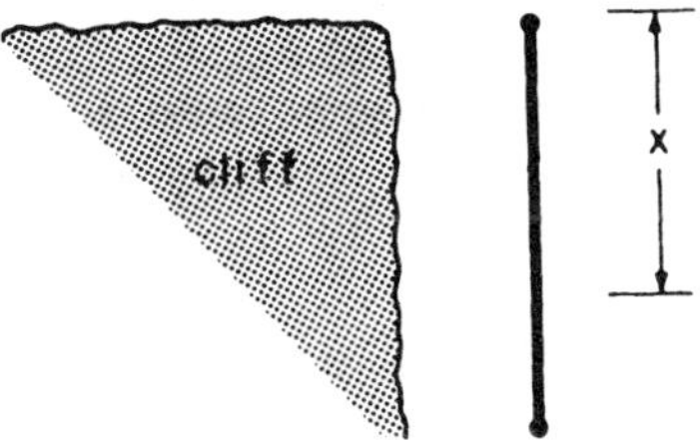

Figure S.42

43. The shape of the reservoir shown in Fig. S.43 is obtained by rotating the graph of $y = x^2$, between 0 and 4, about the y-axis. What is the work done in pumping it full of water from a lake 4 feet below the bottom? Assume water weighs w pounds per cubic foot.

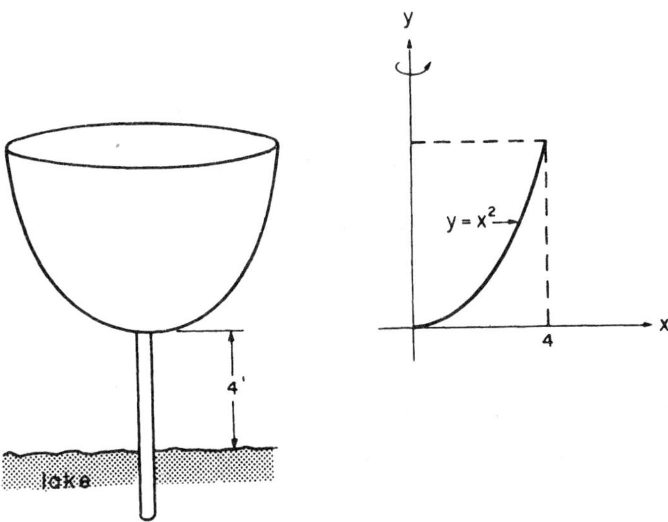

Figure S.43

44. The force on a charged object at a distance of x feet from a small sphere is $\dfrac{k}{x^2}$ lb (towards the sphere) where k is a constant. A force of 3 lb will hold it 6 feet away from the sphere. Find the work done in moving the object from 1 foot to 4 feet from the sphere.

45. A spring with no forces acting on it has length 0.6 metres. A force of 10 N is required to stretch it to a length of one metre. Using Hooke's law, and using integral calculus, determine the amount of work required to stretch the spring from 0.9 to 1.1 metres in length.

46. Say a spring has natural length $L=10$ in, and a force of 10 lb is sufficient to extend it to a length of 16 in.

(a) What is the value of the spring constant K for this spring?

(b) How much work is done in stretching it from a length of 12 in to 16 in?

47. A spiral spring stretches 1.8 inches when a force of 9 pounds is applied. The natural length of the spring is 14 inches. Find the work done in stretching the spring from 16 inches to 20 inches.

48. The force required for Mr. Peabody and his boy Sherman to travel through time satisfies the exponential law $F(t)=e^{Kt}$, where t is measured in years from 1979 (the pivot year of Mr. Peabody's machine) and K is Kingsbury's constant.

If a force of $\frac{\pi}{2}$ Einsteins is required to move from 1979 forward exactly one year to 1980, what is the work done in travelling from the Battle of Hastings (1066 A.D.) to the Roman Year of Confusion (40 A.D.)?

49. If $F(x)$ is the resultant force acting on a particle of fixed mass m located at position $x(t)$ at time t, then according to Newton's second law of motion, $F=ma=m\dfrac{d^2x}{dt^2}$. The kinetic energy of this particle is $\frac{1}{2}\,mv^2$ where $v=\dfrac{dx}{dt}$. Show that the work done by the resultant force in moving a particle from x_1 to x_2 is equal to its change in kinetic energy during the motion.

50. Newton's law of cooling states that the rate at which an object changes temperature is proportional to the difference between its temperature and the temperature of the surrounding medium. If an object placed in a medium whose temperature is 60° cools from 100° to 90° in 10 minutes, how many more minutes will it take for the object to cool to 80°?

51. If a warm body at temperature T is immersed in surroundings of temperature T_0, $T>T_0$, then $x=Ae^{-kt}$, where $x=T-T_0$, t is the time in hours, and A and k are constants. In a house heated by electricity to 70° Fahrenheit, there is a power failure while the outside temperature is 10°. Assuming that for that particular house $k=0.10$, after how many hours will the temperature be down to 32°?

(Note: The purpose of insulating a house is to reduce the magnitude of k.)

(*)52. Suppose a mathematics professor was murdered while lying on his bed. The coroner gathered the following information:

Time	Professor's body temperature
Midnight	94.6°F
1:00 a.m.	93.4°F

The room temperature was found to be 70°F during this period. Assume a reasonable body temperature for the professor before death, and use Newton's law of cooling to estimate when the professor expired.

53. A flywheel spinning on a shaft is slowed down by friction at a rate proportional to the speed of rotation so that $\frac{dp}{dt} = -kp$, where p is the speed of rotation. At $t=0$, the initial speed is $p(0)=1600$ revolutions per minute (rpm) and two minutes later we have $p(2)=800$ rpm. Find

(a) the speed ten minutes later, $p(10)$;

(b) the time, T, at which $p(T)=100$ rpm.

54. When pumping water into a reservoir, the rate of change (increase) of the water content is inversely proportional to the amount of water in the reservoir. If the reservoir had contained 1 ton of water at the beginning and if after 1 hour's pumping the reservoir contains 2 tons of water, when will the reservoir contain 3 tons of water?

55. A particle that is displaced 6 units from the origin and given an initial velocity of 3 units/sec undergoes simple harmonic motion with period equal to $4\sqrt{3}\,\pi$ units. Find the amplitude of the motion.

(*)56. The cross-section of a concave parabolic mirror has the equation $y^2 = kx$ ($k>0$ is constant). Parallel rays of light, represented by horizontal lines $y=$ const., come in from the right and hit the mirror. Prove that the reflections of all such rays pass through a single point (called the *focus* of the parabola $y^2 = kx$). Sketch.

57. In a far-off land, all the elevator repairmen go on strike. Stair climbers discover, by sad experience, that as they climb stairs they lose energy at a rate proportional to the amount present. At the end of 1 minute they lose 20% of their original energy. When will one-half of their energy be used up?

58. The length of an arc $\{(x, y)|y=f(x),\ a \le x \le b\}$ is given by $L = \int_a^b (1+(y')^2)^{1/2}\,dx$. A heavy chain hangs between two concrete piers, 20 metres apart, in the form of a curve $y = 10\cosh(x/10)$, called a *catenary*, the centre of the chain being on the y-axis. Sketch the curve and find its length. Show that the area under the catenary and above the x-axis, between the lines $x=-10$ and $x=10$, is $100(e-e^{-1})$.

59. The separation distance, l ft, between automobiles travelling in a single lane tunnel is related to their average velocity, v ft/sec, by $l=18+v+v^2/32$.

(a) How many cars pass a given point in an hour?

(b) For what speed is the traffic flow a maximum?

Chapter IV

Business and Economics

Recently differential and integral calculus techniques have become important tools for studying problems in business and economics. The major areas covered by the problems in this chapter include the following.

1. *Maxima and Minima*
   (For these problems there is no need for special knowledge of business and economics terms: see Chapter II.)
2. *Interest*
   (a) *principal, balance.*
   The *principal*, $P(t)$, is the amount of money invested at time t.
   The *balance* is $B(t) = P(t) + I(t)$, where $I(t)$ is the total interest earned up to time t.
   (b) *simple interest, compound interest.*
   Let $r\%$ be the annual interest rate, let T be the time in years, and let $P(t) = P = \text{const.}$ be the principal, i.e., the principal is fixed.
   In the case of *simple interest*, the balance after T years is $B(T) = P\left(1 + \frac{rT}{100}\right)$.
   In the case of *compound interest*, if the interest rate is compounded n times over T years, then at time $\frac{T}{n}$, the balance is $B\left(\frac{T}{n}\right) = P\left(1 + \frac{rT}{100n}\right)$; at time $\frac{2T}{n}$, the balance is $B\left(\frac{2T}{n}\right) = P\left(1 + \frac{rT}{100n}\right)^2$; ...; at time T, the balance is $B(T) = P\left(1 + \frac{rT}{100n}\right)^n$.
   In particular, if $T = 1$ year and the annual interest rate is compounded quarterly, then at the end of 1 year the balance is $B(1) =$

$P\left(1+\frac{r}{400}\right)^4$. In this case the *effective annual interest rate* is $100\left[\left(1+\frac{r}{400}\right)^4-1\right]\%$.

(c) *continuously compounded interest.*
In this case $n \to \infty$, where n is the number of compounding periods over the time period T. Here at time T, the balance is

$$B(T) = P \lim_{n \to \infty} \left(1+\frac{rT}{100n}\right)^n = Pe^{rT/100}.$$

(See the Solved Problems on indeterminate forms in Chapter VIII, Section 11, for the evaluation of such a limit.)

(d) *present value*
Say one invests a fixed sum P at time $t = 0$ years with interest earned up to time t according to some interest rule $f(t)$.
In the case of simple interest, $f(t) = 1 + \frac{rt}{100}$.
In the case of continuously compounded interest, $f(t) = e^{rt/100}$.
In the case of interest compounded semi-annually,

$$f(t) = \left(1+\frac{r}{200}\right)^{\tau(t)}, \text{ where } \tau(t) = n \text{ for } \frac{n}{2} \le t < \frac{n+1}{2}, n = 0,1,2,\ldots.$$

For any interest rule, at time t the balance is $B(t) = Pf(t)$. The *present value* for a balance $B(t)$ is said to be $P = \frac{B(t)}{f(t)}$.
For example the present value P of the given sum of \$1000 available in five years time with a continuously compounding interest rate of $r\%$ is:

$$P = \frac{B(5)}{f(5)} = \frac{1000}{e^{5r/100}} = 1000e^{-r/20} \text{ dollars.}$$

(This means that a sum of P dollars invested at time $t = 0$ is worth \$1000 after 5 years if the interest rate of $r\%$ is compounded continuously.)

3. *Economics*

(a) *demand, supply and equilibrium.*
The *demand curve* or *demand equation*, $p = D(q)$ (see Fig. IV.0a), relates the price per unit of a given product p to the quantity of the product q which the consumer is willing to buy. $D(q)$ (or its inverse) is called the *demand function*. The "law of demand" states that the lower the price of a product, the greater is the demand for the product, i.e., the slope of $D(q)$ is negative $\left[\frac{dD}{dq} < 0\right]$, assuming $D(q)$ is differentiable. (This guarantees the existence of its inverse.) Moreover $D(q)$ is usually concave upward, i.e., $\frac{d^2D}{dq^2} > 0$.
The *supply curve* or *supply equation* $p = S(q)$ relates the price per unit of a given product p to the quantity q which a producer is willing to supply.

$S(q)$ is called the *supply function*. If the price of a product decreases, a producer is less willing to produce it, i.e., the slope of $S(q)$ is positive. Moreover $S(q)$ is usually concave upward. Hence, normally, $\frac{dS}{dq} > 0$, $\frac{d^2S}{dq^2} > 0$.

The curves $p = S(q)$, $p = D(q)$ intersect at the *equilibrium point* $(\bar{q}, \bar{p})$. At this point the supply just equals the demand. $\bar{q}$ is called the *equilibrium demand*; $\bar{p}$, the *equilibrium price*.

(b) (*Marshallian*) *consumer's and producer's surplus*.

Theoretically the price per unit of a certain commodity in the market place is the equilibrium price $\bar{p}$. The theoretical market value of a commodity is (price per unit) $\times$ (quantity consumed) $= (\bar{p})(\bar{q})$.

However the consumer is willing to pay a higher price $D(Q) > \bar{p}$ if the quantity supplied is $Q < \bar{q}$. If a consumer buys dq units in the production interval $[Q, Q + dq]$ then he is willing to pay a price of $D(Q)$ per unit. If such a consumer pays the equilibrium price $\bar{p}$ then his saving is $[D(Q) - \bar{p}]\,dq$. The total amount of money that a consumer can save is the sum over all such intervals, namely, the *consumer's surplus* $= \int_0^{\bar{q}} [D(q) - \bar{p}]\,dq$, corresponding to area A in Fig. IV.0*a*.

Similarly the producer is willing to charge a lower price $S(Q) < \bar{p}$ if the quantity supplied is $Q < \bar{q}$. The total amount of money that a producer gains by selling at the equilibrium price is the *producer's surplus* $= \int_0^{\bar{q}} [\bar{p} - S(q)]\,dq$, corresponding to area B in Fig. IV.0*a*.

(c) *total cost, revenue, profit*.

The *total cost function*, $C(q)$, represents the total cost $C(q)$ for a producer to produce a quantity of q units of a given commodity. The

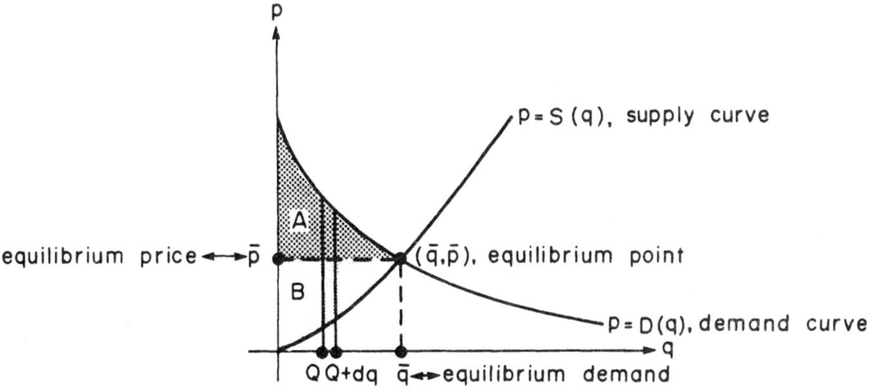

Figure IV.0*a*. Typical demand and supply curves.

revenue function, $R(q)$, is the total income $R(q)$ received by a producer for selling a quantity of q units of a given commodity. If $D(q)$ is the demand function for the commodity, then $R(q) = qD(q)$. The *profit*, $\pi(q)$, for a producer selling q units is $\pi(q) = R(q) - C(q)$ (Fig. IV.0*b*).

(d) *elasticity of demand.*

The *elasticity of demand*, η, is the response of quantity demanded to a change in price. It corresponds to the ratio of the percentage increase in the quantity demanded to the percentage drop in its price. Let Δq be the change in the quantity demanded with change in price $\Delta D(q)$, where the demand is $D(q)$ units. Then $\eta = \dfrac{\dfrac{\Delta q}{q}}{\dfrac{-\Delta D(q)}{D(q)}} = \dfrac{\dfrac{-D(q)}{q}}{\dfrac{\Delta D(q)}{\Delta q}}$. Letting $\Delta q \to 0$, one finds the (instantaneous) elasticity of demand

$$\eta = \frac{-D(q)}{qD'(q)}.$$

If $\eta > 1$, the demand is said to be *elastic.*

If $\eta = 1$, the demand is said to be at *unit elasticity.*

If $\eta < 1$, the demand is said to be *inelastic.*

Revenue is maximized at a value of q corresponding to unit elasticity of demand (see Solved Problem 16).

(e) *marginal analysis.*

In economics a derivative is called a marginal. Hence if $C(q)$ is the cost function, then $C'(q)$ is called the *marginal cost*. Similarly one can speak of marginal revenue, marginal profit, etc. The marginal of a dependent variable is used to estimate its change produced by a change of 1 unit in its independent variable, q.

For example the change in cost in producing $Q+1$ units instead of Q units is $C(Q+1) - C(Q)$. Marginal analysis estimates $C(Q+1) - C(Q) \approx C'(Q)$ (Fig. IV.0*c*).

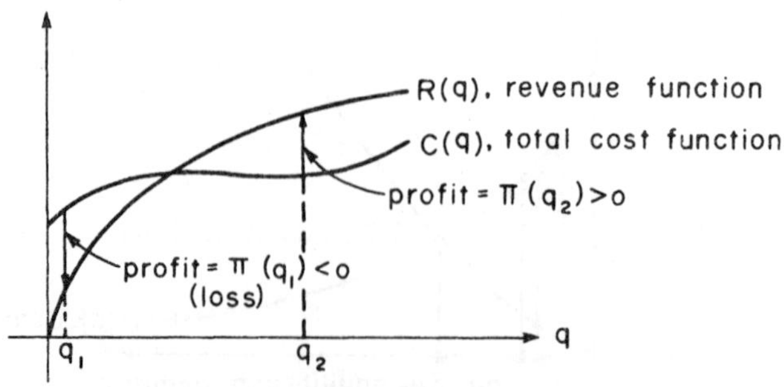

Figure IV.0*b*

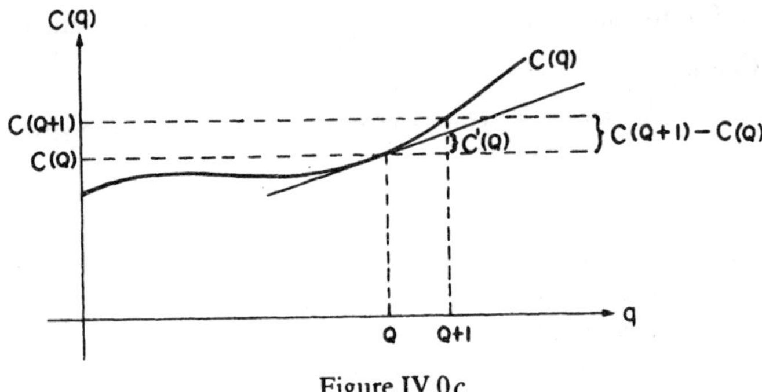

Figure IV.0c

Solved Problems

1. It is required to build a fence enclosing a rectangle of given area A, with one end facing a highway (Fig. IV.1).

The end facing the highway needs a decorative fence costing \$15 per metre, while the other boundaries can have a cheaper fence costing \$5 per metre. What should the proportions be (i.e. what should be the ratio of l to w) in order to minimize the cost of the fence?

$$A = lw = \text{constant}. \tag{1}$$

Let C be the cost of the fence

$$= \text{cost of fence along the highway} + \text{cost of other 3 sides}$$
$$= 15w + 5(w + 2l).$$

Equation (1) leads to $l = \dfrac{A}{w}$. Hence $C = C(w) = 20w + \dfrac{10A}{w}$.

Now one must find the value of w minimizing $C(w)$ where the domain of w is $w > 0$: $C'(w) = 20 - \dfrac{10A}{w^2}$. Hence $C'(w) = 0 \Rightarrow w = \sqrt{\dfrac{A}{2}}$; $C''(w)$

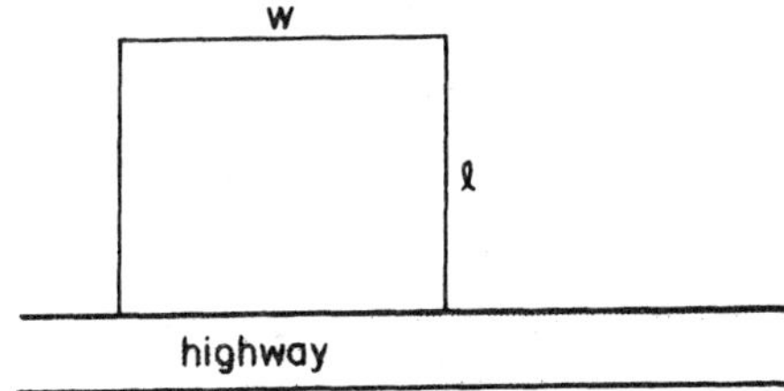

Figure IV.1

$= \dfrac{20A}{w^3} > 0$ for all values of $w > 0 \Rightarrow w = \sqrt{\dfrac{A}{2}}$ corresponds to the minimum value of $C(w)$.

When $w = \sqrt{\dfrac{A}{2}}$, $l = \sqrt{2A}$. Hence the ratio of l to w minimizing the cost is $\dfrac{l}{w} = \boxed{2}$.

(*)2. A river is 300 metres wide. A power line must be installed from a power house on one side to a dwelling on the other side which is 400 metres downstream from the point directly opposite the power house. If it costs \$6 per metre to lay the cable under water and \$2 per metre to lay it on land, find the least cost to install the cable.

From symmetry, for a particular cost to install the cable, there are an infinite number of paths for laying it. Two such equivalent paths are shown in Figure IV.2. (What are other equivalent paths for laying the cable?) Let:

> $P \leftrightarrow$ location of power house, $D \leftrightarrow$ location of dwelling, A or equivalently $A' \leftrightarrow$ node where the underwater cable meets the land cable, $O \leftrightarrow$ point directly across the river from P, ($O' \leftrightarrow$ point directly across the river from D).

The cable follows the path $PA \cup AD$ or, equivalently, $PA' \cup A'D$ (Fig. IV.2). Let x be the distance from O to A. Then the distance from P to A is $l_1 = \sqrt{(300)^2 + x^2}$; the distance from A to D is $l_2 = 400 - x$. Let C be the total cost to lay the cable. Then

$$C = C(x) = 6l_1 + 2l_2$$
$$= 6\sqrt{(300)^2 + x^2} + 2(400 - x).$$

$$\frac{dC}{dx} = \frac{6x}{\sqrt{(300)^2 + x^2}} - 2; \quad \frac{dC}{dx} = 0 \Rightarrow 36x^2 = 4\left((300)^2 + x^2\right)$$

$\leftrightarrow 8x^2 = (300)^2 \Rightarrow x = 75\sqrt{2}$ since it is necessary that $x > 0$ where $\dfrac{dC}{dx} = 0$. As $x \to -\infty$, $\dfrac{dC}{dx} \to -8$; as $x \to +\infty$, $\dfrac{dC}{dx} \to 4$. Hence $x = 75\sqrt{2}$ corre-

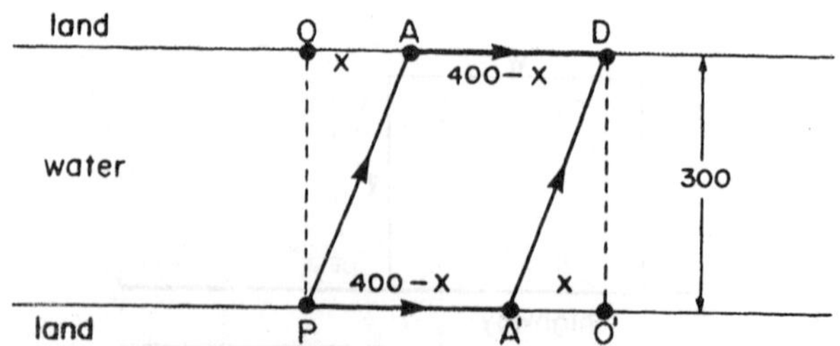

Figure IV.2

sponds to a minimum for C. The minimum cost is

$$\$(800+1200\sqrt{2}) = \boxed{\$2497.06}.$$

3. An underdeveloped country, whose only export is coffee, can sell x tons of coffee per month on the international market at the price of $(300-x/1000)$ dollars per ton. The cost of shipping x tons is 10 dollars per ton plus 1000 dollars overhead. What level of export will maximize the dollar income of the country?

Let $\pi(x)$ be the dollar income of the country if x tons of coffee are exported per month. $\pi(x)$ is the country's profit.

$\pi(x)=R(x)-C(x)$ where $R(x)$ is the revenue and $C(x)$ is the cost of shipping for x tons of coffee. The aim is to find the value of $x=x^*$ maximizing $\pi(x)$ (Fig. IV.3).

$$R(x) = (\text{price per ton})\times(\text{number of tons exported})$$

$$= \left(300-\frac{x}{1000}\right)x.$$

$$C(x)=10x+1000.$$

Hence

$$\pi(x) = -\left(\frac{x^2}{1000}-290x+1000\right) \leftrightarrow \text{concave down parabola.}$$

$$\frac{d\pi}{dx} = 290-\frac{x}{500};\ \frac{d\pi}{dx}=0 \leftrightarrow x=x^*=(290)(500)=145{,}000.$$

The underdeveloped country should export $\boxed{145{,}000 \text{ tons}}$ of coffee every month.

(*)4. Lethbridge City Transit, the bus company, is concerned about losing money next year. It estimates that if the fare is 30¢ it will carry 5000 passengers per day. For each increase of 10¢ beyond 30¢ it will lose 1000 passengers per day. What fare will maximize the revenue per day, how many passengers will be carried, and what is the daily revenue which will result?

Let the fare charged be $(30+x)$¢, $x \geq 0$.

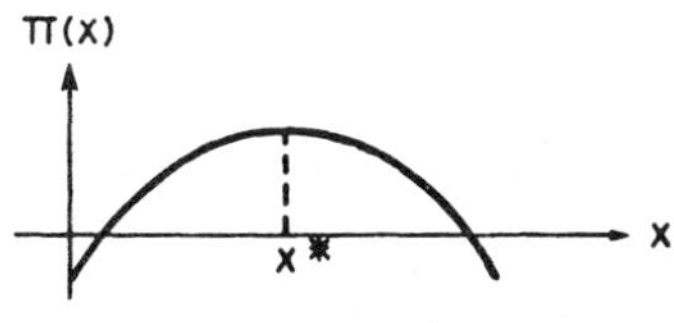

Figure IV.3

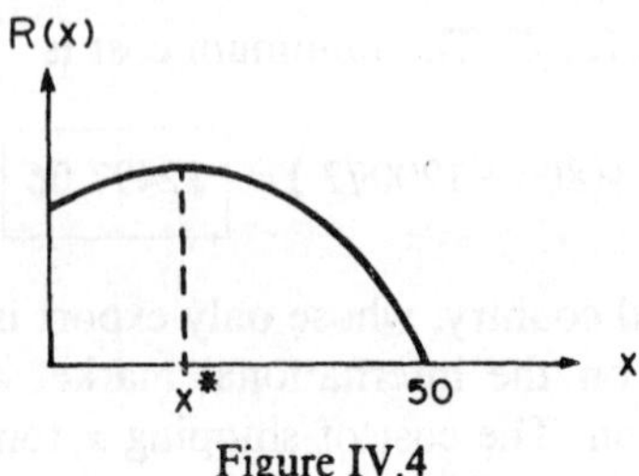

Figure IV.4

The number of passengers carried per day, if the fare is $(30+x)$¢, is $5000-(1000)\frac{x}{10}$, assuming x is a continuous variable. The revenue per day is $R(x)=(30+x)(5000-100x)$ in cents or $(30+x)(50-x)$ in dollars.

$R(x)$ describes a parabola which is concave downward (Fig. IV.4).

$$R'(x)=(50-x)-(30+x)=20-2x.$$

$$R'(x)=0 \leftrightarrow x=x^*=10 \leftrightarrow \text{maximum}.$$

Hence a fare of $\boxed{40¢}$ maximizes the revenue per day. In this case $\boxed{4000}$ passengers per day are carried and the daily revenue is $\boxed{\$1600}$.

5. A truck is to be driven 130 km at a constant speed of x km/hr. Speed laws require that $50 \le x \le 100$. Assume that gasoline costs 90 cents/gallon and is consumed at the rate of $2+\frac{x^2}{360}$ gallons/hr. If the driver is paid 15 dollars/hr, find the most economical speed and the total cost for the trip.

Let x be the speed at which the truck is driven. Then the domain of x is $50 \le x \le 100$. Let $C(x)$ be the total cost for the trip in dollars (Fig. IV.5). At speed x the trip will take $\frac{130}{x}$ hr.

$$C(x)=C_1(x)+C_2(x)$$

where $C_1(x)$ is the cost of gasoline $=\left(2+\frac{x^2}{360}\right)\left(\frac{130}{x}\right)(0.9)$ and $C_2(x)$ is the labour cost $=(15)\left(\frac{130}{x}\right)$. Hence

$$C(x)=\frac{130}{x}\left[15+(0.9)\left(2+\frac{x^2}{360}\right)\right]=130\left[\frac{16.8}{x}+\left(\frac{0.9}{360}\right)x\right].$$

The aim is to minimize $C(x)$ where $50 \le x \le 100$.

$$C'(x)=130\left[\frac{-16.8}{x^2}+\frac{0.9}{360}\right];$$

$$C'(x)=0 \text{ at } x=x^*=8\sqrt{105}\simeq 82.0.$$

As $x \to 0^+$, $C'(x) \to -\infty$; as $x \to +\infty$, $C'(x) \to$ a positive number. Hence $x=x^*$ corresponds to a minimum since $C(x)$ is continuous for $0<x<\infty$

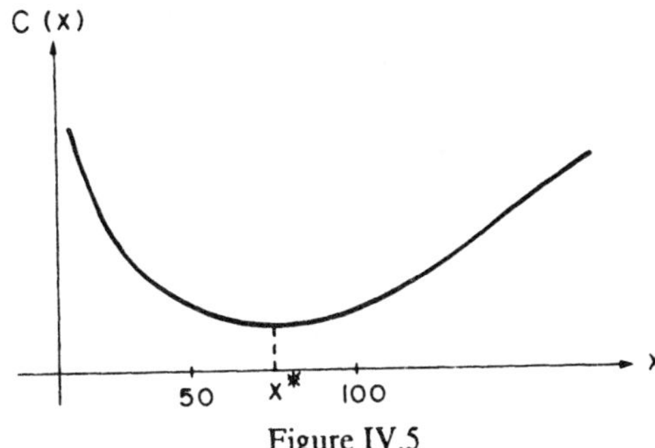

Figure IV.5

and one is interested in the region $50 \le x \le 100$. (Alternatively, $C''(x) > 0$ for $x > 0$ shows that $x = x^*$ is a minimum.)

Hence the most economical speed is $\boxed{82.0 \text{ km/hr}}$ and the corresponding total cost for the trip is $\boxed{\$53.28}$.

6. If \$100 is invested at 8% interest compounded quarterly, what is the balance after 4.5 years?

In this compound interest problem, $r = 8$, $T = 4.5$, the number of compounding periods in 4.5 years is $n = (4)(4.5) = 18$, and the principal $P = 100$. Hence after 4.5 years the balance is

$$B(4.5) = P\left(1 + \frac{rT}{100n}\right)^n = 100\left(1 + \frac{8(4.5)}{100(18)}\right)^{18} = 100(1.02)^{18} = \boxed{\$142.82}.$$

(*)7. \$1000 is invested in an account which pays 10% interest annually.

(a) How many years does the investor have to wait to double his investment?

(b) How many years will it take for the account to grow to one million dollars? (Assume in both cases that the interest rate is unchanged.)

(a) In this problem $r = 10$, $P = 1000$, and there is one compounding period annually. Hence the balance after t years is

$$B(t) = P\left(1 + \frac{r}{100}\right)^{\tau(t)}$$

where $\tau(t) = n$ for $n \le t < n+1$, $n = 0, 1, 2, \ldots$.

The aim is to find the smallest value of $t = t^*$ such that $B(t^*) \ge 2P$. This is equivalent to finding the smallest *integer* n such that $\left(1 + \frac{r}{100}\right)^n \ge 2$, i.e. $(1.1)^n \ge 2 \leftrightarrow n \log 1.1 \ge \log 2 \leftrightarrow n \ge \dfrac{\log 2}{\log 1.1} \simeq 7.27$.

The smallest integer with this property is $n = 8$. Hence the investor has to wait $\boxed{8 \text{ years}}$ in order to double his investment.

(b) Here the aim is to find the smallest integer n such that $B(n) \geq 10^6$. Thus n is the smallest integer such that

$$B(n) = 10^3\left(1 + \frac{r}{100}\right)^n \geq 10^6 \leftrightarrow n \geq \frac{\log 10^3}{\log 1.1} \simeq 72.5.$$

Hence it takes $\boxed{\text{73 years}}$ for the account to grow to one million dollars.

(*)8. At a bank, interest is compounded continuously. The interest rate rises, however, and is given at time t years by the formula $r = 0.08 + (0.015)\sqrt{t}$. In particular, the interest rate now ($t = 0$) is 0.08 (that is, 8%) while the interest rate a year from now ($t = 1$) will be 0.095. If a person deposits \$1000 in an account now and leaves it alone, then the amount of money y in the account at time t satisfies the differential equation

$$\frac{1}{y}\frac{dy}{dt} = 0.08 + (0.015)\sqrt{t}$$

with $y = 1000$ when $t = 0$.

(a) Solve this differential equation to obtain a formula for the amount of money y in the account at any time t.

(b) Find the amount of money in the account after two years (that is, at time $t = 2$).

(c) Find the average interest rate R in the two year period, $0 \leq t \leq 2$, and check that the formula $Y = Pe^{RT}$ gives the same answer as (b) if R is the average interest rate over the time period $T = 2$ and $P = 1000$ is the principal.

(d) Suppose that as well as the fixed deposit of \$1000 at time $t = 0$, the person also deposits money in a steady stream at a rate of \$500 per year (so that at the end of year one, \$1500 will have been deposited, while in the middle of year two, \$1750 will have been deposited, etc.). Assume as before that interest is compounded continuously at a rate $r = 0.08 + (0.015)\sqrt{t}$ at time t. Let y be the amount of money in the account at time t (including interest). Write down, but do not solve, a differential equation and initial condition that y must satisfy.

(a) $\frac{1}{y}\frac{dy}{dt} = 0.08 + (0.015)\sqrt{t}$ where $y(0) = 1000$. Thus $\frac{dy}{y} = (0.08 + (0.015)\sqrt{t})\,dt$, after separation of variables. Hence

$$\int_{1000}^{y} \frac{dy'}{y'} = \int_0^t [0.08 + (0.015)\sqrt{t'}]\,dt'$$

$$\leftrightarrow \log y - \log 1000 = 0.08t + 0.01t^{3/2}$$

$$\leftrightarrow y = \boxed{1000e^{[0.08t + 0.01t^{3/2}]}}.$$

(b) When $t = 2$, $y = 1000e^{[0.16 + 0.02\sqrt{2}]} = \boxed{\$1207.18}$.

(c) If $r(t)$ is the interest rate in decimal form and the interest is compounded continuously, then $\frac{1}{y}\frac{dy}{dt}=r(t)$ where $y(0)=P$ is the principal and $y(t)$ is the balance after t years. Then $\int_P^y \frac{dy'}{y'}=\int_0^t r(t')\,dt' \leftrightarrow$ $\log y-\log P=\int_0^t r(t')\,dt' \leftrightarrow y(T)=P\exp\left[\int_0^T r(t)\,dt\right]$ over a period of T years. By definition, the average interest rate over T years is

$$R=\frac{\int_0^T r(t)\,dt}{T} \leftrightarrow \int_0^T r(t)\,dt=RT.$$

Hence $y(T)=Pe^{RT}$, in general. Note that R is a function of T. When

$$T=2,\ R=\frac{\int_0^T r(t)\,dt}{T}=\frac{0.16+0.02\sqrt{2}}{2}$$
$$=0.094.$$

Hence the average interest rate over the two year period $0\le t\le 2$ is $\boxed{9.4\%}$.

(d) $y(t)$ is the balance at time t; the rate of change of the balance at time t is $\frac{dy}{dt}=$ (balance at time t)×(interest rate at time t)+deposit rate at time t. Hence

$$\frac{dy}{dt}=r(t)\,y(t)+500,\text{ i.e.,}$$

$$\boxed{\frac{dy}{dt}=\left[0.08+(0.015)\sqrt{t}\right]y+500},$$

with the initial condition $y(0)=P=1000$.

(*)9. A person owns a property which has a present worth of \$1000 and expects it to increase in value at the rate of \$150 per year.

(a) If he can invest money at 10% compounded semi-annually, when should he sell the property in order to maximize the present value of the sale price?

(b) Another person wishes to buy this property some time in the future. To provide for the purchase price, he continuously deposits money into a fund at the rate of \$200 per year which earns 10% continuously compounded interest. Will he have enough money to meet the purchase price after 7 years assuming the property is still available?

(a) Say the property is sold after T years. Let $P(T)$ be the present value of the sale price. The sale price of the property will be $B(T)=\$(1000+150T)$. The interest rule is $f(T)=(1.05)^{\tau(T)}$ where

$$\tau(T)=n \text{ for } \frac{n}{2}\le T<\frac{n+1}{2},\qquad n=0,1,2,\ldots.$$

Hence $P(T)=\frac{B(T)}{f(T)}$. The aim is to find the value of T maximizing $P(T)$.

In a fixed compounding time interval $f(T)$ remains fixed whereas $B(T)$ increases. Thus the maximum value of $P(T)$ is attained at the integer n maximizing

$$F(n)=\frac{B\left(\frac{n+1}{2}\right)}{f\left(\frac{n}{2}\right)}=\frac{1075+75n}{(1.05)^n}.$$

Let

$$F(x)=\frac{1075+75x}{(1.05)^x},\text{ on the domain } x\geq 0;$$

$$F'(x)=\frac{75-\log(1.05)(1075+75x)}{(1.05)^x};$$

$$F'(x)=0\leftrightarrow x=x^*=\frac{1}{\log(1.05)}-\frac{43}{3}\simeq 6.16.$$

Since $F'(0)>0$, $F'(x)\to-\infty$ as $x\to+\infty$. Hence $x=x^*$ leads to the maximum value of $F(x)$ on the domain $x\geq 0$. In particular the maximum value of $F(n)$ is either $F(6)$ or $F(7)$. $F(6)\simeq 1138$; $F(7)\simeq 1137$.

Thus the property should be sold after $T=\frac{7+1}{2}=\boxed{4\text{ years}}$.

(b) Let $B(t)$ be the second person's balance at time t. Then $\frac{dB}{dt}=rB+200$ where $r=0.1$ and $B(0)=0$. The aim is to find $B(7)$.

$$\int_0^{B(7)}\frac{dB}{rB+200}=\int_0^7 dt\leftrightarrow\left.\frac{\log(rB+200)}{r}\right|_{B=0}^{B=B(7)}=7.$$

Hence $\log(rB(7)+200)-\log 200=7r\leftrightarrow rB(7)+200=200e^{7r}\leftrightarrow B(7)=\frac{200[e^{7r}-1]}{r}\simeq 2027.51$. Hence the second person's balance after 7 years is \$2027.51. If the property is still available after 7 years, its value then will be $1000+(150)(7)=\$2050$. Thus the second person will not be able to purchase the property.

10. Given that the demand equation is $p+3x=21$ and the supply equation is $x=\frac{4}{3}\sqrt{p}$, find the equilibrium price p_0 and the equilibrium demand x_0, and evaluate the consumer's surplus and the producer's surplus.

Here p is the price and x is the number of units of a commodity.

The demand equation is $p=-3x+21$ and the supply equation is $p=\frac{9x^2}{16}$, $x\geq 0$ (Fig. IV.10).

At equilibrium, the curves intersect. This corresponds to $-3x+21=\frac{9x^2}{16}\leftrightarrow 3x^2+16x-(16)(7)=0\leftrightarrow(3x+28)(x-4)=0$. Hence equilibrium occurs

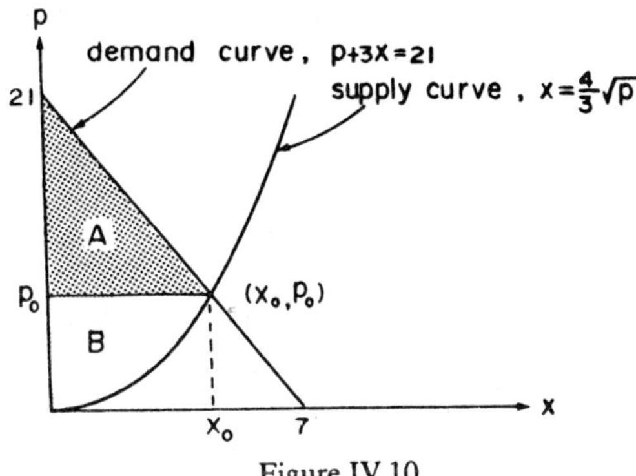

Figure IV.10

when $x = \boxed{x_0 = 4}$. The corresponding equilibrium price is $p = p_0 = -12 + 21 = \boxed{9}$. The consumer's surplus is area $A = \frac{(21-9)}{2}(4) = \boxed{24}$; the producer's surplus is area $B = (9)(4) - \int_0^4 \frac{9x^2}{16}dx = 36 - \frac{9(4)^3}{3(16)} = \boxed{24}$.

(*)11. A company is planning to phase out a product because of falling demand. Its current inventory level is 1680 items and it is producing at the rate of 900 items/month. Demand is currently at the rate of 800 items/month and is dropping at the rate of 10 per month. The company would like to reduce production at a rate of R items/month, with R chosen so as to reduce its inventory to zero at the end of 12 months.

(a) Give the expression for the demand and production rates as functions of time t (taking $t = 0$ as the present).

(b) Give the expression for the inventory at time t.

(c) What should R be to reduce the inventory to 0 at $t = 12$?

Let t be the time in months.

(a) The demand rate at time t is $D(t) = \boxed{800 - 10t \text{ items/month}}$. The production rate at time t is $S(t) = \boxed{900 - Rt \text{ items/month}}$.

(b) Let $I(t)$ be the inventory at time t. Then $\frac{dI}{dt} = S(t) - D(t)$ where $I(0) = 1680$. Hence

$$\frac{dI}{dt} = 100 + (10 - R)t$$

$$\leftrightarrow I(t) - I(0) = \int_0^t [100 + (10-R)t']\, dt'$$

$$= 100t + \frac{(10-R)}{2}t^2 \leftrightarrow I(t) = \boxed{1680 + 100t + \frac{(10-R)}{2}t^2}.$$

(c) $I(12)=0 \leftrightarrow 1680+1200+(10-R)72=0 \leftrightarrow R=\boxed{50}$.

12. A manufacturer has determined that, for a given product, the average cost $\bar{c}$ (in dollars per unit) is given by

$$\bar{c}=2x^2-36x+210-\frac{200}{x},$$

this relation being valid for values of x in the closed interval $[2,10]$.

(a) Write a formula for the *total cost function*, c.

(b) For which values of x in the interval $[2,10]$ is the total cost function increasing?

(c) For which values of x in the interval $[2,10]$ is the graph of the total cost function concave down?

(d) At what level within the interval $[2,10]$ should production be fixed in order to minimize total cost? What is the minimum total cost?

(e) If production were required to lie within the interval $[5,10]$, what value of x would minimize total cost?

x is understood to represent the number of units produced.

(a) $\bar{c}(x)=\dfrac{c(x)}{x}$ is the average cost to produce x units. Hence $c(x)=x\bar{c}(x)=\boxed{2x^3-36x^2+210x-200}$, $2\le x\le 10$.

(b) $\dfrac{dc}{dx}=6x^2-72x+210=6(x-5)(x-7)$. Hence for:

$$2\le x<5,\ \frac{dc}{dx}>0 \leftrightarrow c(x) \text{ increasing};$$

$$5<x<7,\ \frac{dc}{dx}<0 \leftrightarrow c(x) \text{ decreasing};$$

$$7<x\le 10,\ \frac{dc}{dx}>0 \leftrightarrow c(x) \text{ increasing (Fig. IV.12)}.$$

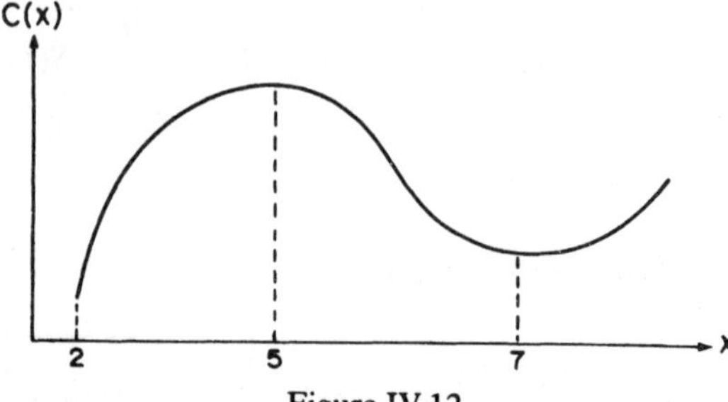

Figure IV.12

The total cost function is increasing if $\boxed{2 \le x < 5}$ and $\boxed{7 < x \le 10}$.

(c) $\dfrac{d^2c}{dx^2} = 12x - 72 = 12(x-6)$. Hence for:

$$2 \le x < 6,\ \frac{d^2c}{dx^2} < 0 \leftrightarrow c(x) \text{ concave down;}$$

$$6 < x \le 10,\ \frac{d^2c}{dx^2} > 0 \leftrightarrow c(x) \text{ concave up.}$$

The total cost function is concave down if $\boxed{2 \le x < 6}$.

(d) $\dfrac{dc}{dx} = 0 \leftrightarrow x = 5, 7.$ $\dfrac{d^2c}{dx^2} < 0$ when $x = 5 \leftrightarrow$ local maximum;

$$\frac{d^2c}{dx^2} > 0 \text{ when } x = 7 \leftrightarrow \text{local minimum.}$$

Clearly $x = 10$ cannot correspond to a global minimum. Hence one need only compare the values of $c(2)$ and $c(7)$:

$c(2) = 92$, $c(7) = 192$. Thus $\boxed{x = 2 \leftrightarrow \text{minimal cost of 92 dollars}}$.

(e) Here minimal cost would occur at $\boxed{x = 7}$.

(*)13. A manufacturer estimates that he can sell 2000 toys per month if he sets the unit price at \$5.00. Furthermore he estimates that for each \$0.20 decrease in price his sales will increase by 200 per month.

(a) Find the demand and revenue functions.

(b) Find the number of toys that he should sell each month in order to maximize the monthly revenue.

(c) What is the maximum monthly revenue?

(a) Let q be the total number of toys sold per month. Let x be the total number of toys sold per month in excess of 2000. Then $q = 2000 + x$. Let $D(q)$ be the demand function. Then

$$D(q) = 5 - (0.2)\frac{x}{200} = 5 - \frac{(q-2000)}{1000}.$$

Hence $D(q) = \boxed{7 - \dfrac{q}{1000}}$. The revenue function is $R = qD(q)$ $= \boxed{7q - \dfrac{q^2}{1000}}$.

(b) $\dfrac{dR}{dq} = 7 - \dfrac{q}{500}$. Hence $q = \boxed{q^* = 3500}$ for the maximum monthly revenue. ($R(q)$ is a parabola, concave downward.)

(c) The maximum monthly revenue is $R(q^*) = \boxed{\$12{,}250}$.

(*)**14.** The demand for an item is 800 units/month. The item is ordered in batches (of say Q units) which are received and put into storage; demand is met by withdrawals from storage (at a uniform rate). Four types of costs are incurred in the operation of this system:

(i) Purchase price of the item ($50.00/unit).

(ii) Fixed order and receiving costs ($500.00 each time an order is placed and delivered).

(iii) Storage costs ($2.00/unit/month, charged on the maximum inventory during an order-reorder period).

(iv) Taxes and insurance ($0.50/unit/month, charged on the average inventory during an order-reorder period).

Let T be the number of months between receiving one order and receiving the next order.

(a) Give an expression (in terms of Q and T) for the total cost (TC) of operating the above system for T months.

(b) Give an expression (in terms of Q) for the average cost/month C of operating the above system.

(c) What order quantity Q^* minimizes C and what is the minimum average cost?

(a) Let T be the number of months in one order-reorder period and let Q be the number of units of the item ordered at the beginning of this period. Then the total cost (TC) = purchase price + fixed costs + storage costs + taxes and insurance;

$$\text{purchase price} = \$50Q;$$

$$\text{fixed costs} = \$500;$$

$$\text{storage costs} = \$2QT.$$

The average inventory over the period is $\frac{Q}{2}$ since the demand is met by withdrawals at a uniform rate. Thus the taxes and insurance costs are $\frac{Q}{2}(0.50)T = \frac{QT}{4}$. Hence $(TC) = \boxed{50Q + 500 + \frac{9}{4}QT}$.

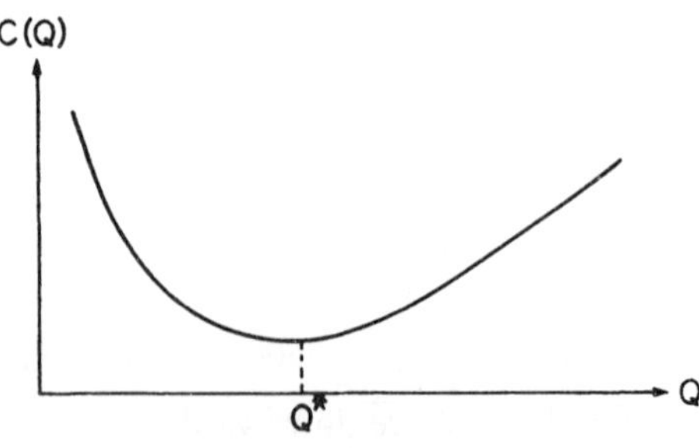

Figure IV.14

(b) Since the demand is 800 units/month, $Q = 800T \leftrightarrow T = \dfrac{Q}{800}$.

$$C = \frac{(TC)}{T} = 50\frac{Q}{T} + \frac{500}{T} + \frac{9}{4}Q.$$

Hence the average cost/month to operate the system is

$$\boxed{C = \frac{(500)(800)}{Q} + \frac{9}{4}Q + (50)(800)}.$$

(c) To find the optimal order quantity Q^*, compute

$$\frac{dC}{dQ} = -\frac{(500)(800)}{Q^2} + \frac{9}{4}.$$

$$\frac{dC}{dQ} = 0 \text{ at } Q = Q^* = \frac{400}{3}\sqrt{10} \simeq 422.$$

$$\frac{d^2C}{dQ^2} = \frac{(1000)(800)}{Q^3} > 0 \text{ if } Q > 0.$$

Hence $Q = Q^*$ leads to the minimum for C (Fig. IV.14). Hence the order quantity $Q^* = \boxed{422}$ minimizes the average cost/month and the minimum average cost/month is $C(Q^*) = \boxed{\$41{,}897}$.

(*)15. A monopolist determines that if $C(x)$ cents is the total cost of producing x units per week of a certain commodity, then

$$C(x) = 25x + 2000.$$

The demand equation is $x + 50p = 5000$, where x units are demanded each week when the unit price is p cents.

(a) How many units should be produced each week in order to maximize the profit?

(b) If the government levies a tax of 10 cents per unit produced, how many units should be produced each week in order to maximize the profit?

(c) What tax should be levied by the government on each unit produced in order that the weekly tax revenue received by the government be maximized?

(a) Let $\pi(x)$ be the profit in producing x units/week.

$\pi(x) = R(x) - C(x)$ where $R(x)$ is the revenue function;

$R(x) = xD(x)$ where the demand function is $D(x) = \dfrac{5000}{50} - \dfrac{x}{50} = 100 - \dfrac{x}{50}$.

Hence $\pi(x) = 100x - \dfrac{x^2}{50} - 25x - 2000 = 75x - \dfrac{x^2}{50} - 2000$; $\dfrac{d\pi}{dx} = 75 - \dfrac{x}{25}$;

$\dfrac{d\pi}{dx} = 0$ at $x = 1875$, corresponding to a maximum since $\pi(x)$ represents a parabola, concave downward. Thus $\boxed{1875}$ units should be produced each week to maximize the profit.

(b) A tax of 10 cents per unit raises the cost function to $C(x) = 25x + 2000 + 10x = 35x + 2000$. Hence $\pi(x) = 65x - \frac{x^2}{50} - 2000$; $\frac{d\pi}{dx} = 65 - \frac{x}{25}$; $\frac{d\pi}{dx} = 0$ at $x = 1625$. In this case $\boxed{1625}$ units/week maximizes the profit.

(c) Let t be the tax/unit. Now one assumes that for a given value of t a monopolist aims to maximize his profit. In this case $C(x) = (25 + t)x + 2000$ and $\pi(x) = (75 - t)x - \frac{x^2}{50} - 2000$; $\frac{d\pi}{dx} = 75 - t - \frac{x}{25}$; $\frac{d\pi}{dx} = 0$ at $x = 1875 - 25t$. Hence the weekly tax revenue to the government is $\rho(t) = (1875 - 25t)t$. The government's aim is to maximize $\rho(t)$ whose graph is a parabola, concave downward. $\frac{d\rho}{dt} = 1875 - 50t$; $\frac{d\rho}{dt} = 0$ at $t = \boxed{37.5 \text{ cents}}$.

(*)16. Suppose you have a product for which the demand function is $p = f(q)$. Show mathematically that if you set the price at the level where revenue is maximized, then one has unit elasticity of demand.

The revenue function $R(q) = qf(q)$; $\frac{dR}{dq} = f(q) + qf'(q)$.

$$\frac{dR}{dq} = 0 \text{ at } q = q^* \text{ where } f(q^*) + q^*f'(q^*) = 0. \quad \frac{d^2R}{dq^2} = 2f'(q) + qf''(q).$$

Assuming $2f'(q^*) + q^*f''(q^*) < 0$, one sees that $q = q^*$ corresponds to maximum revenue (locally).

By definition the elasticity of demand $\eta = \frac{-f(q)}{qf'(q)}$. Hence $\eta(q^*) = 1 \leftrightarrow f(q^*) + q^*f'(q^*) = 0$. Thus it has been shown that if $q = q^*$ maximizes revenue then $\eta(q^*) = 1$.

17. If c is a total cost function and if $e^c = \log(x^2 + 1)$ then what is the marginal cost function?

Use implicit differentiation:

$$e^c = \log(x^2 + 1);$$

$$\frac{d}{dx}e^c = \frac{de^c}{dc}\frac{dc}{dx} = e^c\frac{dc}{dx} = \frac{d}{dx}\log(x^2 + 1) = \frac{2x}{x^2 + 1}.$$

Hence the marginal cost function is $\frac{dc}{dx} = \boxed{\frac{2x}{(x^2 + 1)[\log(x^2 + 1)]}}$.

18. A manufacturer's demand equation is

$$p = \frac{100}{\sqrt{x+4}}$$

(where p is price per unit, expressed in dollars). Use differentials to estimate the price per unit when the level of sales is 12.5 units. Express your answer to the nearest cent.

The demand equation is $p = D(x) = \frac{100}{\sqrt{x+4}}$. The marginal demand is $\frac{dD(x)}{dx} = \frac{-50}{(x+4)^{3/2}}$. The aim is to estimate $D(12.5)$.
Note that $D(12) = \frac{100}{\sqrt{16}} = 25$; $D'(12) = \frac{-50}{(16)(4)} = \frac{-25}{32}$.
$D(12.5) \simeq D(12) + D'(12)(12.5-12)$. Thus the differential $\Delta D \simeq D'(12)\,\Delta x$ where $\Delta x = 0.5$. Hence $D(12.5) \simeq 25 - \left(\frac{25}{32}\right)\left(\frac{1}{2}\right) \simeq 24.61$. Hence, using differentials, one finds that the estimated price/unit is $\boxed{\$24.61}$.

19. Suppose that a monopolist's marginal revenue function is given, in terms of total revenue r and quantity x, by the differential equation $\frac{dr}{dx} = (50-4x)e^{-r/5}$.

Find the demand equation for the monopolist's product.

Let $D(x)$ be the demand function. Then $D(x) = \frac{r(x)}{x}$. Using separation of variables, one sees that

$$e^{r/5}\,dr = (50-4x)\,dx. \text{ Note that } r = 0 \text{ when } x = 0.$$

Hence

$$\int_0^r e^{r'/5}\,dr' = \int_0^x (50-4x')\,dx'$$

$$\leftrightarrow 5[e^{r/5} - 1] = 50x - 2x^2 \qquad (r > 0 \text{ if } 0 < x < 25)$$

$$\leftrightarrow e^{r/5} = \frac{50x - 2x^2 + 5}{5}$$

$$\leftrightarrow r = 5\log\left[\frac{50x - 2x^2 + 5}{5}\right]$$

$$\leftrightarrow \boxed{D(x) = \frac{5}{x}\log\left[\frac{50x - 2x^2 + 5}{5}\right], \qquad 0 < x < 25}.$$

20. From observations of the market over a lengthy period it was determined that the marginal revenue function was that given in the graph in Fig. IV.20. Use both the trapezoidal rule and Simpson's rule to determine

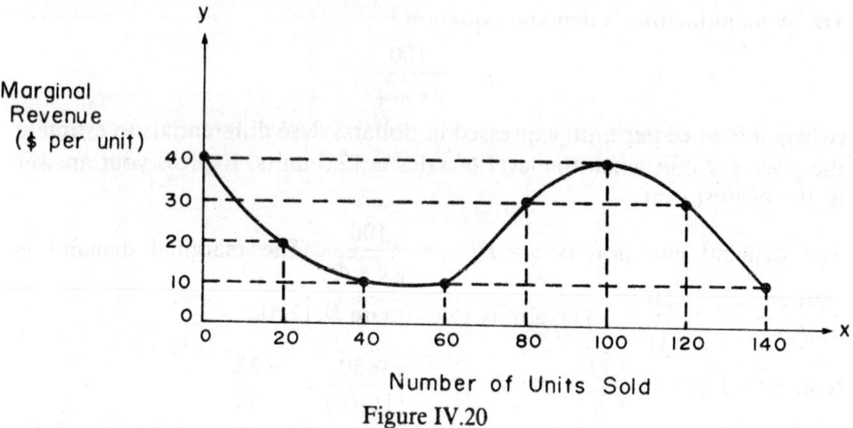

Figure IV.20

the total revenue obtainable from the production and sale of 120 units of the product. (Use $n = 6$.)

Let the marginal revenue function be $y = f(x)$. Then

$$\text{the revenue function is } R(x) = \int_0^x f(x')\,dx'.$$

Hence the total revenue obtainable from the production and sale of 120 units is $R(120) = \int_0^{120} f(x)\,dx$ corresponding to the area under the curve $y = f(x)$ for $0 \le x \le 120$.

The aim is to estimate $R(120)$ from the given data. By the trapezoidal rule with $\Delta x = 20$ $[n = 6]$,

$$R(120) \simeq \frac{\Delta x}{2}[f(0)+2f(20)+2f(40)+2f(60)+2f(80)+2f(100)+f(120)]$$

$$= 10[40+(2)(20)+(2)(10)+(2)(10)+(2)(30)+(2)(40)+30]$$

$$= \boxed{\$2900}.$$

By Simpson's rule with $\Delta x = 20$,

$$R(120) \simeq \frac{\Delta x}{3}[f(0)+4f(20)+2f(40)+4f(60)+2f(80)+4f(100)+f(120)]$$

$$= \frac{20}{3}[40+(4)(20)+(2)(10)+(4)(10)+(2)(30)+(4)(40)+30]$$

$$= \boxed{\$2866.67}.$$

(See Chapter VI on Numerical Methods for discussions of the trapezoidal rule and Simpson's rule.)

Supplementary Problems

1. An open box (no top) with square base is to have a volume of 54 cm^3. If the material for the base costs \$4 per cm^2, and the material for the sides costs \$1 per cm^2, what should be the dimensions of the box in order that the cost be minimized?

2. Plans for a new rectangular building require a floor area of 14,400 square metres. The walls are 10 metres high. Three walls are made of brick and the fourth wall is made of glass. Glass costs 1.88 times as much as brick per square metre. What should the dimensions of the building be so that the cost for the walls is minimum?

3. A company wishes to construct a storage tank in the form of a rectangular parallelepiped with a square horizontal cross-sectional area, as shown in Fig. S.3. In addition, the volume of the tank is required to be 86.4 cubic metres.

(a) If the material for the sides and top costs \$1.25 per square metre, and the material for the bottom costs \$4.75 per square metre, and the 12 welds cost \$7.50 per metre, find the total cost of construction as a function of x.

(b) Show that for a relative extreme value of this function, x must satisfy $x^4 + 5x^3 - 36x - 432 = 0$.

(c) Given that the only real roots of the polynomial equation in (b) are $x = 4$ and $x = -6$, find the dimensions of the most economical tank.

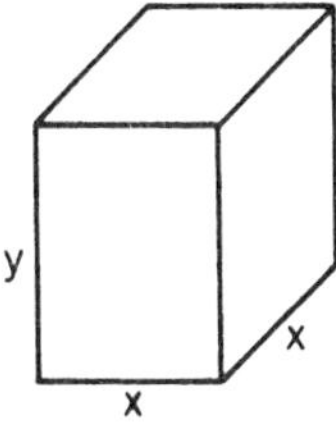

Figure S.3

4. A gardener wishes to enclose a rectangular plot that has one side along a neighbour's property. The fencing costs \$4 per metre. The gardener is to pay for the fence along three sides on his own ground and half of that along the property line with the neighbour. What dimensions would give him the least cost if the plot is to contain 1200 square metres?

5. Supertankers carrying oil to a refinery will discharge their oil to a deep water pumping station 1 km off-shore. The refinery is located 2 km along the shore (which is straight). If an underwater pipeline costs twice as much per km as one on land, how should the pipeline from the pumping station to the refinery be laid in order to minimize the cost?

6. Suppose that one has a power house P situated on one bank of a straight river α metres wide and a factory F situated on the other bank β metres downstream, $\beta = 2\alpha$. One wants to lay a cable from P to F at minimum cost. If underwater cable costs \$1.00 per metre and land cable costs \$0.50 per metre, what path should be chosen for the cable?

(*)7. The locations of two cities on a map scaled in km are $(2,0)$ and $(1,-5)$. Find the cheapest route for a pipeline between these cities if the construction cost for the region $y<0$ is a dollars per km whereas the cost for the region $y \geq 0$ is b dollars per km. Note that you must distinguish between the two cases $a^2/b^2 \leq 26$ and $a^2/b^2 > 26$.

8. The cost of producing x tons of cement per week is

$$c = \tfrac{1}{20}x^2 + 70x + 50 \text{ dollars.}$$

When x tons per week are sold, the manufacturer can sell the cement at a price of

$$p = 270 - \tfrac{3}{20}x \text{ dollars per ton.}$$

If profit is defined by:

$$\pi = \text{profit} = \text{revenue minus cost},$$

find the value of x which will yield the maximum profit.

9. A company can sell $\sqrt{100-x}$ tons of produce when the price is set at \$$x$ per ton. If each ton costs \$10 to produce, what price will maximize the profit? Explain why your answer gives a maximum.

10. A company finds that it can sell x widgets per hour at a price of \$$p$ per widget where x and p are related by $p = 5e^{-x/6}$. If it costs \$$x^2e^{-x/6}$ to produce x widgets, find the number of widgets per hour the company should produce to obtain a maximum profit. Make sure that you explain why your answer gives a maximum.

11. At a price of \$1.00, a dealer can sell 1000 articles that cost him 60¢ each. For each 1¢ that he lowers the price, he can increase the number sold by 50. What price will maximize the profit?

12. A cable T.V. company has 20,000 subscribers and charges an \$8/month rental fee. The sales manager discovers that for each 1¢ reduction he can attract 50 new customers. What reduction will maximize the rental income?

13. A steamship company running an excursion normally charges each passenger \$15. If more than 200 tickets are sold each passenger receives a reduction of 5 cents for every ticket sold in excess of 200. How many passengers will give the company the maximum revenue and what is the maximum revenue?

14. An executive of a commuter railway company is considering raising the fare, so he does a survey and discovers that:

(a) 600 passengers currently ride his train every day at a fare of \$1.00 each;

(b) for each 1¢ he adds to the fare, he will lose 5 passengers. How much should he raise the fare if he wants to maximize his income?

15. A man rents a large motor boat to make a 400 km trip. He has to pay at the rate of \$15 per hour for the boat (so he pays \$5 for 20 minutes, \$1 for four minutes and so on) and he must pay for the gasoline at \$1 per gallon. When the boat is travelling at x km per hour, it burns gasoline at the rate of $10+\frac{x^2}{16}$ gallons per hour. What is the most economical speed?

(*)16. A printer is to produce 100,000 identical posters by using printing blocks that each print 100 posters per hour (so that, for example, if he uses only 2 blocks it would take 500 hours to do the job). Each block costs \$2.00 to make, and no matter how many blocks he uses at the same time, his overhead expenses are \$5.00 per hour. How many blocks should he use to produce the posters at minimum cost? Show, by using the second derivative or otherwise, that you have found a minimum and not a maximum.

17. For a given principal P, what annual interest rate compounded once a year gives the same amount as 10% compounded twice a year?

18. Ms. Jones bought \$50,000 worth of bank bonds, paying 8% interest, 5 months ago, and now wants to sell them to buy a Ferrari. (The bonds cannot be redeemed for another 7 months.) Assuming that the bonds have been increasing in value continuously and that the rate of change is proportional to the value, find the current value of the bonds.

19. A bank advertises that by increasing money proportional to the amount present, it will double invested money in 10 years time. What is the annual interest rate?

20. A man has the current value of a rare painting appraised at \$10,000.00, and the appraiser feels that the value of the painting will increase at a rate of \$1,000.00 per year. If the rate of interest is 8% compounded quarterly, find the best time for the owner to sell the painting.

(*)21. Say bank B_1 pays an interest rate I_1 of $r\%$ annually to its depositors. Say bank B_2 compounds the annual interest rate of $r\%$ semi-annually, i.e. B_2 pays an interest rate of $\frac{r}{2}\%$ every 6 months to depositors, which results in an $\frac{r}{2}\%$ increase in the balance every 6 months. Thus the *effective* annual interest rate I_2 on the *original* principal paid by bank B_2 is $100\left[\left(1+\frac{r}{200}\right)^2-1\right]\%$. Let bank B_n compound the annual interest rate of $r\%$ n times yearly.

(a) What is the effective annual interest rate I_n which B_n pays to depositors?

(b) Let bank $B_\infty\left(=\lim_{n\to\infty} B_n\right)$ compound the annual interest rate of $r\%$ "instantly". What effective annual interest rate I_∞ does B_∞ pay to depositors?

(c) Prove that if $n_1 < n_2$, then $I_{n_1} < I_{n_2}$.

(d) Find the effective annual interest rate which B_∞ pays (accurate to 2 decimals) if (i) $r = 2$, (ii) $r = 10$, (iii) $r = 20$.

(*)**22.** I open a bank account with 1 dollar and henceforth do not touch the account. The bank pays interest at a rate of 8% per annum compounded semi-annually.

(a) How much is in the account after N years?

(b) Suppose I arrange that every time an interest payment is made, half of it goes into my account and the other half is given directly to me to put in an old sock under my pillow. How much is in the account after N years?

(c) How much is in the sock after N years?

(*)**23.** The profit rate of the Acme Novelty Company t years after the company was first formed ($t = 0$) is given by the formula:

$$10^5(t+1)^{1/2} \text{ dollars/year.}$$

(a) In particular, when $t = 1$ (the beginning of the second year of operation) the profit rate is $10^5\sqrt{2}$ dollars/year. If the profit rate had remained constant at this value from then on (instead of increasing as it really did) the company would have received $10^5\sqrt{2}$ dollars profit during the second year $1 \le t \le 2$. Write down and evaluate an integral which gives the amount of profit that the company actually receives during the second year $1 \le t \le 2$. (Your answer should be less than $10^5\sqrt{3}$, the profit rate when $t = 2$. Why?)

(b) What is the average profit rate in the period $1 \le t \le 2$?

(c) If interest rates are 8% per annum compounded continuously, how much more money could the company have made in the second year by depositing its profits continuously during that time $1 \le t \le 2$? (Do not attempt to evaluate the integral which arises in your answer.)

24. Suppose $x = \dfrac{12}{\sqrt{p}}$ is the demand function and $x = \sqrt{p}(1+p)$ is the supply function for a commodity where x is quantity and p is price.

(a) Compute the equilibrium price $\bar{p}$ and equilibrium demand $\bar{x}$.

(b) Compute the producer's surplus at $(\bar{x}, \bar{p})$.

25. The demand curve $p = f(x)$ and the supply curve $p = g(x)$ for a monopolist's product are not known exactly, but the information in the following table is presumed to be reliable.

x	0	2	4	6	8	10	12	14
$f(x)$ (dollars)	68	60	53	46	40	35	32	30
$g(x)$ (dollars)	12	14	17	20	24	28	32	37

(a) Find the best estimate, using the trapezoidal rule, for the consumer's surplus at market equilibrium.

(b) Use Simpson's rule to estimate the producer's surplus at market equilibrium.

26. Suppose the cost in dollars of making x items of a product is given by $C(x)=10x^2+200x+5000$. If the demand function is given by $p=f(x)=2000-5x$, how many items should be produced to maximize the profit? What is the maximum profit?

27. Suppose the demand and cost functions for a commodity are given by $p=1000-10x$ and $C(x)=10x^2+200x+6000$ (dollars). For what values of x will the profit be positive? What value of x gives the maximum profit?

28. (a) Sketch a graph of the equation $y=15e^{-x^2/18}$, indicating locations of relative extrema, points of inflection, intervals of increase and decrease, and intervals of upward and downward concavity.

(b) If the demand equation for a certain commodity is $p=15e^{-x^2/18}$, where p is price per unit and x is quantity demanded, find the demand x that maximizes total revenue.

29. A manufacturer has determined that for a certain product the average cost function is $\bar{c}=x^2+30x-400+\dfrac{200}{x}$ and the demand equation is $p=2x^2-15x+200+\dfrac{300}{x}$, both these relations being valid for values of x in the closed interval $[1,21]$. At what level within this interval should production be fixed to maximize profit? What is the maximum profit obtainable? At what price per unit is this profit obtained?

(*)30. Suppose that the revenue received from selling a certain item is $R=x^2+3xy-6y+100$ where x is the amount sold of the given item and y is the amount sold of its competitor.

(a) Find the maximum revenue if the total amount sold of the item and its competitor will be 50.

(b) When $x=y=25$, $R=2450$. Use differentials to estimate x and y when $R=2500$. (Still assume $x+y=50$, so $\Delta x+\Delta y=0$.)

31. The demand equation for a certain commodity is $p=a-bx$, where x units per year are produced. If dollar production costs for x units are $C(x)=c+dx$, and tax is t dollars per unit, find the following:

(Note: a, b, c, d, t, are positive constants.)

(a) The production level x_0 and price p_0 that yield maximum profit.

(b) The consumer's surplus at price level p_0.

32. Mitchell Mailorder receives a shipment of 600 cases of athletic supporters every 60 days. The number of cases on hand t days after the shipment arrives is $I(t) = 600 - 20\sqrt{15t}$. Find the average daily inventory. If the holding cost for one case is $\frac{1}{2}$ cent per day, find the total daily holding cost.

(*)33. Demand for two products (product 1 and product 2) is 100 pieces/week and 200 pieces/week, respectively. Storage costs for the two products are $0.60/piece/week and $0.20/piece/week, respectively, based on average inventory. Every T weeks, an order is placed for both products, calling for delivery of Q_1 pieces of product 1 and Q_2 pieces of product 2; it costs $800 to place the order, receive the goods, pay the invoice, etc.

(a) Derive an expression for the average cost per week, including order and holding cost, as a function only of T. (Show *how* you obtain the average cost function.)

(b) How often should orders be placed in order to minimize the average cost per week?

(c) What is the minimum average cost per week?

34. Find a manufacturer's cost function, given that his fixed cost is 50 and his cost function satisfies the differential equation $\dfrac{dc}{dx} = (x+1)e^{50-c}$.

35. The demand equation for a certain product is $p = 100 - 0.01x$ (dollars), and the total cost function is $c(x) = 50x + 10{,}000$ (dollars), where x is the number of units produced.

(a) Find the value of x that maximizes the profit and determine the corresponding price and the total profit for this level of production.

(b) Answer the questions in part (a) if the government imposes a tax of $10.00 per unit.

36. The demand equation for deep freeze display units is $x + 4p = 100$, where x units are demanded when the price is p thousand dollars. The cost in thousands of dollars to produce x units is $C(x) = 3x + 400$. A tax t thousand dollars per unit is introduced.

(a) Show that to maximize profit, the manufacturer should pass on half the tax to the purchasers.

(b) At what level of tax does the manufacturer's profit fall to zero?

37. The demand equation for a product is $x^2 = 400 - p$ and the supply equation is $x = \dfrac{p}{20} - 5$. Market equilibrium is established when 10 units are sold.

(a) Find the producer's and consumer's surpluses at market equilibrium.

(b) Express the elasticity of demand η for the above product as a function of x.

(c) For what values of x is demand *elastic*, for the above product?

38. The demand equation for a product is $x = 500 - 40p + p^2$. Find the elasticity of demand with respect to price when $p = 10$. If this price of 10 is increased by $\frac{1}{2}$ percent, what is the approximate change in demand?

39. If $c = x^3 - 6x^2 + 12x + 18$ is a total cost function, for what values of x is *marginal cost* increasing?

40. Find the marginal cost function of a manufacturer given that his average cost function is

$$\bar{c} = 34 - \frac{\log\left[(5x+4)^{1/3}\right]}{x}.$$

41. Suppose that the demand equation for a monopolist is

$$p = 4 + \frac{2}{x} - \frac{10}{x^2 + 5x} \quad \text{where } x > 0.$$

(a) For what values of x is the marginal revenue function increasing? For what values is it decreasing?

(b) Show that the marginal revenue function is concave up for all $x > 0$.

42. The quantity q sold per week for a particular product is related to its unit price p in dollars by $q = 6.6 - 2\log p$.

(a) Find the expression for p as a function of q.

(b) What is the marginal revenue as a function of price (i.e., what is the rate of change in revenue per unit increase in price)?

(c) What is the marginal revenue as a function of quantity (i.e., what is the rate of change of revenue per unit increase in quantity sold)?

(d) What price and quantity yield maximum revenue?

43. A manufacturer's total cost c (in dollars) of producing x units of a certain commodity is given by

$$c = \frac{3x^2}{\sqrt{x^2 + 500}} + 2000.$$

Use differentials to estimate the total cost of producing 19 units of the commodity. Express your answer to the nearest dollar.

44. Suppose a manufacturer's total cost (in dollars) is given by

$$c = \frac{3x^2}{\sqrt{x^2 + 300}} + 2000.$$

(a) Find the total cost when 10 units are produced.

(b) Find the marginal cost when 10 units are produced.

(c) Use differentials and the results of (a) and (b) to approximate the total cost when 11 units are produced.

45. Find the demand equation for a monopolist's product if his marginal revenue function is

$$\frac{dr}{dx}=\frac{\log(x+1)}{\sqrt{x+1}}.$$

Given $\log 2 \simeq 0.693$, find the price per unit for the above product, correct to 2 decimal places, when $x=15$.

(*)46. A new town is to be built in a circular shape with City Hall at the centre. It is planned to have a population density of 10^4 persons per square mile. It is estimated that property taxes will produce an annual revenue of \$200 per person. It is further estimated that running a city costs \$100r per person where r is the distance in miles between City Hall and the person's house. What (non-zero) radius, R, should the city have in order that the revenue matches the expenditure? Does the result depend on the population density?

Chapter V

Biology and Chemistry

This chapter considers elementary calculus problems related to the fields of biology and chemistry. Most of these problems are concerned with growth, decay and chemical reactions.

1. Malthusian Growth Law, Natural Decay (Exponential Growth and Decay)

In the case of *unrestricted* growth or decay of a quantity, the rate of change of the amount of the quantity, denoted by $\frac{dp}{dt}$, is proportional to the amount of the quantity, p, i.e.,

$$\frac{dp}{dt} = kp \tag{1}$$

for some proportionality constant k.

$$k > 0 \leftrightarrow \textit{unrestricted growth}.$$

$$k < 0 \leftrightarrow \textit{unrestricted decay}.$$

If p is the population size of some species and $k > 0$ then equation (1) corresponds to a *Malthusian growth law*. If p represents the amount of a substance undergoing radioactive decay where $k < 0$ then equation (1) corresponds to the *law of natural decay*.

$$\frac{dp}{dt} = kp \leftrightarrow \frac{dp}{p} = k\,dt,$$

after separation of variables. Hence

$$\int \frac{dp}{p} = \int k\, dt$$

$$\leftrightarrow \log p = kt + c \quad \text{for some constant } c \tag{1'}$$

$$\leftrightarrow p = p_0 e^{kt} \quad \text{for some constant } p_0 \tag{1''}$$

where $p = p_0$ when $t = 0$. (Note that $c = \log p_0 \leftrightarrow p_0 = e^c$.) One says that p *grows exponentially* if $k > 0$; p *decays exponentially* if $k < 0$ (see Figs. V.0a–d).

In the case of exponential growth ($k > 0$) the *doubling time* τ is the time it takes for the population size to double, i.e.,

$$p(\tau) = 2p_0 \leftrightarrow 2 = e^{k\tau} \leftrightarrow k\tau = \log 2 \leftrightarrow \tau = \frac{\log 2}{k}.$$

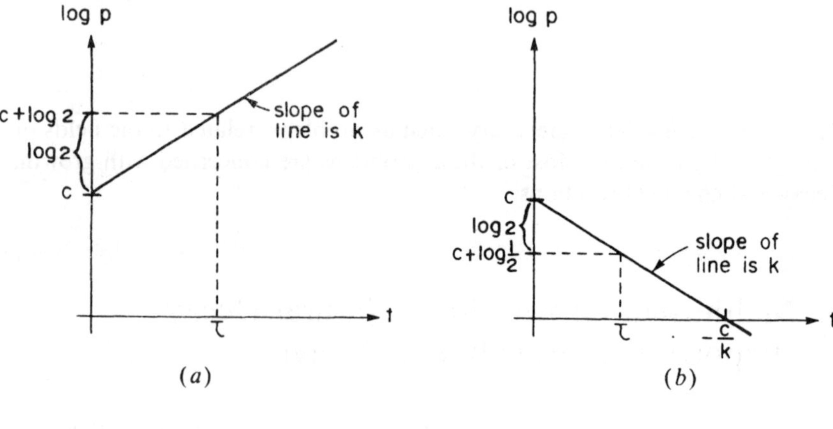

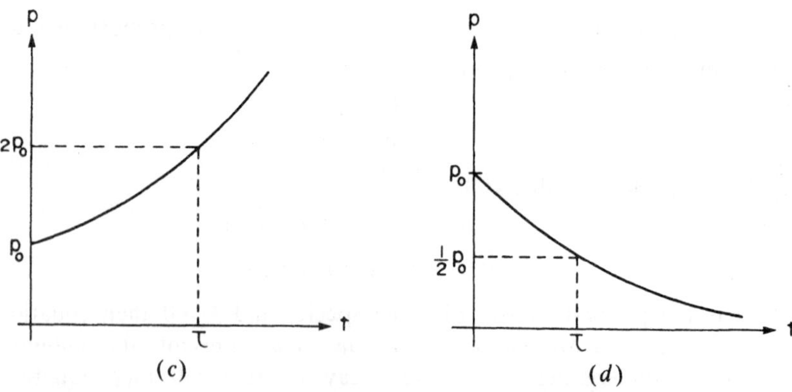

Figure V.0a. Graph of equation (1′) when $k > 0$ (τ = doubling time). b. Graph of equation (1′) when $k < 0$ (τ = half-life). c. Graph of equation (1″) when $k > 0$ (τ = doubling time). d. Graph of equation (1″) when $k < 0$ (τ = half-life).

In the case of exponential decay ($k < 0$) the *half-life* τ is the time it takes for half of the decaying substance to decay, i.e.,

$$p(\tau) = \frac{1}{2}p_0 \leftrightarrow \frac{1}{2} = e^{k\tau} \leftrightarrow k\tau = -\log 2 \leftrightarrow \tau = \frac{-\log 2}{k}.$$

It is easy to show that the values of p form a geometric sequence at equally spaced times if and only if p grows or decays exponentially.

2. Logistic Growth

The *logistic growth curve* or *Pearl-Verhulst curve* is a *restricted* growth curve. It represents a model for the spread of diseases (epidemics), the growth of bacteria, and certain chemical reactions where the amount of a quantity p eventually stabilizes to a level p^*. If $p(0) > p^*$ then the curve decays to p^*, and if $p(0) < p^*$ then the curve grows to p^* (Figs. V.0f, g). In any case $\lim_{t \to +\infty} p(t) = p^*$. The logistic growth curve satisfies the differential equation

$$\frac{dp}{dt} = kp(p^* - p) \tag{2}$$

where $k > 0$. The aim is to graph p as a function of t. After separation of variables, equation (2) becomes

$$\frac{dp}{p(p^* - p)} = k\,dt \leftrightarrow \frac{1}{p^*}\left[\frac{1}{p} - \frac{1}{p - p^*}\right] dp = k\,dt$$

$$\leftrightarrow \frac{1}{p^*}\{\log p - \log|p - p^*|\} = kt + \text{const.} \tag{2'}$$

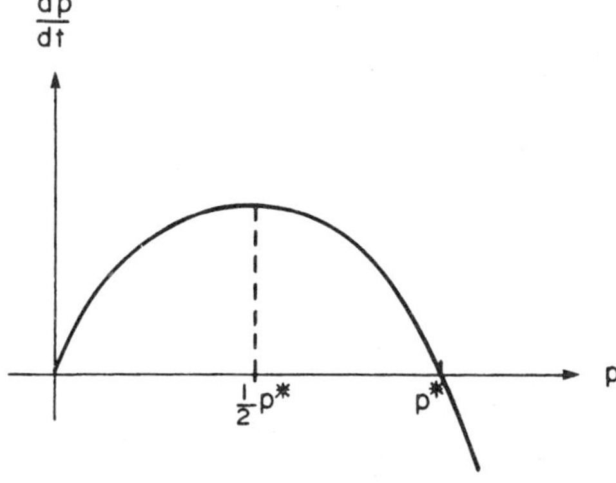

Figure V.0e. Graph of $\frac{dp}{dt}$ vs. p.

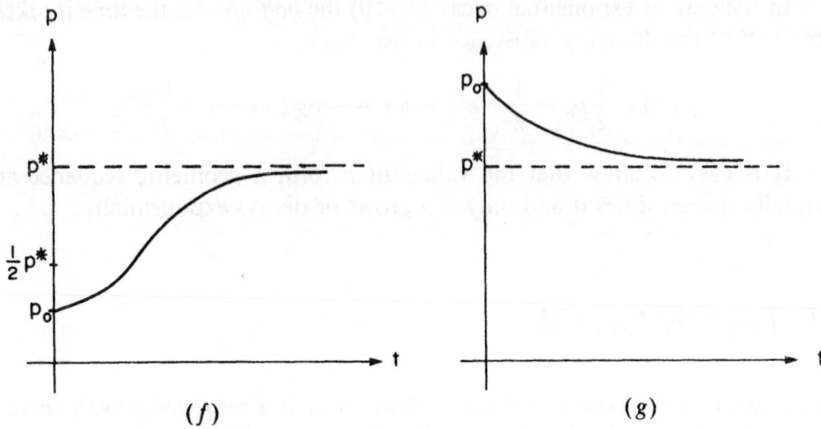

Figure V.0*f*. Graph of equation (2″) for $c > 0 \leftrightarrow p_0 < p^*$. *g*. Graph of equation (2″) for $c < 0 \leftrightarrow p_0 > p^*$.

Say $p(0) = p_0$. Then solving equation (2′) for p one finds that

$$p = \frac{p^*}{1 + ce^{-p^*kt}} \tag{2''}$$

where

$$c = \frac{p^* - p_0}{p_0}.$$

From equation (2) one sees that $\frac{d^2p}{dt^2} = k[p^* - 2p]\frac{dp}{dt}$. Thus $\frac{d^2p}{dt^2} = 0$ when $p = \frac{p^*}{2}$, i.e., $p = \frac{p^*}{2}$ is an inflection point (Fig. V.0*e*). Hence if $p_0 < \frac{p^*}{2}$, the rate of population increase achieves a maximum when the population reaches $\frac{p^*}{2}$ and then it slows down.

3. Chemical Reactions

(See Solved Problem 10)

Solved Problems

1. Population tends to grow with time at a rate roughly proportional to the population present. According to the Bureau of the Census, the population of the United States in 1960 was approximately 179 million and in 1970 it was 205 million.

(a) Use this information to estimate the population in 1940.
(b) Predict the population for the year 2000.

Let $P(t)$ be the estimated population in millions of the United States at time t years where $t=0$ corresponds to the year 1960. $P(t)=P_0e^{kt}$ for some constant k where $P_0=179$, $P(10)=205$. Hence $205=179e^{10k} \leftrightarrow k = \frac{1}{10}\log\left(\frac{205}{179}\right)$.

(a) The year 1940 $\leftrightarrow t=-20$: $P(-20)=179e^{-20k}=179\left(\frac{179}{205}\right)^2 \simeq 136.5$.

The estimated population in the year 1940 is $\boxed{136.5 \text{ million}}$.

(b) The year 2000 $\leftrightarrow t=40$: $P(40)=179e^{40k}=179\left(\frac{205}{179}\right)^4 \simeq 307.9$. The estimated population in the year 2000 would be $\boxed{307.9 \text{ million}}$ assuming unrestricted growth at a constant rate.

2. A pesticide sprayed onto tomatoes decomposes into a harmless substance at a rate proportional to the amount $M(t)$ still unchanged at time t. Write down a differential equation which describes this process and solve it for $M(t)$. If an initial amount of 10 pounds sprayed onto an acre reduces to 5 pounds in 6 days, when will 80% of the pesticide be decomposed?

This is an unrestricted decay problem. $M(t)$ satisfies the differential equation $\frac{dM}{dt}=-kM$ where $k>0$ is a constant. The solution of this differential equation is $M(t)=M_0e^{-kt}$ where $M_0=M(0)=10$ and t is time in days. One is given that $M(6)=5$ and the aim is to find the time $t=T$ such that $M(T)=(0.2)M(0)=(0.2)M_0 \leftrightarrow 0.2=e^{-kT} \leftrightarrow -kT=\log 0.2$

$$\leftrightarrow T=\frac{\log 0.2}{-k}. \; M(6)=5 \leftrightarrow 5=10e^{-6k} \leftrightarrow 0.5=e^{-6k} \leftrightarrow -6k=\log 0.5$$

$$\leftrightarrow -k=\frac{\log 0.5}{6}.$$

Hence

$$T=6\frac{\log 0.2}{\log 0.5}=6\frac{\log 5}{\log 2} \simeq \boxed{13.9 \text{ days}}.$$

Note that for a fixed value of k, T is *independent* of the value of M_0.

3. Professor Willard Libby of U.C.L.A. was awarded the Nobel prize in chemistry for discovering a method of determining the date of death of a once-living object. Professor Libby made use of the fact that the tissue of a living organism is composed of two kinds of carbon, a radioactive carbon A and a stable carbon B, in which the ratio of the amount of A to the amount of B is approximately constant. When the organism dies, the law of natural decay applies to A. If it is determined that the amount of A in a piece of charcoal is only 16% of its original amount and the half-life of A is 5500 years, when did the tree from which the charcoal came die?

Let $C(t)$ be the amount of carbon A in the charcoal at a time t years after the tree died. The law of natural decay corresponds to $\frac{dC}{dt} = -kC$ for some constant $k > 0$. Hence $C(t) = C_0 e^{-kt}$ where $C(0) = C_0$ is the amount of carbon A in the charcoal when the tree died. One is given the half-life $\tau = 5500$ for carbon A. Let $t = T$ be the number of years since the tree died; $C(T) = (0.16)C_0 = C_0 e^{-kT} \leftrightarrow -kT = \log 0.16 \leftrightarrow T = \frac{-\log 0.16}{k}$. But

$$k = \frac{\log 2}{\tau}.$$

Thus

$$\begin{aligned} T &= -\tau\frac{\log 0.16}{\log 2} = -\tau\frac{[\log 16 - \log 100]}{\log 2} \\ &= \tau\frac{[2\log 10 - 4\log 2]}{\log 2} \approx 14{,}500. \end{aligned}$$

Hence the tree died approximately **14,500 years ago**.

(*)4. Assume that the population of Canada satisfies an equation of the form $N_1(t) = C_1 e^{k_1 t}$ and that of B.C. an equation of the form $N_2(t) = C_2 e^{k_2 t}$ where $(N_1(t), N_2(t))$ are the respective populations at time t. Here k_1 and k_2 are constants to be determined for the purpose of "extrapolation". For the given data extrapolate the populations of B.C. and Canada to the years 2001, 2101, 2501, 3001 and comment on the answers.

Data(*)

Year	1901	1911	1921	1931	1941	1951	1961
$N_1(t)$	5.4	7.2	8.8	10.4	11.5	14.0	18.2
$N_2(t)$	0.18	0.39	0.52	0.69	0.81	1.17	1.62

(*)Population in millions.

From Fig. V.0a one sees that $N(t) = Ce^{kt}$ corresponds to the graph of $\log N(t)$ vs. t being a straight line with slope k. For the given data let the year 1901 be $t = 0$. Assuming exponential population growth, for the data points one first computes $\log N_i(t)$, $i = 1, 2$. After plotting $\log N_i(t)$ vs. t for $i = 1, 2$, one can estimate k_i by a least squares fit (Figs. V.4a, b).

t in years after 1901	0	10	20	30	40	50	60
$\log N_1(t)$	1.69	1.97	2.17	2.34	2.44	2.64	2.90
$\log N_2(t)$	−1.71	−0.94	−0.65	−0.37	−0.21	0.16	0.48

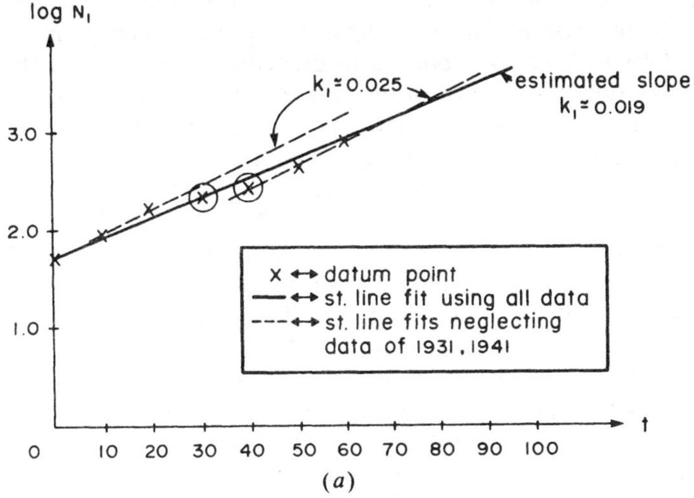

(a)

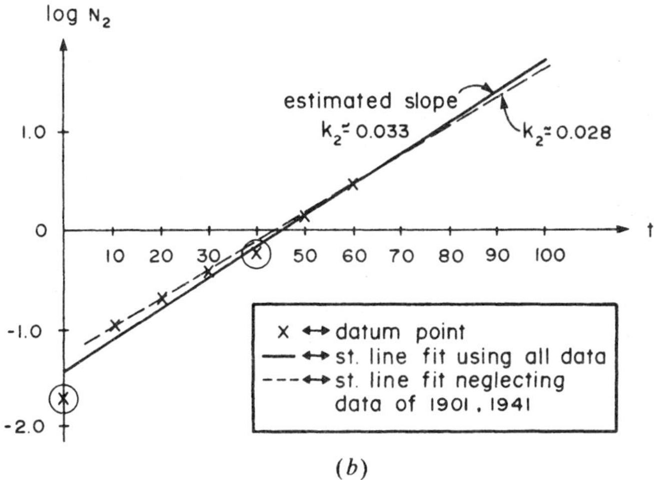

(b)

Figure V.4*a*. $\log N_1(t)$ vs. t (Canada). *b*. $\log N_2(t)$ vs. t (B.C.).

After a least squares fit, using *all* data points, one estimates that $k_1 = 0.019$, $k_2 = 0.033$ and that $C_1 = 5.73$, $C_2 = 0.23$. The years 2001, 2101, 2501 and 3001 correspond to the values of $t = 100$, 200, 600 and 1100 respectively. Based on such exponential growths one estimates (in millions) the populations in the year 2001: $N_1(100) = 38$, $N_2(100) = 6.3$; in the year 2101: $N_1(200) = 256$, $N_2(200) = 170$; in the year 2501: $N_1(600) = 5 \times 10^5$, $N_2(600) = 9 \times 10^7$; in the year 3001: $N_1(1100) = 10^{10}$, $N_2(1100) = 10^{15}$.

Since $k_2 > k_1$, the population of B.C. eventually exceeds that of Canada! In particular one must build in the constraint that B.C. is a province of Canada, i.e., $N_2(t) < N_1(t)$.

One can get "better" extrapolations by neglecting "ridiculous" data points. In the case of B.C. one might neglect the datum point 1901 (the period 1901 to 1911 corresponded to dramatic expansion) and the datum point 1941 (depression era); in the case of Canada one might neglect the data points 1931 (lower immigration) and 1941 (depression era). For Canada one finds that the straight lines fitting $\{1901, 1911, 1921\}$ and $\{1951, 1961\}$ have approximately the same slopes. These better extrapolations correspond to $k_1 \simeq 0.025$ and $k_2 \simeq 0.028$ and are indicated by dotted lines in Figures V.4*a*, *b*. Letting $t = 0$ be the year 1961 one finds $C_1 = 18.2$, $C_2 = 1.62$; consequently, one estimates (in millions) the populations in the year $2001 \leftrightarrow t = 40$: $N_1 = 49$, $N_2 = 5.0$; in the year $2101 \leftrightarrow t = 140$: $N_1 = 600$, $N_2 = 82$; in the year $2501 \leftrightarrow t = 540$: $N_1 = 1.3 \times 10^7$, $N_2 = 6 \times 10^6$; in the year $3001 \leftrightarrow t = 1040$: $N_1 = 3.6 \times 10^{12}$, $N_2 = 7 \times 10^{12}$.

A population of 1.3×10^7 for Canada corresponds to a population density of approximately 8000 ft^2/capita; a population of 6×10^6 for B.C. corresponds to a population density of approximately 2000 ft^2/capita. (In the year 1980, the population density of Hong Kong was approximately 2000 ft^2/capita.) Clearly the extrapolation of population data must take into account historical and geographical factors, and restrictions on growth.

(*)5. The curve marked *b* in Fig. V.5*a* gives the birth-rate in a population as a function of time. That is, at any time t the number of individuals born per unit time is the height of the curve. The curve marked *d* is the death-rate in the same population. Suppose at time $t = 0$ the population has size zero. Indicate geometrically how the following numbers can be found (measured) from the graph.

(a) The total number born by time t.

(b) The number alive at time t.

(c) The time t^* at which the population size is a maximum.

(d) The time $\hat{t}$ at which the net rate of increase of the population is a maximum.

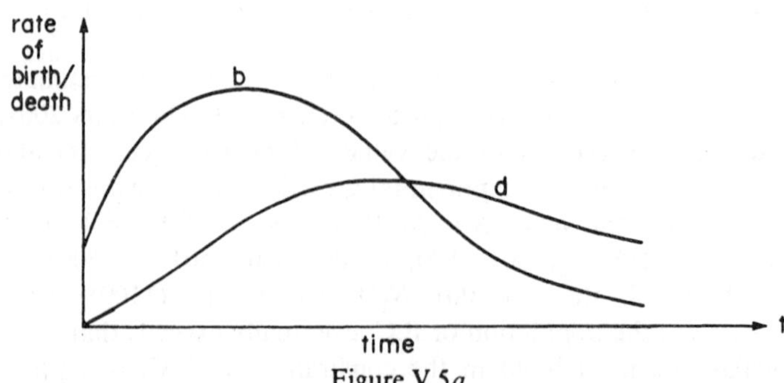

Figure V.5*a*

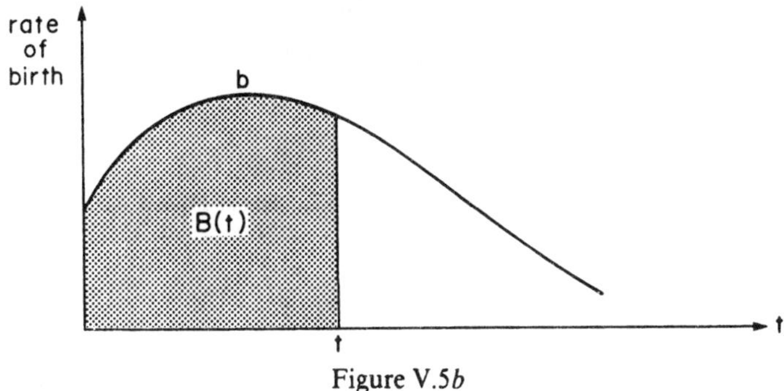

Figure V.5b

(a) Let $B(t)$ be the total number born at time t. Then $\frac{dB}{dt} = b(t)$, $B(0) = 0$. Hence $B(t) = \int_0^t b(t')\,dt'$ is the area under the b curve from 0 to t (Fig. V.5b).

(b) Let $L(t)$ be the number alive at time t and let $D(t)$ be the number who have died by time t. Then $L(t) = B(t) - D(t)$. Since $D(t)$ corresponds to the area under the d curve from 0 to t, $L(t)$ represents the area between the b curve and the d curve from 0 to t. Let t^* be the time when $b(t^*) = d(t^*)$. If $t > t^*$, the area between the curves (cf. Fig. V.5c) to the right of t^* is considered as negative.

(c) The population size is an extremum at the time t^* when

$$\frac{dL}{dt} = 0 \leftrightarrow b(t^*) = d(t^*).$$

This is the time t^* indicated in Fig. V.5c. t^* corresponds to a maximum since $\frac{dL}{dt} = b - d > 0$ if $t < t^*$, and $\frac{dL}{dt} < 0$ if $t > t^*$.

(d) Here the aim is to find the time $\hat{t}$ when $\frac{dL}{dt}$ is a maximum. Thus $\frac{d^2L}{dt^2} = 0$ when $t = \hat{t}$. $\frac{d^2L}{dt^2} = \frac{d}{dt}(b - d)$. Hence $\hat{t}$ is a time when the b and d

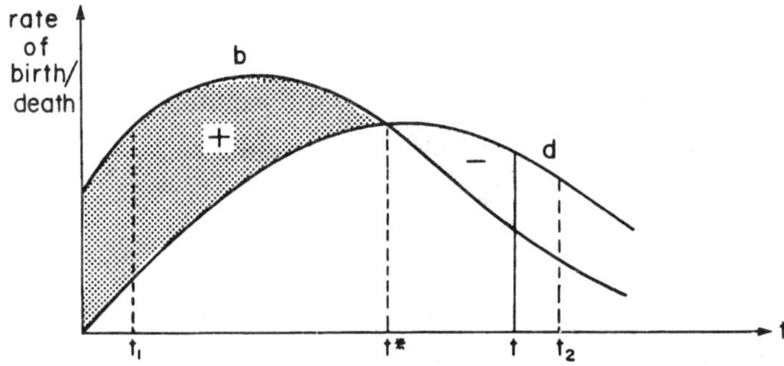

Figure V.5c

curves have the same slope. Hence $\hat{t}$ is either t_1 or t_2 in Fig. V.5*c*. If $t < t_1$, $\dfrac{db}{dt} > \dfrac{d(d)}{dt} \leftrightarrow \dfrac{d^2L}{dt^2} > 0$; if $t_1 < t < t_2$, $\dfrac{d^2L}{dt^2} < 0$; if $t > t_2$, $\dfrac{d^2L}{dt^2} > 0$. Hence $\boxed{\hat{t} = t_1}$.

6. (a) A salmon population living off the B.C. coast grows according to the Malthusian law:

$$\frac{dP(t)}{dt} = 0.03P(t)$$

where t is the time measured in years and $P(t)$ is the number of fish at time t. How long does it take for the number of salmon to double?

(b) Suppose at time $t = 0$, a group of predators moves into the home waters of the salmon, and kills the salmon at a rate of $0.0001[P(t)]^2$ per year. Write down the new law of growth for the salmon population. What happens to the population as $t \to \infty$?

(a) Let τ be the doubling time. The growth constant is $k = 0.03$.

$$\tau = \frac{\log 2}{k} = \frac{\log 2}{0.03} \simeq \boxed{23.1 \text{ years}}.$$

(b) Here $\dfrac{dP}{dt} = 0.03P - 0.0001P^2 = kP(P^* - P)$ where $k = 0.0001$ and $P^* = 300$.

This corresponds to logistic growth. Hence for some constant C

$$\boxed{P = \frac{P^*}{1 + Ce^{-kP^*t}}}.$$

Therefore $$\lim_{t \to \infty} P(t) = P^* = \boxed{300 \text{ salmon}}.$$

(*)7. A contagious disease, say smallpox, begins to spread in a community of 1000 people. This disease has the property that it spreads by contact and that each person who has it immediately and forever infects others. Initially, one person has it, and the speed of the epidemic appears to lessen after a month. Find how many people have had the disease at any time.

Let $P(t)$ be the infected population at time t. Then the epidemic satisfies the logistic equation $\dfrac{dP}{dt} = kP(P^* - P)$ whose solution is

$$P(t) = \frac{P^*}{1 + ce^{-P^*kt}}$$

where t is time in months. The aim is to find P^*, c and k. $P^* = 1000$ since eventually everyone must have had the disease. $P(0) = 1$ leads to $1 = \dfrac{1000}{1 + c}$. Hence $c = 999$.

One is given that $\frac{d^2P}{dt^2} > 0$ if $t < 1$ and $\frac{d^2P}{dt^2} < 0$ if $t > 1$. Hence $\frac{d^2P}{dt^2} = 0$ at $t = 1$, leading to $P(1) = \frac{P^*}{2}$. Thus

$$500 = \frac{1000}{1 + 999e^{-1000k}} \leftrightarrow k = \frac{\log 999}{1000} \simeq 0.0069. \text{ Therefore}$$

$$P(t) = \boxed{\frac{1000}{1 + 999e^{-6.9t}}}.$$

(*)8. There exist a great number of living species for which the birth-rate is not proportional to the size of the population. Suppose, for example, that for the reproduction of a species each member of the population must encounter a partner and that such an encounter is not purely the result of chance. It is natural to suppose that the expected number of encounters is proportional to the product of the number of males and of females making up this population. If it is supposed further that the males and females are equally distributed in the population, then the number of encounters, and consequently the number of births, is proportional to p^2, p designating the number of individuals making up this population. As for the rate of mortality, it is proportional to p. Under these hypotheses one concludes that $p(t)$ satisfies the differential equation

$$\frac{dp}{dt} = bp^2 - ap \quad \text{where} \quad a > 0, b > 0. \tag{1}$$

(a) Comment on the difference between equation (1) and the Verhulst equation.

(b) Suppose that at an instant t_0, one has $p(t_0) = p_0 < a/b$. In this case show that if one lets $y = \frac{a}{b} - p$, the function y satisfies the Verhulst equation.

(c) Supposing always that $p_0 < a/b$, conclude that $p(t) \to 0$ when $t \to \infty$. Furthermore give the precise behaviour of the graph of $p(t)$.

The preceding analysis shows that if some such population falls below the critical size a/b, then this population tends toward extinction. Thus a species is in danger if its population is in the neighbourhood of this critical value a/b. In this case, special measures should command attention.

(a) The Verhulst equation is $\frac{dp}{dt} = kp(p^* - p)$ where $k > 0$ and $p^* > 0$ are constants. Equation (1) can be rewritten as $\frac{dp}{dt} = -bp(p^* - p)$ where $p^* = \frac{a}{b} > 0$. Hence equation (1) is a Verhulst equation where $k < 0$.

(b) $y = \frac{a}{b} - p \leftrightarrow p = \frac{a}{b} - y$; $\frac{dp}{dt} = -\frac{dy}{dt}$; $bp^2 - ap = bp\left[p - \frac{a}{b}\right] = -b\left[\frac{a}{b} - y\right]y$. Hence equation (1) becomes

$$\frac{dy}{dt} = ky(y^* - y), \tag{2}$$

where $k = b > 0$, $y^* = \frac{a}{b} > 0$; $y(t_0) = \frac{a}{b} - p(t_0) > 0$. Moreover $y(t_0) < y^*$.

(c) The solution of equation (2) is (Fig. V.8a)

$$y(t) = \frac{y^*}{1 + ce^{-y^*kt}},$$

where $k = b$, $y^* = \frac{a}{b}$ and c is a constant satisfying the equation $\frac{a}{b} - p_0$ $= \dfrac{\frac{a}{b}}{1 + ce^{-at_0}} \leftrightarrow c = \dfrac{p_0 e^{at_0}}{\frac{a}{b} - p_0} > 0.$

Since $p(t) = \frac{a}{b} - y(t)$, it follows that $\lim_{t \to \infty} p(t) = \frac{a}{b} - \lim_{t \to \infty} y(t) = \frac{a}{b} - y^* = 0$ (Figs. V.8a–c).

(Note that if $p_0 > \frac{a}{b}$, then $c < 0$ and at some time $t^* > t_0$, $p(t^*) = \infty$.)

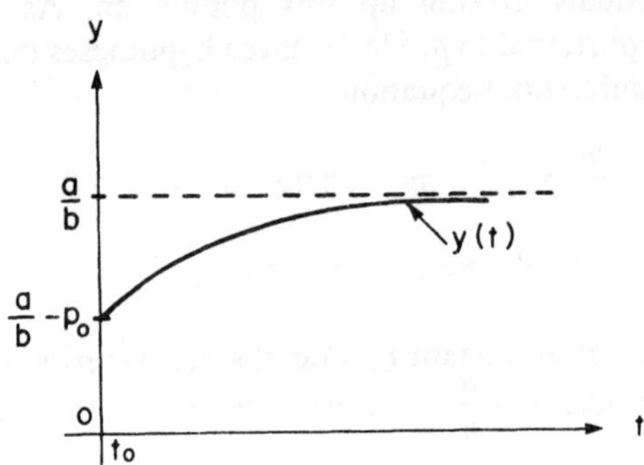

Figure V.8a. Graph of $y(t)$.

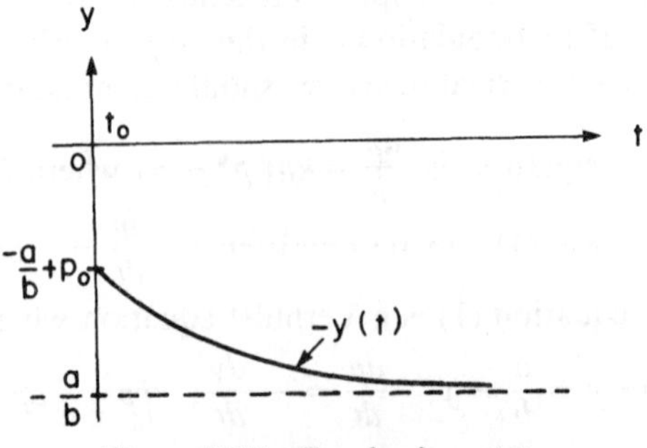

Figure V.8b. Graph of $-y(t)$.

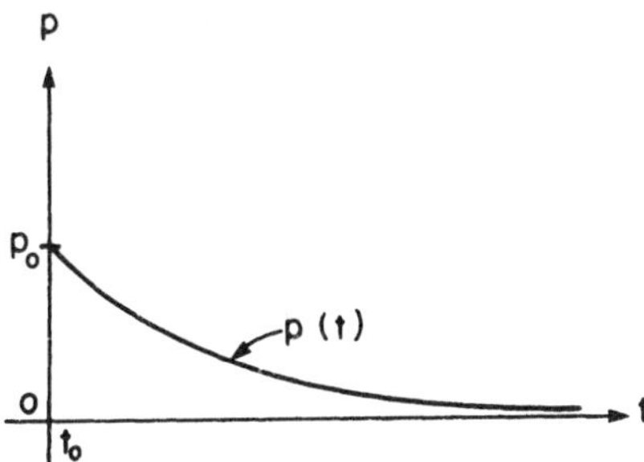

Figure V.8*c*. Graph of $p(t) = -y(t) + \frac{a}{b}$.

(*)**9.** Toxins in the medium of a bacterial culture kill bacteria at a rate proportional to the product of the number of bacteria present and the concentration of toxins. In the absence of toxins, the number of bacteria would grow at a rate proportional to the number of bacteria present. One can hope to control the number of bacteria by controlling the concentration of toxins; for example, by introducing toxins into the medium, or by removing some of the toxins from the medium.

Suppose the concentration of toxins varies with time at a constant rate c. At time t let $y(t) > 0$ be the number of bacteria and $T(t)$ the concentration of toxins. Let $y(0) = y_0$ and $T(0) = T_0$.

(a) Write and solve a differential equation satisfied by y.

(b) What happens when $t \to \infty$? Discuss according to the sign of c, indicating what it signifies concretely.

(c) Show, according to the values of $c \neq 0$, that the population of bacteria passes a maximum or a minimum. Find the time of this extremum. Sketch a graph indicating the behaviour of $y(t)$. Is it true that if $c = 0$, the population is Malthusian? If yes, give its rate of growth.

(a) From the data, bacteria grow at a rate of $k_1 y$ and are destroyed at a rate of $k_2 Ty$ where $k_1, k_2 > 0$ are constants. Hence $\frac{dy}{dt} = k_1 y - k_2 Ty$, $y(0) = y_0$. Moreover $\frac{dT}{dt} = c$, $T(0) = T_0$, and hence $T(t) = T_0 + ct$. Thus

$$\frac{dy}{dt} = y[k_1 - k_2 T_0 - k_2 ct] \leftrightarrow \frac{dy}{y} = [k_1 - k_2 T_0 - k_2 ct]\, dt = -k_2 c[t - \alpha]\, dt,$$

where $\alpha = \frac{k_1 - k_2 T_0}{k_2 c}$. Thus $\log y = -k_2 c \frac{(t-\alpha)^2}{2} + \text{constant}$. Hence

$$\boxed{y = \beta \exp\left[-\frac{k_2 c (t-\alpha)^2}{2}\right]}, \tag{1}$$

where $\beta = y_0 \exp\left[\frac{k_2 c \alpha^2}{2}\right]$.

(b) $\boxed{c>0}$: $\lim_{t\to\infty} y(t)=0$. Hence if $c>0$, the concentration of toxins grows indefinitely which leads to the total destruction of all bacteria.

$$\boxed{c<0}: \qquad T(t)=\begin{cases} T_0+ct & \text{if} \quad 0\le t\le \dfrac{-T_0}{c} \\ 0 & \text{if} \quad t\ge \dfrac{-T_0}{c} \end{cases}.$$

Here $y(t)$ is given by equation (1) if $t\le \frac{-T_0}{c}$; if $t>\frac{-T_0}{c}$, then $\frac{dy}{dt}=k_1 y$, $y\left(\frac{-T_0}{c}\right)>0$. Thus $y(t)=y\left(\frac{-T_0}{c}\right)e^{k_1(t+T_0/c)}$ and hence $\lim_{t\to\infty} y(t)=\infty$. In this case the toxins disappear after a finite amount of time and thereafter the bacteria have unrestricted growth.

(c) At an extremum, $\frac{dy}{dt}=0$, $k_1-k_2T_0=k_2ct\leftrightarrow t=\alpha$. Hence an extremum exists if and only if $c(k_1-k_2T_0)>0$. At $t=\alpha$, $\frac{d^2y}{dt^2}=-k_2cy$. Thus

$$t=\alpha\leftrightarrow\begin{cases}\text{maximum if } c>0\\ \text{minimum if } c<0.\end{cases}$$

Graphs for various cases of extrema are shown in Figs. V.9*a*, *b*.

If $c=0$: $T(t)=T_0\Rightarrow\frac{dy}{dt}=[k_1-k_2T_0]y$. This corresponds to Malthusian growth if $k_1-k_2T_0>0$; Malthusian decay if $k_1-k_2T_0<0$; $y=y_0$ if $k_1=k_2T_0$.

(*)10. (a) Say a, α and β are three positive constants, $\alpha\ne\beta$. Solve the equation $\frac{dy}{dx}=a(\alpha-y)(\beta-y)$; $y(0)=0$, $y<\min\{\alpha,\beta\}$.

(b) *Application.* Analyze a chemical reaction such that one molecule of a reactant A combines with one molecule of a reactant B to form one

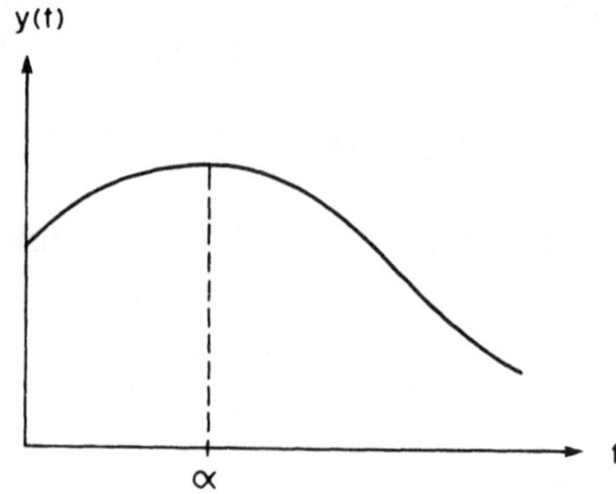

Figure V.9*a*. $k_1-k_2T_0>0$, $c>0$.

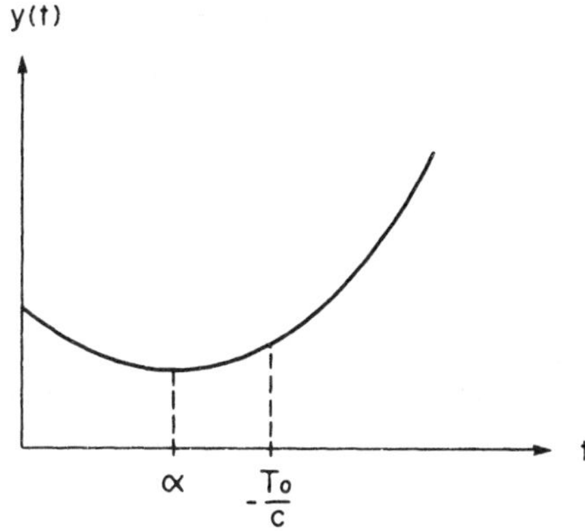

Figure V.9*b*. $k_1 - k_2T_0 < 0, c < 0$.

molecule of a product X and one or more other products:

$$A + B \to X + Y + \dots$$

$$(\text{example: } NaOH + HCl \to NaCl + H_2O).$$

A law of chemistry says that at each instant t, the rate of formation of product X is proportional to the *product* of the amount of reactant A and the amount of reactant B present at the time t. Designate by $k > 0$ this constant of proportionality, by a and b the initial quantities of reactants A and B respectively and by $x(t)$ the amount of product X at time t. Suppose that $x(0) = 0$.

(i) Write (with explanation) the differential equation satisfied by $x(t)$.

(ii) If $a \neq b$, show that the quantity $x(t)$ is given by:

$$x(t) = a\left[1 + \frac{b-a}{a - be^{(b-a)kt}}\right] = b\left[1 + \frac{a-b}{b - ae^{(a-b)kt}}\right].$$

(iii) For each case $a > b$ and $a < b$ discuss what happens when $t \to \infty$ and represent graphically the behaviour of the amount of product X. Is there a similarity to the logistic curve?

(iv) What happens in the case $a = b$?

(a) Let $u = -y + \alpha$. Then the equation becomes the logistic equation $\frac{du}{dx} = au(\alpha - \beta - u)$;

$$y < \alpha \leftrightarrow u > 0;$$

$$y < \beta \leftrightarrow \alpha - \beta - u < 0;$$

$$y(0) = 0 \leftrightarrow u(0) = \alpha.$$

Hence from the solution of the logistic equation, one sees that

$$u = \frac{\alpha - \beta}{1 + ce^{-a(\alpha-\beta)x}}, \quad \text{where} \quad c = -\frac{\beta}{\alpha} < 0, \; x \geq 0.$$

Note that $1 + ce^{-a(\alpha-\beta)x} \neq 0$ for any $x > 0$. Hence

$$y = -u + \alpha = \boxed{\alpha\left[1 + \frac{\beta - \alpha}{\alpha - \beta e^{-a(\alpha-\beta)x}}\right]}.$$

(b)

(i) If $x(t)$ molecules of X are produced then there are $x(t)$ fewer molecules left of each of the reactants A and B. Hence if $x(t)$ is the amount of product X at time t, then $a - x(t)$, $b - x(t)$ are the respective amounts of reactants A and B at time t. Thus $x(t)$ satisfies the differential equation

$$\frac{dx}{dt} = k(a - x)(b - x) \tag{1}$$

where $x(0) = 0$, $x(t) > 0$, $x(t) < \min\{a, b\}$.

(ii) From part (a), the result follows by appropriate substitutions.

(iii) *Case I*: $a > b$ (Fig. V.10a). This corresponds to the situation for $u = -x + a$ illustrated by Fig. V.0g where $p = u$, $p_0 = a$ and $p^* = a - b$. $\lim_{t\to\infty} x(t) = -\lim_{t\to\infty} u(t) + a = -(a - b) + a = b$.
Case II: $a < b$ (Fig. V.10b). This corresponds to interchanging a and b in equation (1) and consequently in Fig. V.10a. Thus $\lim_{t\to\infty} x(t) = a$.

(iv) Let $b = a + \varepsilon$ in the equation for $x(t)$ in part (ii) and then let $\varepsilon \to 0$. Hence, using a Taylor series expansion of $e^{\varepsilon kt}$ about $\varepsilon = 0$,

$$x(t) = \lim_{\varepsilon\to 0}\left\{a\left[1 + \frac{\varepsilon}{a - (a+\varepsilon)(e^{\varepsilon kt})}\right]\right\}$$

$$= a \lim_{\varepsilon\to 0}\left[1 + \frac{\varepsilon}{a - (a+\varepsilon)(1 + \varepsilon kt + \ldots)}\right]$$

$$= a \lim_{\varepsilon\to 0}\left[1 + \frac{\varepsilon}{-\varepsilon[1 + akt + \ldots]}\right] = a\left[1 - \frac{1}{1 + akt}\right] = \boxed{a - \frac{a}{1 + akt}}.$$

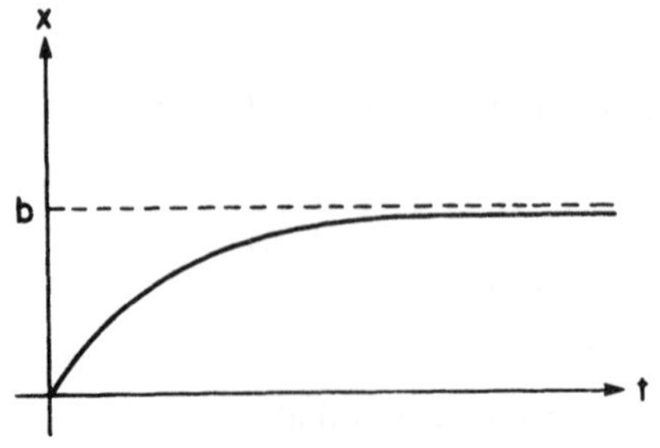

Figure V.10a. *Case I*: $a > b$.

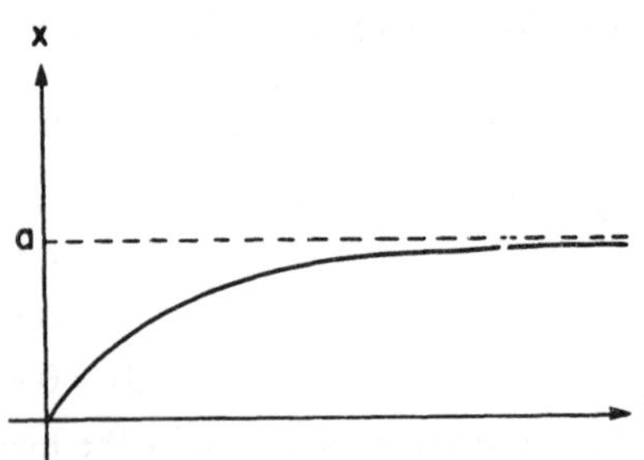

Figure V.10b. *Case II*: $a < b$.

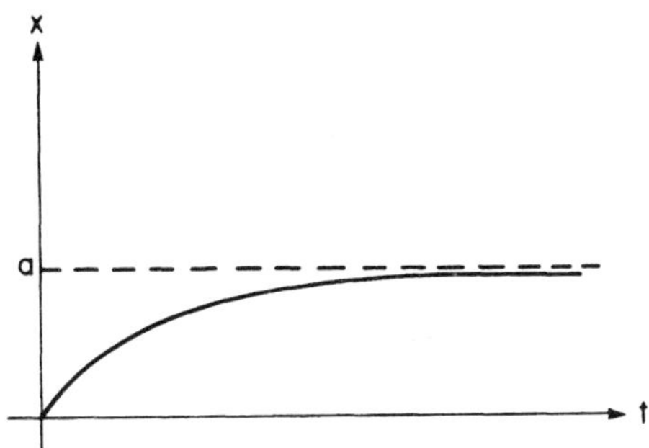

Figure V.10c. *Case III*: $a = b$.

Alternatively, note that $\lim_{t \to 0} \varepsilon \left[\frac{(a+\varepsilon)e^{\varepsilon k t} - a}{\varepsilon} \right] = \left(\frac{d}{d\varepsilon} \left[(a+\varepsilon)e^{\varepsilon k t} \right] \right) \Big|_{\varepsilon = 0}$

$= 1 + akt$. Hence $x(t) = a - \dfrac{a}{1+akt}$ (Fig. V.10c).

In this case $x(t)$ tends to the value a as $t \to \infty$. The approach to the equilibrium value is algebraic rather than exponential.

11. In the reaction 'trypsinogen → trypsin' the end product, trypsin, catalyzes the reaction. Let $\alpha(t)$ be the concentration of trypsinogen at time t (expressed as 'no. of molecules/unit volume'), and let B be the initial concentration of trypsin. Each molecule of trypsinogen yields just one molecule of trypsin. If $B = \alpha(0) = 1$, and $\alpha(1) = 1/2$, find $\alpha(4)$. Roughly plot the curve $y = \alpha(t)$.

If $\alpha(t)$ is the concentration of trypsinogen at time t, then $1 - \alpha(t)$ molecules of trypsinogen per unit volume have been converted to trypsin. At time t the number of molecules of trypsin per unit volume is

$$\beta(t) = 1 + 1 - \alpha(t) = 2 - \alpha(t).$$

$$\frac{d\beta}{dt} = k\beta(t)\alpha(t) \text{ where } k = \text{constant} > 0.$$

$$\frac{d\alpha}{dt} = -\frac{d\beta}{dt} = -k\alpha[2-\alpha] \quad \text{where } \alpha(0) = 1,\ \alpha(1) = 1/2.$$

From the solution of the logistic equation

$$\alpha(t) = \frac{2}{1 + ce^{2kt}} \text{ where } c = 1;$$

$$\alpha(1) = 1/2 = \frac{2}{1+e^{2k}} \Rightarrow e^{2k} = 3.$$

$$\alpha(4) = \frac{2}{1+e^{8k}} = \frac{2}{1+3^4} = \frac{1}{41} \simeq \boxed{0.024}.$$

The curve $y = \alpha(t)$ is plotted in Fig. V.11.

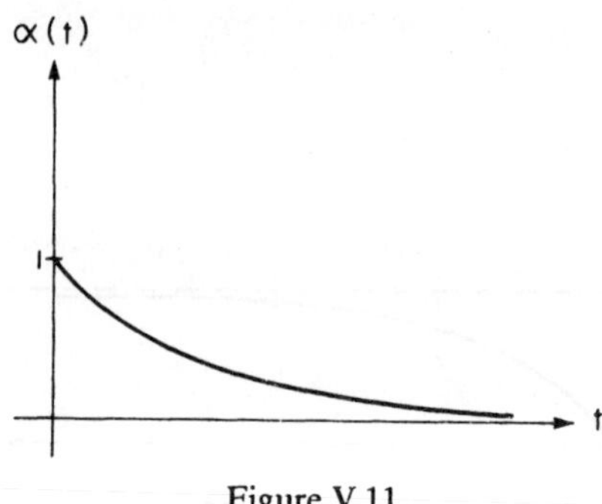

Figure V.11

Supplementary Problems

1. The population of a certain bacteria is known to double every 3 hours. If there are 1,000 bacteria present at 12 noon, what is the population at 7 p.m.?

2. The rate of decay of radium is proportional to the amount present at any time. If 60 mg of radium are present now and the half-life of radium (the time required for half of the substance to decay) is 1690 years, how much radium will be present 100 years from now?

3. The population of a city grows at a rate proportional to the population of the city. It takes 20 years for the population to grow from 30,000 to 90,000 people.

(a) How many years would it take for the population to grow from 30,000 to 150,000 people?

(b) What would be the average rate of change of population over the period of time in part (a)?

4. In a western town during the gold rush it was noted that the rate of decrease of population $p(t)$ at any instant (due to 'natural' causes) was proportional to the population at that instant. If the populations in 1834 and 1840 were 512 and 256 respectively, in what year did the town become a ghost town?

5. A country is growing in population at a rate proportional to its size. In August 1975 it had a population of 10 million, and the population is now 10.5 million. The government plans to introduce immigration restrictions when the population reaches 15 million. In what year will these restrictions go into effect?

6. Let P be the world's population. If the rate of increase is proportional to the population, then $P = P_0 e^{kt}$. In 1965 P was approximately 3,000,000,000. Let $t = 0$ represent the year 1965, and let $k = 0.018$ (a percentage increase of

1.8% per year, which actually occurred in 1965). Assuming that one-quarter acre is required to provide food for 1 person, and that the world contains 10 billion acres of arable land, when will the world reach its saturation point?

7. The total land suitable for farming on this planet is estimated to be 10^9 acres. (There are 36.8×10^9 acres of land on the earth.) At the present level of food production, one acre will support four people, thereby limiting the population which can be sustained by present standards to 4×10^9 people. The 1967 world population was estimated at 3.4×10^9 people, and the rate of increase, $\frac{p'}{p}$, at $\frac{19}{1000}$ per year. If this rate is constant, when will the world population reach 4×10^9 people?

8. (a) The population of town A grows at the rate of 2% per year. How long will it take to double itself?

(b) The population of town B doubles itself in one-third of the time required by that of town A to double itself. Assuming B grows at a rate proportional to its population, what is the percentage growth of B per year?

(c) By what factor will the population of B increase in the time taken by the population of A to double itself?

9. The rate of growth of a population of size P is proportional to P but also, because of a deteriorating environment, inversely proportional to the square root of the time $\left(\text{i.e., the rate of growth is proportional to } \frac{P}{\sqrt{t}}\right)$. If the initial population was 10^6 and it doubled in 4 years, find the population after 100 years.

10. A mycologist believes that the rate of increase of the weight of a certain colony of mould is proportional to its weight and inversely proportional to the square of $t+4$, where t is the time in weeks (i.e., proportional to $(\text{weight})/(t+4)^2$). If he is correct, and if the colony doubles its weight in the first week, what will have happened to its weight after 16 weeks, and what will its weight approach as $t\to\infty$?

(*)11. There are two countries, A and B. At time $t=0$ there are 10 million people in country A and no people in country B. Then emigration begins. If the rate of emigration is proportional to the difference between the population of A and the population of B, and if at time $t=10$ there are 2.5 million people in B and 7.5 million in A, what will the value of t be when the population of B reaches 4 million?

(*)12. Initially (at $t=0$) a tank contains 1,000 gallons of brine containing 50 lb of salt. Fresh water is added to the tank at the rate of 20 gal/min and the fluid is drawn off at the same rate (the concentration is maintained uniform by stirring). Write a differential equation for $S(t)$, the number of lb of salt in the tank at time t. *Do not solve.*

13. A tank initially has 600 gal of brine containing 450 lb of salt dissolved. Pure water is run into the tank at the rate of 15 gal/min, and the mixture, kept uniform by stirring, is withdrawn at the same rate. How many pounds of salt remain after 20 minutes?

14. A tank contains 200 gallons of water and 10 pounds of salt. Water is pumped in at 3 gal/min and the mixture pumped out at 1 gal/min.

(a) Find how much salt $S(t)$ is in the tank at any time t and graph this function.

(b) When will there be 5 pounds of salt in the tank?

(*)15. A population is changing due to births (at 0.02 births/person/year) and immigration (at 5 people/year). Initially the population is 100. Find how many descendants the immigrants have had after 10 years have elapsed.

(*)16. A population is changing due to deaths (at 0.01 deaths/person/year) and immigration (at 10 people/year). Initially the population was 2,000.

(a) Find the population after 10 years.

(b) How many died in the first 10 years?

17. An epidemic of falling ague spreads through a town of 10,000. Originally one person has it, and in two weeks ten people have it.

(a) When will the spread be most rapid?

(b) When will 900 people be falling about?

(*)18. A rumour is spreading in a population of 500. Each person meets four people each day. Initially one person knew the rumour.

(a) When will 250 people know it?

(b) When will 499 people know it?

(*)19. Two chemicals A, B react together, one molecule of A combining with one of B. Initially their concentrations are equal and in one hour they are halved. When will they be one-quarter of their initial value?

20. The rate of formation of a certain chemical in a reaction is known to be governed by the equation

$$\frac{dx}{dt} = (a - x)(b - x)$$

where x is the amount (mass) of the chemical present at time t and a, b are the amounts of other chemicals present when $t = 0$, with $0 < b < a$. For $t = 0$, $x = \frac{1}{2}(a + b)$. Find x as a function of time and determine $\lim_{t \to +\infty} x(t)$.

21. Bacteria are introduced at a constant rate α^2 as food into a population of protozoans. It is observed that the bacteria are consumed at a rate proportional to the square of their number. If $p(t)$ designates the population of bacteria in the medium at time t, then:

$$\frac{dp}{dt} = \alpha^2 - \beta^2 p^2 \text{ where } \alpha > 0 \text{ and } \beta > 0.$$

Suppose that at any time the population of bacteria is less than α/β.

(a) Find $p(t)$, knowing that $p(0)=0$, and show that $p(t)$ is of the form $p(t)=c_1-\dfrac{c_2}{1+e^{c_3 t}}$ where c_1, c_2 and c_3 are three constants to be identified.

(b) Does the population of bacteria attain a state of equilibrium? Represent graphically the behaviour of the population, indicating if there are any points where the population is at its maximum as well as the inflection points.

(*)22. Chemical reactions are not instantaneous and very many are reversible. Consider one such case. Let A and B be chemical compounds such that under given fixed conditions A transforms into B and B into A. It is reasonable to assume that the rate of transformation from A to B is proportional to the amount of product A present at time t, and similarly the rate of passage of B towards A is proportional to the amount of product B present at time t. This is represented in Fig. S.22.

$$A \underset{k'}{\overset{k}{\rightleftharpoons}} B$$

Figure S.22

The constants of proportionality k and k', sometimes called *specific rates of reaction*, are positive by convention, an appropriate sign preceding them according to the form of the equation. Note that one molecule of A reacts to produce one molecule of B and conversely.
Designate by

a : the initial amount of product A.
0(zero): the initial amount of product B.
$x(t)$: the amount of product A at time t.
$y(t)$: the amount of product B at time t.

(a) Is it true that $x(t)+y(t)=a$ for all time? Give a physical reason and a mathematical reason.

(b) Write a differential equation governing the system and solve it to obtain $y(t)$.

(c) The chemical reaction attains a state of equilibrium, that is, a state in which the amounts of A and B do not vary any more. Show this.

(d) Find the upper limit β of $y(t)$, for $t>0$, and verify that this amount is never attained.

(e) Calculate the half-life of the reaction, that is, the time necessary until the amount of B equals $\beta/2$. What time is required in order to obtain $(3/4)\beta$?

(*)23. *The cigarette kills.*
By 10:00, on a good Friday night, a discotheque of dimensions 30 feet by 50 feet by 10 feet is full of clients. The very great majority of them are smokers: hence cigarette smoke containing 4% carbon monoxide is introduced into the room at a rate of 0.15 cubic feet per minute. Suppose that this rate does not vary significantly during the course of the evening. Fortunately, before 10:00 there is no trace of monoxide in the disco, and furthermore, this disco is equipped with good ventilators. These ventilators allow the formation of a homogeneous smoke-air mixture in the room, and the ejection of this mixture to the outside at the rate of 1.5 cubic feet per minute, that is, at a rate 10 times greater than that of the arrival of pollutant. You want to dance and to socialize but also you want to preserve your health. A prolonged exposure to a concentration of carbon monoxide greater than or equal to 0.012% is considered dangerous by the ministry of health and welfare. Knowing that the disco closes its doors at 3:00 in the morning, will you allow yourself to stay until the end? To be more precise, find the moment when the concentration of monoxide reaches the critical concentration of 0.012%.

Hint: Let $Q(t)$ be the amount of monoxide in the disco at time t and find the equation satisfied by $Q(t)$.

(*)24. For the purification of water, the aerobic reactions (in the presence of oxygen) are more rapid than the anaerobic reactions. In order to favour the former, the medium must contain a sufficient concentration of dissolved oxygen. One must in fact compensate for the oxygen consumed by the mass of organisms in the water with an artificial aeration.

Under normal conditions, the concentration $c(t)$ (in mg/l) of oxygen dissolved in the water may not exceed a certain level c_s, called the saturation concentration. If air is undersupplied in the water, the rate of supply of the concentration of dissolved oxygen is proportional to the difference between the saturation concentration and the concentration at time t. Let k be this constant of proportionality.

In fact, designate by V, the volume of water in the aeration vats, and by Θ, the rate of consumption of oxygen by the microorganisms in the water of the vats (in mg/min). Experimentally it has been determined that this rate of consumption remains relatively constant as long as $c(t) > c_0 = \frac{1}{2}$ mg/l and that otherwise it decreases very rapidly, something one wishes to absolutely avoid. Suppose that $c(t_0) = \frac{1}{2}$ mg/l.

(a) Write the differential equation governing the system.

(b) Show that $c(t) = \left(c_s - \frac{\Theta}{kV}\right)(1 - e^{-k(t-t_0)}) + \frac{1}{2}e^{-k(t-t_0)}$ is the solution to this equation for the given initial condition.

(c) Show that if $t \to \infty$, then $c(t)$ tends towards a constant c_e (to be determined), called the equilibrium concentration.

(d) Show that it is necessary to limit the mass of the microorganisms in the water such that

$$\Theta \le kV(c_s - c_0).$$

(e) Show that $\dfrac{dc}{dt} = k(c_e - c(t))$. From this conclude that if c_e replaces c_s, then the system is entirely analogous to that without microorganisms.

(f) Verify that whatever the constant L, the function $c(t) = c_e + Le^{-kt}$ solves the equation given in (e) and thus show that one can determine k by measuring $c(t)$ at two distinct times t_1 and t_2, assuming that c_e is already known.

25. In chemistry an adiabatic process is one in which there is no gain or loss of heat. During an adiabatic process then the pressure $P = P(t)$ and volume $V = V(t)$ of oxygen in a container are related for all time by the formula

$$PV^{1.5} = C$$

where C is a constant. Suppose that the volume of oxygen in a closed container is increasing at a rate of 2 m^3/s. How fast is the pressure decreasing when the volume is 20 m^3 and the pressure is 0.5 $\mathrm{N/m}^2$?

26. The rate of reaction to a certain dose of a drug at time t hours after administration is given by $r(t) = te^{-t^2}$ (measured in appropriate units). Why is it reasonable to define the *total reaction* as the area under $r(t)$ from $t = 0$ to $t = \infty$? Evaluate the total reaction to the given dose of the drug.

Chapter VI

Numerical Methods

Often the solution of a problem is a complicated function, an integral which is not elementary, or an infinite series. For computational purposes such a solution can be estimated by some *numerical method*. If it is possible an *error analysis* should be given for any estimate. At the very least it is necessary to understand the limitations of an estimate.

1. Linear Approximation—Differential, Error in Measurement, Linear Interpolation

Say $y = f(x)$ is differentiable at $x = x_0$. Then near $x = x_0$, $f(x)$ can be *approximated linearly* by $l(x) = f(x_0) + (x - x_0)f'(x_0)$, i.e., near $x = x_0$,

$$f(x) = f(x_0) + (x - x_0)f'(x_0) + E(x)$$

where $E(x)$ is the *error* in approximating $f(x)$ by $l(x)$ (Fig. VI.0*a*). If $f(x)$ is differentiable at $x = x_0$, then $\lim\limits_{x \to x_0}\left[\dfrac{E(x)}{x - x_0}\right] = 0$, i.e., near $x = x_0$ the error $E(x) = \eta(x)(x - x_0)$ where $\lim\limits_{x \to x_0} \eta(x) = 0$. Note that $y = l(x)$ is the *tangent line* to $y = f(x)$ at $x = x_0$.

If $l(x)$ is simple to compute (in effect meaning that the values $f(x_0)$, $f'(x_0)$ are simple to compute) and if one can estimate that $|E(x)| \ll |l(x)|$, then $l(x)$ is a useful approximation for computing $f(x)$.

Let $\Delta x = x - x_0$, $\Delta y = f(x) - f(x_0) = f(x_0 + \Delta x) - f(x_0)$. In terms of this notation, near $x = x_0$, $\Delta y = f'(x_0)\Delta x + E(x)$ where $f'(x_0)\Delta x$ is called

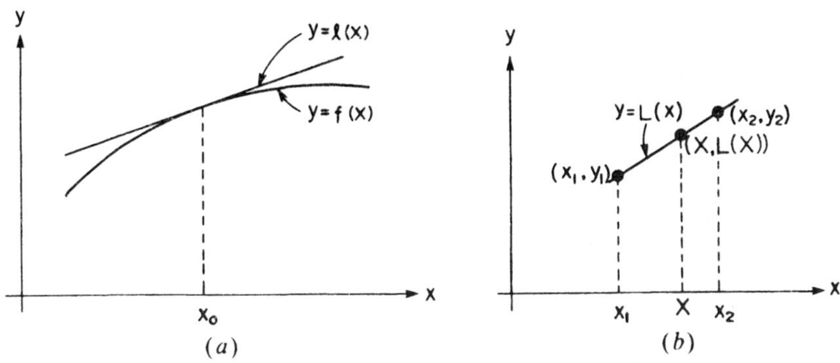

Figure VI.0*a*. Approximation of $y = f(x)$ by the straight line $y = l(x)$ near $x = x_0$. *b*. Approximation by linear interpolation.

the *differential* of $y = f(x)$ *near* $x = x_0$. ($dy = f'(x_0)dx$ is the differential of $f(x)$ *at* $x = x_0$.)

$y = f(x)$ could represent the solution to a problem with an error Δx in the measurement of x, i.e., $x = x_0 + \Delta x$ where Δx is the estimated error (usually expressed as $\pm$ some number) in using the measured value x_0 for x. The resulting error in determining y from such an *error in measurement* is Δy which is estimated by the differential $f'(x_0)\Delta x$, i.e.,

$$f(x) = f(x_0) + \Delta y \simeq f(x_0) + f'(x_0)\Delta x.$$

Given measured values $y_1 = f(x_1)$, $y_2 = f(x_2)$, where $x_2 = x_1 + \Delta x$, $|\Delta x| \ll 1$, one often tries to estimate $Y = f(X)$ for $x_1 < X < x_2$ by using *linear interpolation* (Fig. VI.0*b*). This method is based on the approximation of $y = f(x)$, $x_1 \le x \le x_2$, by the straight line $y = L(x)$ joining (x_1, y_1) to (x_2, y_2), i.e.,

$$f(x) \simeq L(x) = y_1 + \left(\frac{y_2 - y_1}{x_2 - x_1}\right)(x - x_1),$$

so that $Y \simeq L(X)$. Linear interpolation is useful if $f(x)$ can be assumed to be differentiable on the interval $[x_1, x_2]$.

2. Newton's Method

Newton's method (*Newton–Raphson method*) is a method for finding a root (zero) of a function $f(x)$, i.e., it is a method for finding $x = r$ such that $f(r) = 0$. The method succeeds if $f'(x)$ is continuous near $x = r$, or $f(x)f''(x) > 0$ on one side of r. As a first step one must find an interval $I = [a, b]$ containing one and only one root of $f(x)$. (There can be at most one root of $f(x)$ on I if $f'(x) \neq 0$ on I.) The procedure is then as follows:

(a) Let $x_1 \in [a, b]$ be a first guess for the root.

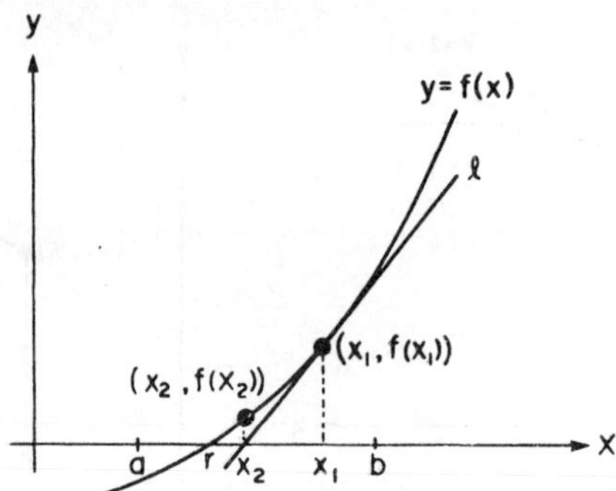

Figure VI.0c. Illustration of Newton's method.

(b) At $(x_1, f(x_1))$, construct the tangent line l to $y = f(x)$. Let $\hat{x}_2$ be the point where this tangent line l cuts the x-axis (Fig. VI.0c).

(c) Assuming $\hat{x}_2 \in [a, b]$, let $x_2 = \hat{x}_2$ (or an appropriate estimate close to $\hat{x}_2$) be a second guess for the root. Repeat the procedure.

(d) Continue iteratively until the desired accuracy is reached for the root.

In general $\hat{x}_{n+1} = x_n - \dfrac{f(x_n)}{f'(x_n)}$, $n = 1, 2, \ldots$.

If $f(x)f''(x) > 0$ on the interval (r, b), then $x_n \in (r, b) \Rightarrow \hat{x}_{n+1} \in (r, x_n)$. Similarly if $f(x)f''(x) > 0$ on the interval (a, r), then $x_n \in (a, r) \Rightarrow \hat{x}_{n+1} \in (x_n, r)$. When Newton's method is succeeding it often happens that if $x_n = r + \alpha\varepsilon$, then $\hat{x}_{n+1} \simeq r + \alpha\varepsilon^2$, i.e., the rate of convergence to the root is geometrical (See Solved Problem 18 in Chapter VII, Theory).

Usually an effective error analysis for an estimate x_n for r is to choose convenient numbers $\varepsilon \geq 0$, $\delta \geq 0$ and check the signs of $f(x_n + \varepsilon)$, $f(x_n - \delta)$. If $f(x_n + \varepsilon)f(x_n - \delta) < 0$ then $x_n - \delta < r < x_n + \varepsilon \leftrightarrow r = \bar{x}_n \pm \left(\dfrac{\varepsilon + \delta}{2}\right)$, where $\bar{x}_n = x_n + \dfrac{\varepsilon - \delta}{2}$.

Of course, one should use a calculator as an aid in computing all iterates.

3. Numerical Integration—Trapezoidal Rule, Simpson's Rule

If a given definite integral $I = \int_a^b f(x)\,dx$ (Fig. VI.0d) is not elementary, then one must resort to a numerical method based on an appropriate approximation of $f(x)$ which leads to an easily computable integral. The simplest effective numerical integration methods are the *trapezoidal rule* (based on approximating the curve $y = f(x)$ by a polygonal (linear) inter-

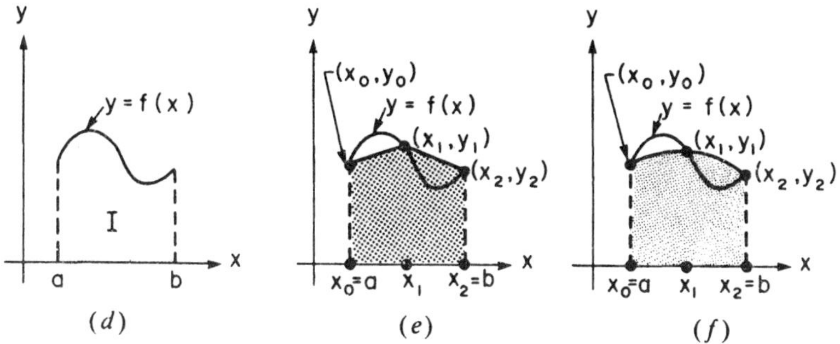

Figure VI.0*d*. $I = \int_a^b f(x)\,dx$. *e*. Trapezoidal rule approximation to I for $n = 2$. *f*. Simpson's rule approximation to I for $n = 2$.

polating curve (Fig. VI.0*e*)) and *Simpson's rule* (based on approximating $y = f(x)$ by interpolating parabolic curves (Fig. VI.0*f*)). I is estimated by computing the area under the approximating curve. The accuracy of the estimation improves as the step size $h = \dfrac{b-a}{n} = \Delta x$ decreases where $n+1$ is the number of equally-spaced interpolating points; n must be even for Simpson's rule. Let $x_0 = a$, $x_n = b$, $y_k = f(x_k)$, $k = 0, 1, 2, \ldots, n$. Then for step size h,

$$T(h) = \frac{h}{2}[y_0 + 2y_1 + 2y_2 + \cdots + 2y_{n-1} + y_n]$$

is the *trapezoidal rule* approximation to I;

$$S(h) = \frac{h}{3}[y_0 + 4y_1 + 2y_2 + 4y_3 + 2y_4 + \cdots + 4y_{n-1} + y_n]$$

is the *Simpson's rule* approximation to I. $I = T(h) + E_T(h) = S(h) + E_s(h)$ where $E_T(h)$, $E_s(h)$ are the respective errors in the trapezoidal rule and Simpson's rule approximations to I. One can show that

$$E_T(h) = -\frac{(b-a)}{12}h^2 f''(c) \text{ for some } c \in (a, b);$$

$$E_s(h) = -\frac{(b-a)}{180}h^4 f^{(4)}(\xi) \text{ for some } \xi \in (a, b).$$

4. Estimation by Taylor's Formula (Taylor Polynomials)

Say $f(x)$ is such that $\{f(x), f'(x), \ldots, f^{(n+1)}(x)\}$ exist for x in the open interval $I = (a, b)$. Let $x_0 \in I$. Then on I, $f(x) \equiv P_n(x) + R_n(x)$ where $R_n(x)$ is the error in approximating $f(x)$ by the *nth degree Taylor poly-*

nomial $P_n(x)$ *about* x_0:

$$P_n(x) = f(x_0) + f'(x_0)(x - x_0) + \cdots + \frac{f^{(n)}(x_0)}{n!}(x - x_0)^n,$$

$$R_n(x) = \frac{f^{(n+1)}(c)}{(n+1)!}(x - x_0)^{n+1} \text{ for some } c \in I.$$

$P_n(x)$ is useful for computing $f(x)$ if

(a) $R_n(x)$ can be estimated to be small for some reasonable integer $n = N$, and

(b) $f(x_0), f'(x_0), f''(x_0), \ldots, f^{(N)}(x_0)$ are simple to compute.

Note that for fixed n, the approximation of $f(x)$ by $P_n(x)$ worsens as $|x - x_0|$ increases.

The *Taylor series* for $f(x)$ about x_0 is $\sum_{n=0}^{\infty} \frac{f^{(n)}(x_0)}{n!}(x - x_0)^n$.

If $x_0 = 0$, then the Taylor series is called a *Maclaurin series.*

5. Estimation of Tails of Infinite Series

(a) *Alternating series.* Consider an *alternating series* $S = \sum_{k=1}^{\infty} (-1)^k a_k$ where $a_k > 0$, $a_k > a_{k+1}$ for $k = 1, 2, \ldots$. Say $\lim_{k \to \infty} a_k = 0$. Then for each n, $S = S_n + R_n$ where the *tail* R_n is the error in approximating S by its *nth partial sum* $S_n = \sum_{k=1}^{n} (-1)^k a_k$. One can show that $|R_n| < a_{n+1}$ and that R_n and a_{n+1} have the same sign, i.e., $R_n a_{n+1} > 0$. Hence $S = S_n + \frac{1}{2}a_{n+1} + E_n$ where $|E_n| < \frac{1}{2}|a_{n+1}|$, after the error is centred.

(b) *Estimating by comparison with elementary series.* Often the tail of a series can be estimated by comparing the tail with a known series (e.g. a geometric series) or dominating it by an elementary improper integral. Here $S = \sum_{k=1}^{\infty} a_k = S_n + R_n$ where $S_n = \sum_{k=1}^{n} a_k$ and the tail is $R_n = \sum_{k=n+1}^{\infty} a_k$. If $|a_k| \le M_k$ for $k = 1, 2, \ldots$ and $E_n = \sum_{k=n+1}^{\infty} M_k$ is easy to compute, then the estimate $|R_n| \le E_n$ could be useful. See Solved Problem 20 for examples.

Solved Problems

1. Find the best linear approximation (i.e. the best approximating line) for x^3 near $x = \frac{1}{2}$.

Here $f(x) = x^3$, $f'(x) = 3x^2$, $x_0 = \frac{1}{2}$. $f'(x_0) = \frac{3}{4}$. The line is

$$y = l(x) = f(x_0) + f'(x_0)(x - x_0) = \tfrac{1}{8} + \tfrac{3}{4}\left(x - \tfrac{1}{2}\right) = \boxed{\tfrac{3}{4}x - \tfrac{1}{4}}.$$

(*)**2.** (a) Find the function which approximates $f(x)=\sqrt{1+\sin(x^2-2)}$ linearly near $x=\sqrt{2}$.

(b) Use your function to calculate $f(1.5)$ approximately.

(c) Can you predict whether your approximate value is too large or too small?

(a) $f'(x)=\dfrac{x\cos(x^2-2)}{\sqrt{1+\sin(x^2-2)}}$, $x_0=\sqrt{2}$, $f(\sqrt{2})=1$, $f'(\sqrt{2})=\sqrt{2}$. Hence the desired function is $l(x)=f(\sqrt{2})+f'(\sqrt{2})(x-\sqrt{2})=\boxed{\sqrt{2}x-1}$.

(b) $f(1.5)\simeq l(1.5)=\frac{3}{2}\sqrt{2}-1\simeq\boxed{1.12}$.

(c) $f''(x)=\dfrac{\cos u-x^2(1+\sin u)}{\sqrt{1+\sin u}}$ where $u=x^2-2$.

If $f''(x)<0$ for $\sqrt{2}\le x\le 1.5$ then $f(x)$ is concave down in this interval and hence $l(x)$ estimates too large a value for $f(1.5)$, i.e. $f(1.5)<l(1.5)$ (see Fig. VI.2). Now it is shown that $f''(x)<0$ if $x\in[\sqrt{2},1.5]$ (i.e., if $u\in[0,\frac{1}{4}]$):

$$f''(x)=\frac{\cos u-(u+2)(1+\sin u)}{\sqrt{1+\sin u}}$$
$$=\sqrt{1-\sin u}-(u+2)\sqrt{1+\sin u}. \qquad \text{(Note that } \cos u\ge 0.)$$

Hence $f''(x)=0$

$$\leftrightarrow \frac{1-\sin u}{1+\sin u}=(u+2)^2 \leftrightarrow \sin u=g(u) \text{ where } g(u)=-1+\frac{2}{1+(u+2)^2}.$$

But on the interval $[0,\frac{1}{4}]$: $\sin u>0$, $g(u)<0$. Thus $f''(x)\ne 0$ on the interval $[\sqrt{2},1.5]$; $f''(\sqrt{2})=-1$. Hence $f''(x)<0$ if $\sqrt{2}\le x\le 1.5$.

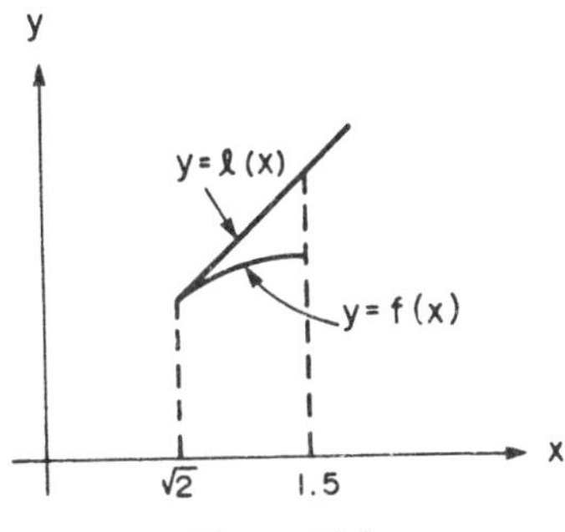

Figure VI.2

3. Use differentials to find an approximate value of arctan (1.04).

Let $y=f(x)=\arctan x$. Then $f'(x)=\dfrac{1}{1+x^2}$. $f(1)=\dfrac{\pi}{4}$, $f'(1)=\dfrac{1}{2}$. Let $x_0=1$, $\Delta x=0.04$; $f(1.04)\simeq f(1)+f'(1)\Delta x=\dfrac{\pi}{4}+0.02\simeq\boxed{0.81}$.

(*)**4.** At a point 180 feet from the base of a building on level ground, the angle of elevation of the top of the building is 30°. Use differentials to find the error in the height of the building due to an error of $\frac{1}{4}°$ in this angle.

The height of the building is $h(\theta) = a\tan\theta$ where $a = 180$, $\theta = \theta_0 + \Delta\theta$, and $\theta_0 = 30° = \frac{\pi}{6}$ radians (Fig. VI.4). The error in measurement $\Delta\theta$ is such that $|\Delta\theta| = \frac{1}{4}° = \frac{\pi}{720}$ radians. $h'(\theta) = a\sec^2\theta$, $h'(\theta_0) = a\sec^2\frac{\pi}{6} = (180)\left(\frac{4}{3}\right) = 240$. Hence the error in estimating the height of the building from the error in measurement is approximately $\pm|h'(\theta_0)\Delta\theta| = \pm\frac{\pi}{3}$ feet $\simeq$ $\boxed{\pm 1.05 \text{ feet}}$.

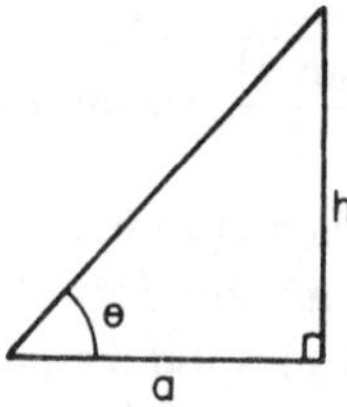

Figure VI.4

(*)**5.** Let $F(a,b) = \int_a^b \cos^6\pi x\,dx$. Given that $F(1,3) = 0.625$:

(a) approximate $F(1.01, 3.02)$;
(b) approximate $F(2.01, 4.02)$.

(a) For any a, b, $F(a,b) = F(0,b) + F(a,0)$. Hence $F(1.01,3.02) = F(0,3.02) + F(1.01,0)$. Let $G(x) = F(0,x) = \int_0^x \cos^6\pi X\,dX$. By the Fundamental Theorem of Calculus, $G'(x) = \cos^6\pi x$. Let $H(x) = F(x,0)$ $= -G(x)$. Thus $H'(x) = -\cos^6\pi x$. Using differential approximations:

$$G(3.02) \simeq G(3) + (0.02)G'(3),$$
$$H(1.01) \simeq H(1) + (0.01)H'(1).$$

Hence

$$\begin{aligned} F(1.01,3.02) &= G(3.02) + H(1.01) \simeq G(3) + H(1) \\ &\quad + (0.02)G'(3) + (0.01)H'(1) \\ &= F(1,3) + (0.02) - (0.01) = \boxed{0.635}. \end{aligned}$$

(b) Note that $f(x) = \cos^6\pi x$ is such that $f(x+1) = f(x)$. Hence $F(a,b) = F(a+1, b+1)$ for all a, b. In particular, $F(2.01, 4.02) =$ $F(1.01, 3.02) \simeq \boxed{0.635}$.

6. Use Newton's method to find, correct to 3 decimals, the root of $x^2 + 10x - 2 = 0$ which is close to 0.

Let $f(x) = x^2 + 10x - 2$. $f(0) < 0$, $f(1) > 0$. Hence there is some root r, $0 < r < 1$, such that $f(r) = 0$. $f(-10) < 0$, $f(-11) > 0$. Thus the other root lies in the interval $(-11,-10)$. $f'(x)=2x+10$, $f''(x)=2>0$. Hence $f(x)f''(x)>0$ if $x>r$.

Newton's method:

$$\hat{x}_{n+1}=x_n-\frac{f(x_n)}{f'(x_n)}=x_n-\frac{\left[(x_n)^2+10x_n-2\right]}{2x_n+10}$$

$$=\frac{(x_n)^2+2}{2x_n+10}.$$

Let $x_1=1$. Then $\hat{x}_2=1-\frac{9}{12}=0.2500$. Letting $x_2=\hat{x}_2\Rightarrow\hat{x}_3<\tilde{x}_3=0.19643$. $f(x)f''(x)>0$ if $x>r\Rightarrow\tilde{x}_3>r$. Hence $f(\tilde{x}_3)>0$. One can check that $f(0.196)<0$. Hence $0.196<r<0.19643$. Thus $r=\boxed{0.1962\pm0.0003}$.

7. Compute (to 3 decimal places) all the positive roots of $x-\dfrac{1}{x}+\sqrt{x}$.

Let $f(x)=x-\dfrac{1}{x}+\sqrt{x}$; $f'(x)=1+\dfrac{1}{x^2}+\dfrac{1}{2x^{1/2}}$; $f''(x)=-\dfrac{2}{x^3}-\dfrac{1}{4x^{3/2}}$. Since $f'(x)>0$ if $x>0$, there is at most one root r where $r>0$. Since $\lim_{x\to0^+}f(x)=-\infty$, $\lim_{x\to+\infty}f(x)=+\infty$, there is precisely one root r of $f(x)$ on the interval $I=(0,\infty)$. $f(1)>0\Rightarrow r\in(0,1)$. Use Newton's method to find r. $f''(x)<0$ if $x>0\Rightarrow f(x)f''(x)>0$ on $(0,r)$. Moreover if $x_n>r$ then $\hat{x}_{n+1}<r$ since $f(x)f''(x)<0$ if $x>r$.

$$\hat{x}_{n+1}=x_n-\frac{f(x_n)}{f'(x_n)}=\frac{2x_n-\frac{1}{2}x_n^{5/2}}{x_n^2+1+\frac{1}{2}x_n^{3/2}}.$$

Let $x_1=1$. Then $\hat{x}_2=0.6$. Let $x_2=\hat{x}_2$. Then $\hat{x}_3>0.666$. Let $x_3=0.666$. Then $\hat{x}_4>0.671$. Clearly $f(0.671)<0$. One can check that $f(0.672)>0$. Hence $r=\boxed{0.6715\pm0.0005}$.

A better way to find r is to let $g(x)=xf(x)$. For $x>0$, $g(x)$ and $f(x)$ have the same roots. Now apply Newton's method to $g(x)$.

(*)8. Consider the equation $\dfrac{2x}{3}-\sin x=0$.

(a) Determine the number of roots of this equation.
(b) Find the largest root to 3 decimal accuracy.

(a) Let $f(x)=\dfrac{2x}{3}-\sin x$. $f(-x)=-f(x)$. Hence $f(r)=0\Rightarrow f(-r)=0$. $f(0)=0$. $\sin x\le1$. Thus $f(x)>0$ if $x\ge\frac{3}{2}$. $f'(x)=\frac{2}{3}-\cos x$, $f'(0)=-\frac{1}{3}<0$.

Hence there is some r^*, $0<r^*<\frac{3}{2}$ such that $f(r^*)=0$. Since $f'(x)=0$ only once on the interval $I=(0,\frac{3}{2})$, $f(x)$ has one and only one root $r^*\in I$. Thus $f(x)$ has 3 roots: $-r^*$, 0, r^*. Alternatively one can see this graphically in Figure VI.8.

(b) Newton's method is used to determine r^*. $f''(x)=\sin x>0$ on I.

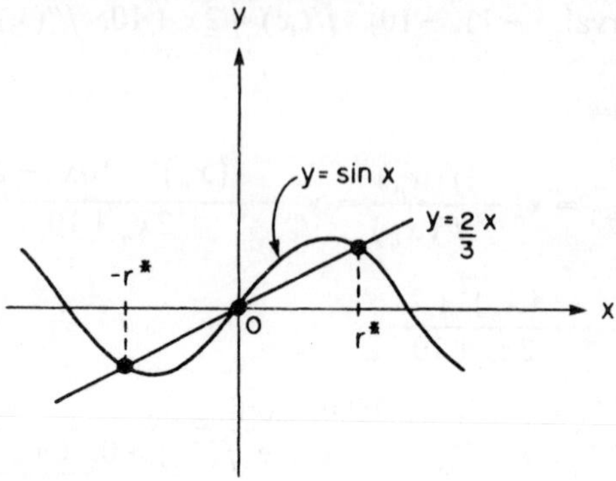

Figure VI.8. Determination of number of roots of $f(x)$ by graphical means. The roots occur where $y = \sin x$ intersects $y = \frac{2}{3}x$.

Hence $f(x)f''(x) > 0$ if $x \in (r^*, \frac{3}{2})$. Let $x_1 = \frac{3}{2}$. In general $\hat{x}_{n+1} = x_n - \dfrac{f(x_n)}{f'(x_n)} = \dfrac{-x_n \cos x_n + \sin x_n}{\frac{2}{3} - \cos x_n}$. $\hat{x}_2 = 1.49579\ldots \Rightarrow f(1.4958) > 0$.

One can check that $f(1.4957) < 0$. Hence $r^* = \boxed{1.49575 \pm 0.00005}$.

(*)9. Evaluate the integral $J = \displaystyle\int_1^3 \frac{x+1}{x}\,dx$ using the trapezoidal rule with $n = 5$. Compute the maximum value of $|E_T|$.

In this problem $a = x_0 = 1$, $b = x_5 = 3$, $h = \dfrac{b-a}{5} = \dfrac{2}{5}$, $f(x) = \dfrac{x+1}{x}$.

n	0	1	2	3	4	5
x_n	1	$\frac{7}{5}$	$\frac{9}{5}$	$\frac{11}{5}$	$\frac{13}{5}$	3
$y_n = f(x_n)$	2	$\frac{12}{7}$	$\frac{14}{9}$	$\frac{16}{11}$	$\frac{18}{13}$	$\frac{4}{3}$

$$T(\tfrac{2}{5}) = (\tfrac{1}{2})(\tfrac{2}{5})[2 + \tfrac{24}{7} + \tfrac{28}{9} + \tfrac{32}{11} + \tfrac{36}{13} + \tfrac{4}{3}] = \boxed{3.110267\ldots}.$$

$$f(x) = 1 + \frac{1}{x};\ f'(x) = \frac{-1}{x^2};\ f''(x) = \frac{2}{x^3};\ f'''(x) = \frac{-6}{x^4}.$$

Since $f'''(x) \neq 0$ for any x, the maximum and minimum values of $f''(x)$ on the interval $[1,3]$ occur at the endpoints. Hence $\frac{2}{27} < f''(c) < 2$. Thus $\dfrac{-h^2}{3} < E_T(h) < \dfrac{-h^2}{81} \Rightarrow |E_T| < \dfrac{h^2}{3}$. Thus the maximum value of $|E_T|$ is $\boxed{\frac{4}{75}}$. By *centring the error*, one can give a better estimate of the integral:

$$J = T(h) + E_T(h) \quad \text{where} \quad \alpha < E_T(h) < \beta \Rightarrow J = T(h) + \frac{\beta + \alpha}{2} + E(h)$$

where $|E(h)| < \frac{\beta - \alpha}{2}$. Thus in this case $J = T\left(\frac{2}{5}\right) - \frac{14h^2}{81} + E(h)$ where $|E(h)| < \frac{13}{81}h^2 = \left(\frac{13}{27}\right)\left(\frac{4}{75}\right)$.

(*)**10.** Say one wishes to evaluate $J = \int_1^3 \frac{1}{x^3}dx$ by Simpson's rule with an error not exceeding 10^{-3}. Obtain a good estimate of the number of subintervals which will be needed.

Here $f(x) = \frac{1}{x^3}$; $f'(x) = -\frac{3}{x^4}$; $f''(x) = \frac{12}{x^5}$; $f'''(x) = -\frac{60}{x^6}$; $f^{(4)}(x) = \frac{360}{x^7}$. Thus the maximum and minimum values of $f^{(4)}(x)$ on the interval $[1,3]$ occur at the endpoints. For step size $h = \frac{b-a}{n} = \frac{2}{n}$, the error in the Simpson's rule approximation to J is $E_s(h) = -\frac{(b-a)}{180}h^4 f^{(4)}(c)$ for some c, $a < c < b$. Thus here $E_s(h) = -\frac{8f^{(4)}(c)}{45n^4}$ where n is even and $1 < c < 3$. $\frac{360}{3^7} < f^{(4)}(c) < 360$. Hence $-\frac{64}{n^4} < E_s(h) < -\frac{64}{3^7 n^4}$. If the error is not centred, $|E_s(h)| < 10^{-3}$ if $\frac{64}{n^4} < 10^{-3} \leftrightarrow n^4 > 64 \times 10^3$; $n = 16$ works, i.e., 16 subintervals are needed to guarantee that $|J - S(h)| < 10^{-3}$.

(*)**11.** Use Taylor's formula up to degree 3 to calculate $e^{-0.1}$. Determine the accuracy of your answer by a careful estimate of the magnitude of the error term.

$f(x) = e^x$ is such that $f^{(n)}(x) = e^x$. Let $x_0 = 0$. Then $f^{(n)}(0) = 1$. Hence $e^x = P_3(x) + R_3(x)$ where $P_3(x) = 1 + x + \frac{x^2}{2} + \frac{x^3}{6}$, and $R_3(x) = \frac{e^c}{24}x^4$ is the error term in approximating e^x by $P_3(x)$. If $x < 0$, then c is some number such that $x < c < 0$. Hence $R_3(-0.1) = \frac{e^c}{24}10^{-4}$ is such that $\frac{(10^{-4})e^{-0.1}}{24} < R_3(-0.1) < \frac{10^{-4}}{24}$. In particular $0 < R_3(-0.1) < \frac{10^{-4}}{24}$. Hence $e^{-0.1} = 1 - \frac{1}{10} + \frac{1}{200} - \frac{1}{6000} + \frac{10^{-4}}{48} + E$ where $|E| < \frac{10^{-4}}{48}$. Thus $e^{-0.1} = 0.904835 \pm 0.000003$, after the error from the remainder term $R_3(x)$ is centred. (Alternatively one can do the error analysis by observing that the Taylor series for $f(x)$ evaluated at $x = -0.1$ is an alternating series.)

12. Find the first two nonzero terms of the Taylor series of $y = \sin^{-1}x$ about $x = 0$. Use your result to calculate an approximate value of $\sin^{-1}(\frac{1}{2})$ and compare it with the exact value which to three decimal places of accuracy is 0.524.

Method 1:

$y(0)=0;\ \ y'=\dfrac{1}{\sqrt{1-x^2}} \Rightarrow y'(0)=1;\ \ y''=\dfrac{x}{(1-x^2)^{3/2}} \Rightarrow y''(0)=0;$

$y'''=\dfrac{1+2x^2}{(1-x^2)^{5/2}} \Rightarrow y'''(0)=1$. Hence

$$y=f(x)=\sin^{-1}(x)=f(0)+f'(0)x$$
$$+\frac{f''(0)x^2}{2!}+\frac{f'''(0)x^3}{3!}+\cdots=x+\frac{x^3}{6}+\cdots.$$

Method 2:

From the binomial theorem,

$$(1+x)^{-1/2}=1+\left(-\frac{1}{2}\right)x+\frac{(-\frac{1}{2})(-\frac{3}{2})x^2}{2!}+\cdots$$
$$\Rightarrow (1-x^2)^{-1/2}=1+\frac{1}{2}x^2+\frac{3}{8}x^4+\cdots \text{ about } x=0. \text{ Hence}$$
$$\sin^{-1}x=\int_0^x \frac{1}{(1-z^2)^{1/2}}dz=\int_0^x\left[1+\frac{1}{2}z^2+\frac{3}{8}z^4+\ldots\right]dz$$
$$=x+\frac{x^3}{6}+\ldots \quad \text{about } x=0.$$

If $\sin^{-1}x$ is approximated by $x+\dfrac{x^3}{6}$ near $x=0$, then.

$\boxed{\sin^{-1}(\tfrac{1}{2})\simeq 0.5208\ldots}$.

13. (a) Find the Taylor polynomial of degree 3, about $x=\pi/4$ (i.e. in powers of $(x-\pi/4)$), for $f(x)=\tan x$.
(b) Use (a) to estimate $\tan(50°)$.

(a) $f\left(\dfrac{\pi}{4}\right)=1;\ f'(x)=\sec^2x \Rightarrow f'\left(\dfrac{\pi}{4}\right)=2;\ f''(x)=2\sec^2x\tan x \Rightarrow f''\left(\dfrac{\pi}{4}\right)$ $=4;\ \ f'''(x)=2[2\sec^2x\tan^2x+\sec^4x] \Rightarrow f'''\left(\dfrac{\pi}{4}\right)=16$. Hence the Taylor polynomial of degree 3 for $f(x)$ about $x=\dfrac{\pi}{4}$ is

$$P_3(x)=\boxed{1+2\left(x-\frac{\pi}{4}\right)+2\left(x-\frac{\pi}{4}\right)^2+\frac{8}{3}\left(x-\frac{\pi}{4}\right)^3}.$$

(b) $50° \leftrightarrow \dfrac{\pi}{4}+\dfrac{\pi}{36}$ radians; $\tan 50° \simeq P_3\left(\dfrac{\pi}{4}+\dfrac{\pi}{36}\right)$ $=1+\dfrac{\pi}{18}+\dfrac{2\pi^2}{(36)^2}+\dfrac{8\pi^3}{3(36)^3}=\boxed{1.1915}$. ($\tan 50°=1.192\ldots$)

(*)14. Using Taylor series with remainder, show how to compute the natural logarithm of 3 to five digits accuracy (don't compute).

Since $e < 3 < e^2$, the first nonzero digit occurs before the decimal point. Thus one wants to show how to compute $\log 3$ to four decimal accuracy.

Method 1:

$$\log 3 = -\log\frac{1}{3} = -\log\left(1-\frac{2}{3}\right) = \int_0^{2/3}\frac{dx}{1-x} = \int_0^{2/3}\left(\sum_{k=0}^{\infty}x^k\right)dx$$

$$= \sum_{k=0}^{\infty}\int_0^{2/3}x^k\,dx = \sum_{k=0}^{\infty}\frac{(2/3)^{k+1}}{k+1} = S.$$

$\log 3 = S = S_n + R_n$ where $S_n = \sum_{k=0}^{n}\frac{(2/3)^{k+1}}{k+1}$ and $R_n = \sum_{k=n+1}^{\infty}\frac{(2/3)^{k+1}}{k+1}$ is the error in approximating $\log 3$ by S_n. $0 < R_n < \frac{1}{n+2}\sum_{k=n+1}^{\infty}(2/3)^{k+1} = \frac{(2/3)^{n+2}}{n+2}\sum_{m=0}^{\infty}(2/3)^m = \frac{(2/3)^{n+2}}{n+2}\left[\frac{1}{1-(2/3)}\right]$ $= 3\frac{(2/3)^{n+2}}{n+2}$. One wants $R_n < 5\times 10^{-5}$; $3\frac{(2/3)^{n+2}}{n+2} < 5\times 10^{-5}$ if $n = 18$.

Thus $\log 3 = \boxed{\sum_{k=0}^{18}\frac{(2/3)^{k+1}}{k+1}}$ to 5 digits accuracy.

Method 2:

$$\log(1+x) = x - \frac{x^2}{2} + \frac{x^3}{3} - \frac{x^4}{4} + \cdots \quad \text{about } x = 0.$$

$$\log(1-x) = -x - \frac{x^2}{2} - \frac{x^3}{3} - \frac{x^4}{4} + \cdots \quad \text{about } x = 0.$$

Hence

$$\log\left(\frac{1+x}{1-x}\right) = 2x + \frac{2x^3}{3} + \frac{2x^5}{5} + \cdots = 2\sum_{k=0}^{\infty}\frac{x^{2k+1}}{2k+1} \quad \text{about } x = 0.$$

$\frac{1+x}{1-x} = \frac{1}{3} \leftrightarrow x = -\frac{1}{2}$. Hence $\log 3 = -\log\frac{1}{3} = \sum_{k=0}^{\infty}\frac{1}{2^{2k}(2k+1)}$.

$\log 3 = s_n + r_n$ where $s_n = \sum_{k=0}^{n}\frac{1}{2^{2k}(2k+1)}$ and $r_n = \sum_{k=n+1}^{\infty}\frac{1}{2^{2k}(2k+1)}$ is the error in approximating $\log 3$ by s_n.

$$0 < r_n < \frac{1}{2n+3} \sum_{k=n+1}^{\infty} \frac{1}{2^{2k}} = \frac{1}{2^{2n+2}(2n+3)} \sum_{m=0}^{\infty} \frac{1}{4^m}$$

$$= \frac{1}{4^{n+1}(2n+3)} \left[\frac{1}{1-\frac{1}{4}} \right] = \frac{1}{3(4^n)(2n+3)}.$$

One wants $\dfrac{1}{3(4^n)(2n+3)} < 5 \times 10^{-5}$; $n = 5$ works. Thus

$$\log 3 = \boxed{\sum_{k=0}^{5} \frac{1}{2^{2k}(2k+1)}}$$ to 5 digits accuracy.

(*)15. Determine the Taylor polynomial for $f(x) = \cos x$ about $x = 0$ in which the error is less than 0.01 for all values of x in the interval $\left[0, \frac{\pi}{4}\right]$.

$f(x) = \cos x = \displaystyle\sum_{k=0}^{\infty} \frac{(-1)^k x^{2k}}{(2k)!}$ is the Taylor series of $\cos x$ about $x = 0$.

$\cos x = P_{2n}(x) + R_{2n}(x)$ where $P_{2n}(x) = \displaystyle\sum_{k=0}^{n} \frac{(-1)^k x^{2k}}{(2k)!}$ is the Taylor polynomial of degree $2n$ for $f(x)$ about $x = 0$ and $R_{2n}(x) = \displaystyle\sum_{k=n+1}^{\infty} \frac{(-1)^k x^{2k}}{(2k)!}$ is the error in approximating $f(x)$ by $P_{2n}(x)$. If $x > 0$ then the Taylor series is alternating. Hence, from the error analysis for an alternating series: $|R_{2n}(x)| < 0.01$ on $\left[0, \frac{\pi}{4}\right]$ if $\dfrac{\left(\frac{\pi}{4}\right)^{2n+2}}{(2n+2)!} < 10^{-2}$; $n = 2$ works. The desired polynomial is $P_4(x) = \boxed{1 - \dfrac{x^2}{2} + \dfrac{x^4}{24}}$.

(*)16. Find a solution of the equation $10x^2 = \cos x$ which is correct to 3 decimals by replacing $\cos x$ with a polynomial approximating $\cos x$ quite accurately near $x = 0$.

Let $x = r$ be a solution. $10r^2 = \cos r \Rightarrow 10r^2 < 1 \leftrightarrow r^2 < \dfrac{1}{10}$.

$$\cos r = 1 - \frac{r^2}{2} + R(r) \text{ where } 0 < R(r) < \frac{r^4}{24}$$

$$\Rightarrow 0 < R(r) < \frac{1}{2400} \text{ since } r^2 < \frac{1}{10}. \text{ Thus}$$

$\dfrac{21}{2} r^2 = 1 + R(r) \leftrightarrow r^2 = \dfrac{2}{21}(1 + R(r))$. Then the root

$$r = \sqrt{\frac{2}{21}} \sqrt{1 + R(r)} = \sqrt{\frac{2}{21}} \left[1 + \frac{R(r)}{2} - \frac{R^2(r)}{8} + \cdots \right].$$ Hence

$r = \sqrt{\dfrac{2}{21}} + E$ where

$0 < E < \frac{1}{\sqrt{42}} R(r) \Rightarrow 0 < E < \frac{1}{\sqrt{42}\,(2400)}$. Thus $r = \boxed{\sqrt{\frac{2}{21}} = 0.3086}$ is accurate to 3 decimals. (Note that $-r$ is also a solution.)

(*)17. Using an appropriate partial sum of an infinite series, compute an approximate value for $J = \int_0^{0.3} e^{-t^2}\,dt$ with an error less than 0.00005.

$$e^x = \sum_{k=0}^{\infty} \frac{x^k}{k!} \quad \text{about } x = 0.$$

Hence

$$e^{-t^2} = \sum_{k=0}^{\infty} \frac{(-1)^k t^{2k}}{k!} \Rightarrow \int_0^{0.3} e^{-t^2}\,dt = \sum_{k=0}^{\infty} \frac{(-1)^k}{k!} \int_0^{0.3} t^{2k}\,dt$$

$$= \sum_{k=0}^{\infty} \frac{(-1)^k (0.3)^{2k+1}}{(k!)(2k+1)} = S_n + R_n$$

where $S_n = \sum_{k=0}^{n} \frac{(-1)^k (0.3)^{2k+1}}{(k!)(2k+1)}$ and R_n is the error in approximating J by the partial sum S_n. Hence $R_n = \sum_{k=n+1}^{\infty} \frac{(-1)^k (0.3)^{2k+1}}{(k!)(2k+1)}$; by using the error analysis for an alternating series, $|R_n| < 5 \times 10^{-5}$ if $n = 2$. Thus $J = \frac{3}{10} - \frac{1}{3}\left(\frac{3}{10}\right)^3 + \frac{1}{10}\left(\frac{3}{10}\right)^5$, accurate to 4 decimals, i.e., $\boxed{J \simeq 0.29124}$.

(*)18. Consider $I = \int_0^1 x^{1/2} \sin(x^2)\,dx$.

(a) Use a series method to evaluate I to 4 decimal places.

(b) With $h = 0.25$ evaluate I by either the trapezoidal rule or Simpson's rule.

(c) Discuss the effectiveness of the methods of (a) and (b) to evaluate $J = \int_0^{0.01} \frac{\sin(x^3)}{x^3}\,dx$.

(a) $\sin x = \sum_{k=0}^{\infty} \frac{(-1)^k x^{2k+1}}{(2k+1)!} \Rightarrow \sin(x^2) = \sum_{k=0}^{\infty} \frac{(-1)^k x^{4k+2}}{(2k+1)!}$.

$I = \int_0^1 x^{1/2} \sin(x^2)\,dx = \sum_{k=0}^{\infty} \frac{(-1)^k}{(2k+1)!} \int_0^1 x^{4k+5/2}\,dx =$ $\sum_{k=0}^{\infty} \frac{(-1)^k}{(4k+\frac{7}{2})(2k+1)!}$, an alternating series. Hence $I = S_n + R_n$ where $S_n = \sum_{k=0}^{n} \frac{(-1)^k}{(4k+\frac{7}{2})(2k+1)!}$ and $|R_n| < \frac{1}{(4n+\frac{15}{2})(2n+3)!}$.

$|R_n| < 5 \times 10^{-5}$ if $n = 2$. Thus $I = \frac{2}{7} - \frac{1}{45} + \frac{1}{1380} + R_2 = \boxed{0.26422} + R_2$ where $-0.00005 < R_2 < 0$.

(b) $T(\frac{1}{4}) = \boxed{0.27220}$; $S(\frac{1}{4}) = \boxed{0.26364}$.

(c) Method (a) is superior to method (b) since the series solution would converge very rapidly, whereas with the use of Simpson's rule or the trapezoidal rule, $f(x) = \dfrac{\sin(x^3)}{x^3}$ involves a small number $\sin(x^3)$ divided by a small number x^3 leading to the possibility of large round-off errors in evaluating $f(x_i)$.

(*)19. Find the first three nonzero terms in the Maclaurin series expansion of $\log(1+\sin x)$. Consequently, obtain an approximate expression for $\int_0^{\frac{1}{4}} \log(1+\sin x)\,dx$. Show that the absolute value of the error in the approximation is less than 2×10^{-5}.

Let $f(x) = \log(1+\sin x)$; $f'(x) = \dfrac{\cos x}{1+\sin x}$; $f''(x) = \dfrac{-1}{1+\sin x}$; $f'''(x) = \dfrac{\cos x}{(1+\sin x)^2}$; $f^{(4)}(x) = -\dfrac{(1+\sin x+\cos^2 x)}{(1+\sin x)^3}$. Hence $f(0)=0$; $f'(0) = 1$; $f''(0) = -1$; $f'''(0) = 1$. Thus the first three nonzero terms in the Maclaurin series of $f(x)$ are represented by the polynomial $P(x) = \boxed{x - \dfrac{x^2}{2} + \dfrac{x^3}{6}}$.

$f(x) = P(x) + R(x)$, where $R(x)$ is the error in approximating $P(x)$ by $R(x)$ on the interval $I = \left[0, \dfrac{1}{4}\right]$; $R(x) = \dfrac{f^{(4)}(c)}{4!}x^4$ for some $c \in I$. Since $0 \le \sin x \le x$ on I, it follows that $|f^{(4)}(c)| \le |1+\sin c+1| \le \dfrac{9}{4}$.

Hence

$$\int_0^{\frac{1}{4}} \log(1+\sin x)\,dx = \int_0^{\frac{1}{4}} [P(x)+R(x)]\,dx = \left[\frac{x^2}{2} - \frac{x^3}{6} + \frac{x^4}{24}\right]\Bigg|_{x=0}^{x=\frac{1}{4}} + E$$

$$= \frac{1}{32} - \frac{1}{384} + \frac{1}{6144} + E = \boxed{0.02881} + E.$$

$$|E| = \left|\int_0^{\frac{1}{4}} R(x)\,dx\right| \le \int_0^{\frac{1}{4}} |R(x)|\,dx \le \frac{9}{4(4!)}\int_0^{\frac{1}{4}} x^4\,dx = \frac{9}{4^6(5!)} < 2\times 10^{-5}.$$

(*)20. (a) Find an expression computing $S = \sum\limits_{n=1}^{\infty} ne^{-n^2}$ to 5 decimal accuracy.

(b) Find an expression computing $S = \sum\limits_{n=1}^{\infty} \dfrac{(-1)^n}{n^3}$ to 3 decimal accuracy.

(c) Find an expression computing $S = \sum\limits_{n=1}^{\infty} \dfrac{1}{4^n+3}$ to 1% accuracy.

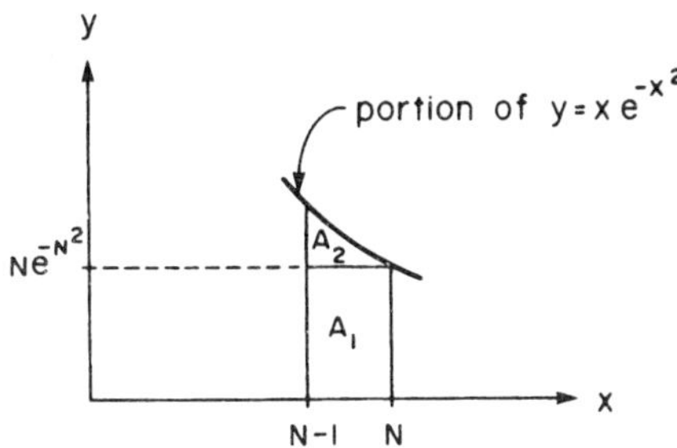

Figure VI.20. Area of $(A_1) = Ne^{-N^2} <$ area of $(A_1 + A_2) = \int_{N-1}^{N} xe^{-x^2}\,dx$ $(N \geq 2)$.

(a) $S = S_n + R_n$ where $S_n = \sum_{k=1}^{n} ke^{-k^2}$ and $R_n = \sum_{k=n+1}^{\infty} ke^{-k^2}$. R_n can be compared with an elementary integral. Note that $Ne^{-N^2} < \int_{N-1}^{N} xe^{-x^2}\,dx$ if $N \geq 2$ (see Fig. VI.20). Hence $ke^{-k^2} < M_k = \int_{k-1}^{k} xe^{-x^2}\,dx$ $\Rightarrow R_n < \sum_{k=n+1}^{\infty} M_k = \int_{n}^{\infty} xe^{-x^2}\,dx = \frac{e^{-n^2}}{2}$. Now the aim is to find n such that $\frac{e^{-n^2}}{2} < 5\times 10^{-6} \leftrightarrow e^{n^2} > 10^5$; $n = 4$ works. Thus

$$S = \boxed{\frac{1}{e} + \frac{2}{e^4} + \frac{3}{e^9} + \frac{4}{e^{16}} + R_4} \quad \text{where} \quad |R_4| < 5\times 10^{-6}.$$

(b) For this alternating series,

$$S = S_n + \tfrac{1}{2}a_{n+1} + E_n$$

where $S_n = \sum_{k=1}^{n} \frac{(-1)^k}{k^3}$, $a_n = \frac{(-1)^n}{n^3}$, $|E_n| < \frac{1}{2}|a_{n+1}| = \frac{1}{2(n+1)^3}$.

$|E_n| \leq 5\times 10^{-4} \leftrightarrow (n+1)^3 \geq 10^3$. Hence $n = 9$ works. Thus

$$S = \boxed{\left(\sum_{k=1}^{9} \frac{(-1)^k}{k^3}\right) + \frac{1}{2(10^3)} \pm 5\times 10^{-4}}.$$

(c) Since each term in S is positive, $S > S_1 = \frac{1}{7}$. Hence 1% of S equals $\frac{S}{100} > \frac{1}{700}$. $S = S_n + R_n$ where $S_n = \sum_{k=1}^{n} \frac{1}{4^k + 3}$ and $R_n = \sum_{k=n+1}^{\infty} \frac{1}{4^k+3}$.

The aim is to find n such that $R_n < \frac{1}{700}$.

$$a_k = \frac{1}{4^k + 3} < M_k = \frac{1}{4^k},\ k = 1, 2, \ldots$$

Then $R_n < E_n = \sum_{k=n+1}^{\infty} M_k = \sum_{k=n+1}^{\infty} \frac{1}{4^k} = \frac{1}{4^{n+1}} \sum_{m=0}^{\infty} \frac{1}{4^m} = \frac{1}{4^{n+1}}\left[\frac{1}{1-\frac{1}{4}}\right] =$

$\frac{1}{(3)(4^n)} \cdot E_n < \frac{1}{700} \leftrightarrow (3)(4^n) > 700$; $n = 4$ works. Hence

$$S = \boxed{\frac{1}{7} + \frac{1}{19} + \frac{1}{67} + \frac{1}{259} + R_4} \text{ where } 0 < R_4 < \frac{S}{100}.$$

(*)21. Derive an approximate value for the sum of one billion terms: $S = 1 + \frac{1}{2} + \frac{1}{3} + \cdots + 10^{-9}$.

$$S = \sum_{k=1}^{n} \frac{1}{k} + R_n \quad \text{where} \quad R_n = \sum_{k=n+1}^{10^9} \frac{1}{k}.$$

Comparing areas, note that $\int_k^{k+1} \frac{1}{x} dx < \frac{1}{k} < \int_{k-1}^{k} \frac{1}{x} dx$. Hence $\int_{n+1}^{10^9+1} \frac{1}{x} dx < R_n < \int_n^{10^9} \frac{1}{x} dx$. Thus $\log(10^9+1) - \log(n+1) < R_n < \log 10^9 - \log n$. Observe that if $\alpha_n < R_n < \beta_n$, $R_n = \frac{\alpha_n + \beta_n}{2} + E_n$ where $|E_n| < \frac{\beta_n - \alpha_n}{2}$, after the error is centred. In this case

$$\frac{\alpha_n + \beta_n}{2} = \frac{\log 10^9 + \log(10^9+1) - [\log(n+1) + \log n]}{2};$$

$$\frac{\beta_n - \alpha_n}{2} = \frac{1}{2}\left[\log\left(1 + \frac{1}{n}\right) - \log(1 + 10^{-9})\right].$$

If $0 < x < 1$, $\log(1+x) < x$. Hence $|E_n| < \frac{1}{2n}$. Thus

$$S = \boxed{\left(\sum_{k=1}^{n} \frac{1}{k}\right) + \frac{1}{2}\left[\log 10^9 + \log(10^9+1) - \log(n+1) - \log n\right] \pm \frac{1}{2n}}.$$

(*)22. Discuss how you could evaluate $I = \int_0^\infty e^{-x^2} dx$ to 2 decimal places.

As a first step one should split up the integral $I = S + T$ where $S = \int_0^X e^{-x^2} dx$ and the tail $T = \int_X^\infty e^{-x^2} dx$, X to be determined. To estimate T, let $x^2 = y$. Then $dx = \frac{dy}{2x} = \frac{dy}{2\sqrt{y}}$. $x = X \leftrightarrow y = X^2$. Hence $T = \frac{1}{2}\int_{X^2}^\infty \frac{e^{-y}}{\sqrt{y}} dy \le \frac{1}{2X}\int_{X^2}^\infty e^{-y} dy \Rightarrow T \le \frac{e^{-X^2}}{2X}$. If $X = 2$, $0 < T \le \frac{e^{-4}}{4} <$ 0.0046. Hence $I = S + 0.0023 \pm 0.0023$ where $S = \int_0^2 e^{-x^2} dx$. Simpson's rule can be used to evaluate S: let $f(x) = e^{-x^2}$. Then $f^{(4)}(x) = 4e^{-z}(4z^2 - 12z + 3)$ where $z = x^2$; $0 \le x \le 2 \leftrightarrow 0 \le z \le 4$. $g(z) = 4z^2 - 12z + 3 = 4\left[z^2 - 3z + \frac{3}{4}\right] = 4\left[\left(z - \frac{3}{2}\right)^2 - \frac{3}{2}\right]$. Thus $-6 \le g(z) \le 19$, and $-24 \le$

$f^{(4)}(x) \leq 76$. Hence for Simpson's rule with spacing $h = \frac{2}{n}$, the error $E_s(h)$ satisfies the inequality:

$$\frac{-(76)(2)(16)}{180n^4} < E_s(h) < \frac{(24)(2)(16)}{180n^4}.$$

After the error is centred one wishes to find n even such that $\frac{80}{9n^4} < 0.0027$; $n = 8$ works, i.e., Simpson's rule with 9 interpolation points on the interval $[0, 2]$ is sufficient to calculate S to the desired accuracy.

(*)23. Consider

$$I = \int_0^\infty \frac{1}{x^{1/2}(1 + x^{50})} dx.$$

(a) Discuss the difficulties in computing I numerically.
(b) Compute I with an error not exceeding 0.01.

(a) I is improper due to its integrand's singular behaviour at $x = 0$ and also improper from its domain of integration.

(b) $I = I_1 + I_2$ where $I_1 = \int_0^1 \frac{1}{x^{1/2}(1 + x^{50})} dx$ and $I_2 = \int_1^\infty \frac{1}{x^{1/2}(1 + x^{50})} dx$. In I_2 make the substitution $y = \frac{1}{x}$. Then $I_2 = \int_0^1 \frac{y^{50-3/2}}{1 + y^{50}} dy$. If $|x| \leq 1$, $\frac{1}{1 + x^{50}} = 1 - x^{50} + E(x)$ where $0 \leq E(x) \leq x^{100}$. Hence $I = \int_0^1 [1 - x^{50}][x^{-1/2} + x^{50-3/2}]\, dx + R$ where $0 \leq R \leq \int_0^1 x^{100}[x^{-1/2} + x^{50-3/2}]\, dx = \frac{2}{201} + \frac{2}{299} \leq 0.018$. Thus
$I = \int_0^1 [1 - x^{50}][x^{-1/2} + x^{50-3/2}]\, dx + 0.009 \pm 0.009 = \boxed{1.999 \pm 0.009}$, after the error is centred.

(*)24. Tabulated below are the values of a function f on $[0, 1]$ with intervals of $1/5$. Suppose $f''(x) > 0$ on $[0, 1]$. Obtain as good an estimate as you can of $f(\frac{1}{2})$ (i.e., find the best possible upper and lower bounds).

x	$f(x)$
0	1.000
0.2	1.018
0.4	1.064
0.6	1.130
0.8	1.212
1.0	1.307

Since $f''(x) > 0$, $f(\frac{1}{2})$ lies below the secant line (Fig. VI.24) joining data

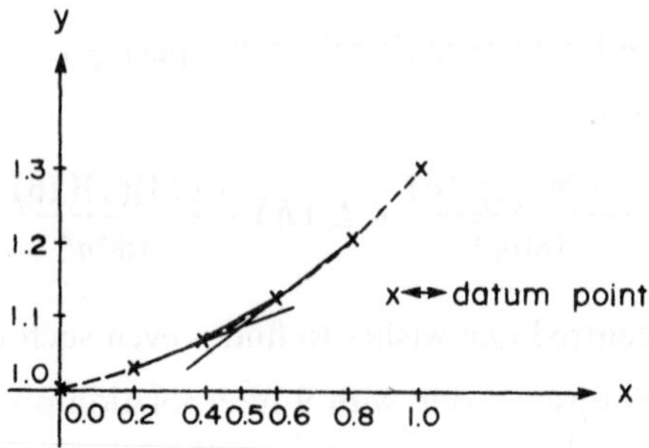

Figure VI.24. $f''(x) > 0$ indicated by dotted line. Secant lines indicated by solid lines.

points at 0.4 and 0.6. Hence

$$f\left(\frac{1}{2}\right) < f(0.4) + (0.5-0.4)\left[\frac{f(0.6)-f(0.4)}{0.6-0.4}\right] = 1.097;$$

moreover $f(\frac{1}{2})$ lies above the secant lines joining data points at 0.2 and 0.4, and data points at 0.6 and 0.8, respectively. Thus

$$f\left(\frac{1}{2}\right) > f(0.4) + (0.5-0.4)\left[\frac{f(0.4)-f(0.2)}{0.4-0.2}\right] = 1.087;$$

$$f\left(\frac{1}{2}\right) > f(0.6) - (0.6-0.5)\left[\frac{f(0.8)-f(0.6)}{0.8-0.6}\right] = 1.089.$$

Hence $1.089 < f(\frac{1}{2}) < 1.097 \leftrightarrow f(\frac{1}{2}) = \boxed{1.093 \pm 0.004}$.

Supplementary Problems

1. Find an approximate value for $\sqrt[3]{1002}$ by using a linear approximation.

(*)2. Use a suitable tangent line approximation to find an approximate value for $\cos 46°$. Is the approximation greater than or less than the actual value of $\cos 46°$? How do you know this?

3. Use differentials to approximate $\dfrac{1}{\sqrt[3]{120}}$ to three significant digits.

4. By means of differentials, estimate $\sqrt{195}$.

5. (a) Find the differential dy of $y = x^3$.

(b) Use differentials to find an approximation for $\sqrt[3]{121}$.

6. Use differentials to find approximations for $\sqrt{82}$ and $\sin(29°)$.

7. Use the differential to estimate $\sqrt{99}$.

8. Using differentials, find an approximate value for $\sqrt[4]{15}$.

9. Calculate the derivative of $(\sin x)^{10}$. Use this result to approximate $\left[\sin\left(\frac{\pi}{4}+0.01\right)\right]^{10}$ by the method of differentials.

10. Use differentials to approximate $(10.135)^4$.

11. Using differentials, find an approximate value of $(65)^{1/2}+(65)^{2/3}+1$.

(*)12. A spherical shell is made of a material having density 2 ounces per cubic inch. It has an inner radius of 3 inches and its thickness is $\frac{3}{32}$ inches. Use differentials to find its approximate weight.

13. Using differentials, find the difference in volume between a sphere of radius 10 and a sphere of radius 10.02.

14. Use the differential to estimate the possible error in calculating the area of a square table, if the sides are 1.250 ± 0.005 metres long.

(*)15. (a) Evaluate $\int_0^1 \frac{1}{1+x^2}\,dx$.

(b) Use the Fundamental Theorem of Calculus as an aid in evaluating, approximately, $I=\int_0^{1.01}\frac{1}{1+t^2}\,dt$.

16. A function has the following values: $f(6.55)=2.315$, $f(6.60)=2.592$. Obtain a linear estimate of $f(6.57)$.

17. Use the Newton–Raphson method to estimate the zero of $f(x)=x^2+x-1$ which lies between 0 and 1. Your answer should be correct to three decimal places.

18. (a) Derive Newton's method for obtaining a root of $f(x)=0$.

(b) Starting with a reasonable guess to the *smaller* positive root of the equation $x^3-4x+2=0$, use Newton's method to obtain the next approximation.

19. Consider the function $f(x)=x^3-3x^2+3x-5=(x-1)^3-4$.

(a) Show that the equation $f(x)=0$ has exactly one real root r.

(b) Find the integer n such that $n-1<r<n$.

(c) Give a second approximation to r by Newton's method, using the integer n found in (b) as a first approximation.

(*)20. Find all the real roots of $x^5+2x-1=0$ to 2 decimal places. Give a careful error analysis.

21. The Newton–Raphson method is to be used to estimate a root of the equation $f(x)=0$.

(a) Assuming one has an estimate, x_n, of a desired root, derive the formula for an improved estimate, x_{n+1}. Illustrate your answer with a well labelled sketch.

(b) The equation $5 \sin x = 4x$ is satisfied by a positive value of x. Make a *reasonable* first estimate of this value that is useful for starting the Newton–Raphson method.

22. Verify that the equation $x^2 + 10 \cos x = 0$, where x is measured in radians, has a root in the interval $\frac{\pi}{2} \leq x \leq \frac{3\pi}{4}$. Then proceed to find this root using Newton's method. The computations may be stopped when the first three decimal digits of two successive approximations to the root remain identical.

(*)23. (a) Show that if u is an approximate solution to the equation $x^5 = 35$, Newton's method yields the further approximation $\frac{4}{5}u + 7u^{-4}$. With the aid of a diagram and Newton's method, show that $2 < (35)^{1/5} < 2.0375$.

(b) Show that an application of the method of differentials with the aid of a graph, or an application of the Mean Value Theorem, to the function $x^{1/5}$, also leads to $2 < (35)^{1/5} < 2.0375$.

(*)24. (a) Calculate the area of the shaded polygon shown in Fig. S.24. The curve is a quarter of the circle $x^2 + y^2 = 1$.

(b) Without using the fact that you know the value of π, show that the difference between the areas of the polygon and quarter-circle is less than 1/8.

(c) The above results give upper and lower estimates of π. What are these estimates?

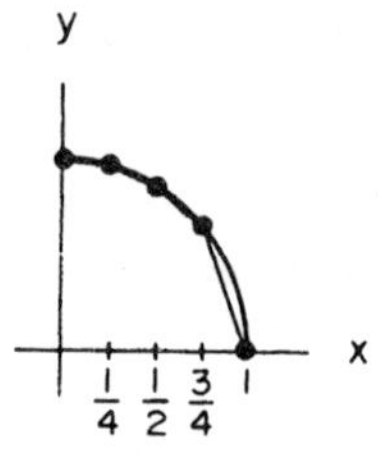

Figure S.24

25. (a) Use the trapezoidal rule with $n = 4$ to approximate $\int_1^3 \frac{1}{x+1} dx$.

(b) Give a bound for the error.

(c) How large should n be taken in order to ensure that the error is less than 0.001?

26. Consider $J = \int_0^2 \frac{1}{x^2+4} dx$.

(a) Approximate the value of this integral by using Simpson's rule with $n = 8$.

(b) Find the exact value of this integral and determine the error made in the approximation.

27. For β use the last digit of your own student number and evaluate the integral $\int_0^{15} \sqrt{x^4 + 10 + \beta}\, dx$ numerically by Simpson's rule with the number of subintervals $n = 6$. In your computation carry out all arithmetic up to 5 decimal places and round off your final answer to three decimal places.

(*)28. (a) Using Simpson's rule, with 4 subintervals, approximate the area between the curve $y = 2 + 4x^2 - \log(2x + 3)$, the x-axis, and lines $x = -1$ and $x = 1$.

(b) Determine the accuracy of this approximation.

x	$\log x$
1	0
2	0.69315
3	1.09861
4	1.38629
5	1.60944

(*)29. Determine the value of $\log 2 = \int_1^2 \frac{1}{x} dx$

(a) using the trapezoidal rule with $n = 4$;

(b) using Simpson's rule with $n = 4$.

(c) Compare with the exact value of $\log 2 = 0.69315$ and explain the difference.

Use at least five decimals for your calculations.

(*)30. In order to compute $\int_0^1 e^{-x^2} dx$ to 5 decimal accuracy ($|\text{Error}| < 5 \times 10^{-6}$) how many equal subintervals are needed for

(a) the trapezoidal rule?

(b) Simpson's rule?

In each case use the appropriate error formula.

31. Approximate $\int_0^4 f(x)\, dx$ when the graph of f is as shown in Figure S.31.

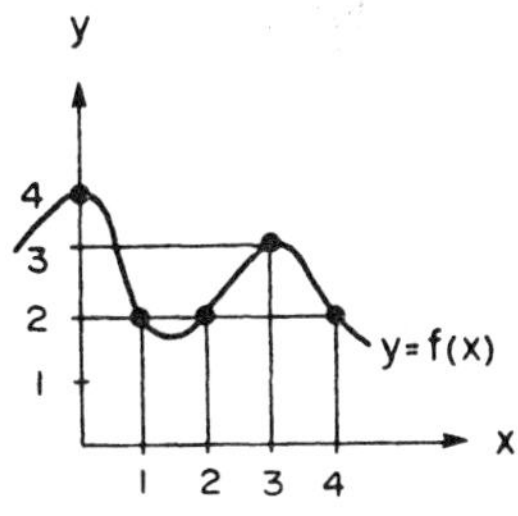

Figure S.31

32. Using either Taylor's formula (with $f(x) = \sqrt{x}$ and $x_0 = 16$) or Newton's method (finding a root of $f(x) = x^2 - 15$), find $\sqrt{15}$ correct to 2 decimal places, indicating why your answer is this accurate.

33. Compute $\sqrt[3]{9}$ to 2 decimal accuracy using
(a) Newton's method, and
(b) an appropriate Taylor series expansion.

34. Using an appropriate partial sum of an infinite series, compute an approximate value for $\dfrac{1}{\sqrt{e}}$ with an error less than 0.005.

35. Using an appropriate partial sum of an infinite series, compute an approximate value for $\cos\frac{1}{10}$ with an error less than 10^{-7}.

36. Use Taylor's formula to approximate $\log(\frac{4}{5})$ with an error less than 0.05.

(*)37. (a) Exhibit the first four terms of the Maclaurin series for

$$f(x) = (1+x)^{-1/2}.$$

(b) Using the series in (a) and $\sqrt{2} = \frac{7}{5}\sqrt{\frac{50}{49}} = \frac{7}{5}(\frac{49}{50})^{-1/2} = \frac{14}{10}(1-\frac{2}{100})^{-1/2}$, compute $\sqrt{2}$ to seven decimal places.

(*)38. Use a Taylor series to evaluate $\int_0^{0.2} xe^{-x^3}\,dx$ with an error less than 10^{-5}. Explain clearly how you decide the error.

(*)39. Evaluate $\int_0^{0.4}(1+t^3)^{1/2}\,dt$ using a Taylor series. (Give a careful error analysis in computing your answer to 3 decimal accuracy.)

(*)40. Find the minimal number of terms in the Taylor expansion of $\log(1+x)$ about $x=0$ needed in order to calculate $\int_0^{1/2}\log(1+x)\,dx$ accurately to 3 decimals.

(*)41. Compute $\int_0^{0.1}\dfrac{\sin^2 x}{x^2}dx$ to 1% accuracy.

(*)42. Consider the integral:

$$I = \int_0^1 \frac{\sin^2 x}{x^2}\,dx.$$

(a) Evaluate I to three decimal places using a series expansion method. Include a careful error analysis.
(b) Taking $h = 0.25$, use Simpson's rule to evaluate I (approximately).

43. (a) Write the Taylor polynomial of degree 5 for $\sin x$ about $x = 0$.
(b) Using (a), compute an approximate value for $\sin 1$ and prove that the absolute error is less than $2(10^{-4})$.
(c) Use the Taylor polynomial from (a) to compute an approximate value for

$$I = \int_0^1 \frac{\sin(t^2)}{t^2}\,dt;$$

give an estimate for the error.

(*)44. (a) Find a power series expansion for the function $f(x) = \cos 2x$ about the point $x = 0$, and determine its radius of convergence.

(b) Use the first 3 terms of this power series to find an approximation to the integral

$$I = \int_0^{\pi/4} \sin^2 x \, dx.$$

(*)45. Evaluate $\int_0^{0.1} \sqrt{1 - \frac{1}{2}\sin^2 t} \, dt$ to 4 decimal accuracy.

46. Find finite series expansions computing to 4 decimals (give careful error analyses) the infinite series

(a) $\sum_{n=1}^{\infty} \frac{(-1)^{n+1}}{n^4}$;

(b) $\sum_{n=1}^{\infty} \frac{1}{n^4}$.

(*)47. Compute $\sum_{n=3}^{100} \frac{1}{n^3}$ to 1% accuracy. Give a careful error analysis.

(*)48. By an appropriate comparison of $\log(n!)$ and $\int_1^n \log x \, dx$, estimate a lower bound for $n!$

(*)49. I want to calculate the average height of the people in the room. So I measure them all, add their heights and divide by the number of people. Suppose all my measurements were accurate to within $\frac{1}{4}''$. What is the maximum possible error in my estimate of the average height? Prove your assertion.

(*)50. Compute approximately

$$I = \int_0^{1000} \frac{e^{-10x} \sin x}{x} \, dx.$$

(*)51. The table below lists values for a function $F(x)$. You are told that $F(x) = \int_0^x f(t) \, dt$, that f is a continuous function, and that f is an even function (i.e. $f(-t) = f(t)$ for all t). Answer the following questions, showing your reasoning.

x	0.0	0.1	0.2	0.3	0.4	0.5	0.6	0.7	0.8
$F(x)$	0.0000	0.3124	0.6147	0.8972	1.1515	1.3708	1.5502	1.6873	1.7817

x	0.9	1.0	1.1	1.2	1.3	1.4	1.5	1.6	1.7
$F(x)$	1.8352	1.8519	1.8373	1.7982	1.7420	1.6764	1.6090	1.5464	1.4949

x	1.8	1.9	2.0	2.1	2.2	2.3	2.4	2.5
$F(x)$	1.4596	1.4460	1.4605	1.5126	1.6192	1.8096	2.1353	2.6829

(a) Is it true that $f(t) \geq 0$ for all t in $[0, 2]$? If this question cannot be answered with the information given, explain why. Remember: the table tabulates the function F; the question here is about f.

(b) Find $\int_{1.7}^{2.4} f(x)\,dx$.

(c) Find $\int_{-0.3}^{0.5} f(t)\,dt$.

(d) Find an approximate value for $f(1.3)$.

(e) Which of the following is more plausible?

(i) $f(t) \leq 3 - 5t^2$ for t in $[0.0, 0.1]$,

(ii) $f(t) \geq 3 - 5t^2$ for t in $[0.0, 0.1]$.

Chapter VII

Theory

In this chapter problems of a more theoretical nature are considered. For background material the reader should consult references such as:

1. C.W. Clark, *Elementary Mathematical Analysis*, 2nd ed., 1982, Wadsworth.
2. C.W. Clark, *The Theoretical Side of Calculus*, 1972, Wadsworth.
3. T.M. Apostol, *Calculus*, Vol. 1, 2nd ed., 1973, Wiley.

Solved Problems

1. Given

$$f(x) = \begin{cases} 2x-2 & \text{if} \quad x<-1, \\ Ax+B & \text{if} \quad -1 \le x \le 1, \\ 5x+7 & \text{if} \quad x>1, \end{cases}$$

determine the constants A and B so that the function f is continuous for all real values of x.

$f(x)$ is obviously continuous for $x<-1$, $-1<x<1$, $x>1$.

$$\lim_{x \to (-1)^-} f(x) = -4; \qquad \lim_{x \to (-1)^+} f(x) = -A+B;$$

$$\lim_{x \to 1^-} f(x) = A+B; \qquad \lim_{x \to 1^+} f(x) = 12.$$

$f(x)$ is continuous for all real x if and only if $-A+B=-4$, $A+B=12$. Hence, solving these equations leads to $\boxed{B=4, A=8}$.

2. Discuss the continuity of the function

$$f(x)=\begin{cases}\dfrac{x^2-x-6}{x-3} & \text{if} \quad x\neq 3,\\ 4 & \text{if} \quad x=3.\end{cases}$$

Can you redefine the function at $x=3$ such that it is continuous for all x?

If $x\neq 3$, $f(x)=\dfrac{x^2-x-6}{x-3}=\dfrac{(x-3)(x+2)}{x-3}=x+2$.
$\lim\limits_{x\to 3} f(x)=5\neq 4=f(3)$. Hence $f(x)$ is discontinuous at $x=3$. $f(x)$ is continuous for all x if one redefines $f(3)=5$.

3. Graph $f(x)=\frac{1}{2}(\cos x+|\cos x|)$. Indicate at what points (if any) $f(x)$ is (a) discontinuous, (b) non-differentiable.

If $\cos x>0$, $f(x)=\cos x$. If $\cos x\leq 0$, $f(x)=0$. $f(x+2\pi)=f(x)$. The graph is as shown in Fig. VII.3.

(a) $f(x)$ is continuous for all x; (b) $f(x)$ is non-differentiable at $x=\dfrac{\pi}{2}+n\pi$ for any integer n.

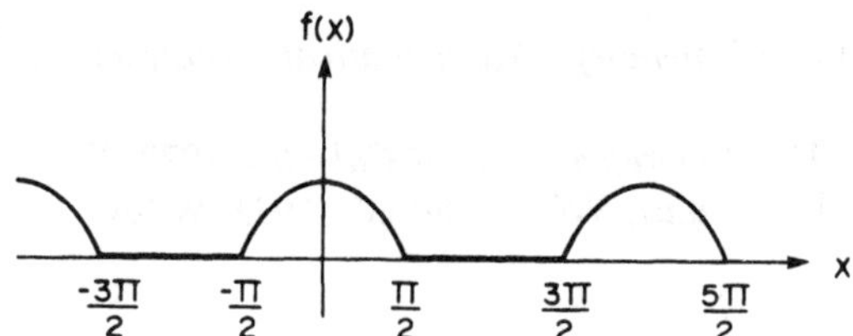

Figure VII.3. Graph of $f(x)=\frac{1}{2}(\cos x+|\cos x|)$.

4. If $f(a)=0$ and $f'(a)=6$ then $\lim\limits_{h\to 0}\dfrac{f(a+h)}{2h}=?$

$f'(a)=6 \leftrightarrow \lim\limits_{h\to 0}\dfrac{f(a+h)-f(a)}{h}=6$. But $f(a)=0$. Hence

$$\lim_{h\to 0}\frac{f(a+h)}{2h}=\frac{1}{2}\lim_{h\to 0}\frac{f(a+h)}{h}=\left(\frac{1}{2}\right)(6)=\boxed{3}.$$

5. If $f(x)=xe^{-1/x^2}$, $x\neq 0$, $f(0)=0$, use the definition of the derivative to evaluate $f'(0)$.

$$f'(0)=\lim_{h\to 0}\frac{f(0+h)-f(0)}{h}=\lim_{h\to 0}\frac{he^{-1/h^2}}{h}=\lim_{h\to 0}e^{-1/h^2}=e^{-\infty}=\boxed{0}.$$

6. Sketch the graph of $f(x)=|x^2-1|$ without computing the derivatives of f. By inspection determine where f is differentiable and compute the derivative of f where possible.

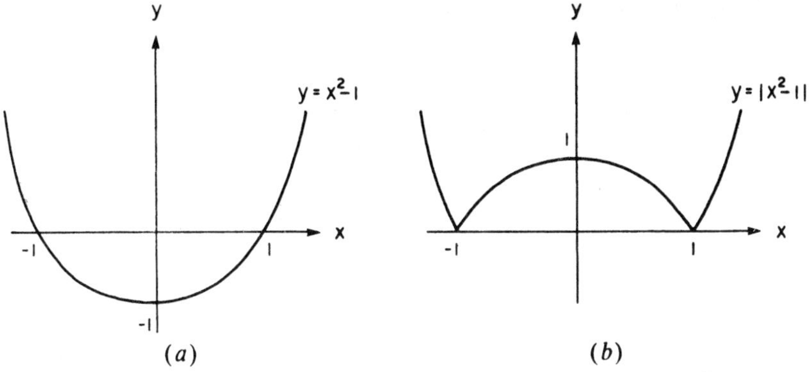

Figure VII.6*a*. Graph of $y = x^2 - 1$. *b*. Graph of $y = f(x) = |x^2 - 1|$.

Observe that

$$f(x) = \begin{cases} x^2 - 1, x \geq 1; \\ -(x^2 - 1), \ -1 < x < 1; \\ x^2 - 1, x \leq -1. \end{cases}$$

Thus the graph of f may be obtained from the graph of $y = x^2 - 1$ by reflecting its negative part about the x-axis. See Figs. VII.6*a*, *b*.

By inspection, $f(x)$ is differentiable at all x except $x = \pm 1$.

$$f'(x) = \begin{cases} 2x, x < -1, x > 1; \\ -2x, \ -1 < x < 1. \end{cases}$$

(*)7. A function f defined for all real x is called *even* if $f(-x) = f(x)$ for all x, and is called *odd* if $f(-x) = -f(x)$ for all x. Prove that, if f is even and f' exists, then f' is odd; and if f is odd and f' exists, then f' is even. Illustrate with examples.

Case I. Say $f(x)$ is even and $f'(x)$ exists. Then

$$\begin{aligned} f'(-x) &= \lim_{h \to 0} \frac{f(-x+h) - f(-x)}{h} \\ &= \lim_{h \to 0} \frac{f(x-h) - f(x)}{h} \\ &= -\lim_{h \to 0} \frac{f(x-h) - f(x)}{(-h)}. \end{aligned}$$

Let $\Delta x = -h$. Then

$$f'(-x) = -\lim_{\Delta x \to 0} \frac{f(x + \Delta x) - f(x)}{\Delta x} = -f'(x).$$

Case II. Say $f(x)$ is odd and $f'(x)$ exists.

$$f'(-x) = \lim_{h \to 0} \frac{f(-x+h)-f(-x)}{h}$$
$$= \lim_{h \to 0} \frac{f(x-h)-f(x)}{(-h)}$$
$$= \lim_{\Delta x \to 0} \frac{f(x+\Delta x)-f(x)}{\Delta x} = f'(x).$$

Examples.

(a) $f(x)=x^2$ is even, $f'(x)=2x$ is odd;
(b) $f(x)=\sin x$ is odd, $f'(x)=\cos x$ is even.

(*)8. Given that f is a function such that
(a) $f(x_1+x_2)=f(x_1)\cdot f(x_2)$ for all x_1, x_2 and
(b) $f(x)=1+xg(x)$ where $\lim_{x \to 0} g(x)=1$, prove that $f'(x)=f(x)$.

From (a), $f(x+\Delta x)=f(x)f(\Delta x)$. Hence $f(x+\Delta x)-f(x)=[f(\Delta x)-1]f(x)$.

From (b), $f(\Delta x)=1+\Delta x g(\Delta x)$. Thus

$$f'(x) = \lim_{\Delta x \to 0} \frac{f(x+\Delta x)-f(x)}{\Delta x} = f(x)\cdot \lim_{\Delta x \to 0}\left[\frac{f(\Delta x)-1}{\Delta x}\right]$$
$$= f(x)\cdot \lim_{\Delta x \to 0} g(\Delta x) = f(x)$$

since $\lim_{\Delta x \to 0} g(\Delta x)=1$ from (b).

(*)9. Prove that $|\sin b - \sin a| \le |b-a|$.

Let $f(x)=\sin x$. $f(x)$ is continuous and differentiable for all x. Hence by the Mean Value Theorem, for any a, b, $a \ne b$, $\dfrac{f(b)-f(a)}{b-a} = f'(c) = \cos c$ for some c between a and b. But $|\cos c| \le 1$. Hence $\left|\dfrac{f(b)-f(a)}{b-a}\right| = |\cos c| \le 1$. It follows that $|f(b)-f(a)| \le |b-a| \Leftrightarrow |\sin b - \sin a| \le |b-a|$, $a \ne b$. Clearly equality holds if $a=b$.

10. (a) State Rolle's Theorem.

(b) Let $f(x)=x^m(x-1)^n$, where m and n are positive integers. For this function, show that the number c guaranteed by Rolle's Theorem is unique and that it divides $[0,1]$ into segments whose lengths have ratio m/n.

(a) If $f(x)$ is continuous on $[a,b]$ and differentiable on (a,b) and $f(a)=f(b)=0$, then $f'(c)=0$ for some $c \in (a,b)$.

(b) $f(x)=x^m(x-1)^n$ is continuous and differentiable on $[0,1]$, $f(0)=f(1)=0$. Hence Rolle's Theorem applies to $f(x)$ on the interval $[0,1]$.

$$f'(x) = mx^{m-1}(x-1)^n + nx^m(x-1)^{n-1}$$
$$= x^{m-1}(x-1)^{n-1}[m(x-1)+nx].$$

On $(0,1)$, $f'(x)=0 \leftrightarrow m(x-1)+nx=0 \leftrightarrow (m+n)x=m \leftrightarrow x=\frac{m}{m+n}$. Hence the c of Rolle's Theorem is unique: $c=\frac{m}{m+n}$. c divides $[0,1]$ into segments of lengths c and $1-c$. The ratio of their lengths is

$$\frac{1-c}{c}=\frac{1-\frac{m}{m+n}}{\frac{m}{m+n}}=\frac{n}{m}.$$

(*)11. What is the smallest value of m that satisfies $mx-1+\frac{1}{x}\geq 0$ for all $x>0$?

Consider the graph of the situation where $mx-1+\frac{1}{x}>0$ for all $x>0$, i.e., the graph of $y=mx$ is above the graph of $y=1-\frac{1}{x}$ for all $x>0$ (cf. Fig. VII.11a). If the graph of $y=mx$ cuts the graph of $y=1-\frac{1}{x}$ precisely once at some point $x=x^*$ for $x>0$, and $mx\geq 1-\frac{1}{x}$ for all $x>0$, then it is necessary that $y=mx$ be tangent to $y=1-\frac{1}{x}$ at x^* (cf. Fig. VII.11b). In this case let $m=m^*$. Clearly m^* is the required value of m. Let $f(x)=m^*x$, $g(x)=1-\frac{1}{x}$. If $f(x)$ is tangent to $g(x)$ at $x=x^*$, then x^* satisfies the following two equations:

$$f(x^*)=g(x^*)\leftrightarrow m^*x^*=1-\frac{1}{x^*}. \tag{1}$$

$$f'(x^*)=g'(x^*)\leftrightarrow m^*=\frac{1}{(x^*)^2}. \tag{2}$$

Equation (1) implies that $m^*(x^*)^2=x^*-1$. Then equation (2) leads to $1=x^*-1\leftrightarrow x^*=2\leftrightarrow \boxed{m^*=\frac{1}{4}}$.

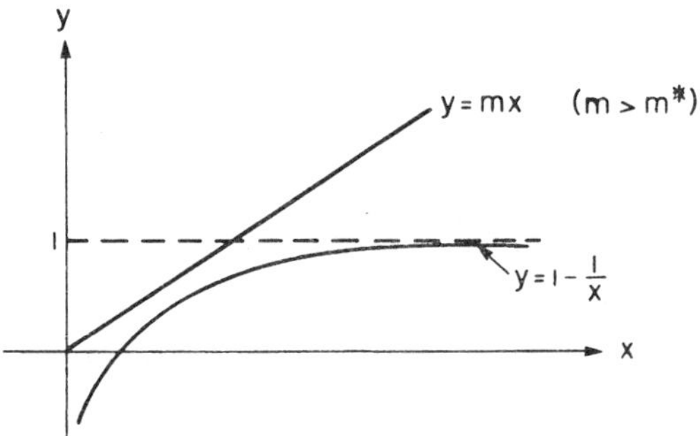

Figure VII.11a. $mx>1-\frac{1}{x}$.

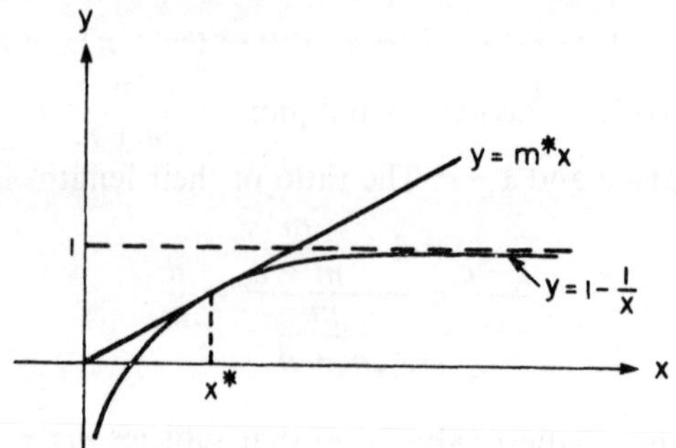

Figure VII.11b. $m^*x \geq 1 - \dfrac{1}{x}$ for smallest value of $m = m^*$.

(*)12. Prove that functions of the form $f(x) = 2x^3 - 3x^2 - a^2x + b$ take the value 0 at most once in the interval $[0, 1]$.

By Rolle's Theorem if $f(x) = 0$ twice in the interval $[0, 1]$, then $f'(c) = 0$ for some $c \in (0, 1)$.

$$f'(x) = 6x^2 - 6x - a^2 = 6\left(x^2 - x - \frac{a^2}{6}\right) = 6\left[\left(x - \frac{1}{2}\right)^2 - \frac{1}{4} - \frac{a^2}{6}\right];$$

$$f'(x) = 0 \leftrightarrow \left(x - \frac{1}{2}\right)^2 = \frac{1}{4} + \frac{a^2}{6} \geq \left(\frac{1}{2}\right)^2 \leftrightarrow \left|x - \frac{1}{2}\right| \geq \frac{1}{2},$$

so if $f'(x^*) = 0$ then $x^* \notin (0, 1)$, since the distance from x^* to $\dfrac{1}{2}$ is at least $\dfrac{1}{2}$. Thus $f(x) = 0$ at most once in $[0, 1]$.

(*)13. Prove that $\log x < \sqrt{x}$ for all $x > 0$. Interpret the result graphically by drawing the curves $\log x$ and $\sqrt{x}$ on the same axes.

Let $f(x) = \log x - \sqrt{x}$; $f'(x) = \dfrac{1}{x} - \dfrac{1}{2\sqrt{x}} = \dfrac{2 - \sqrt{x}}{2x}$. $f'(x) = 0 \leftrightarrow x = 4$. If 0

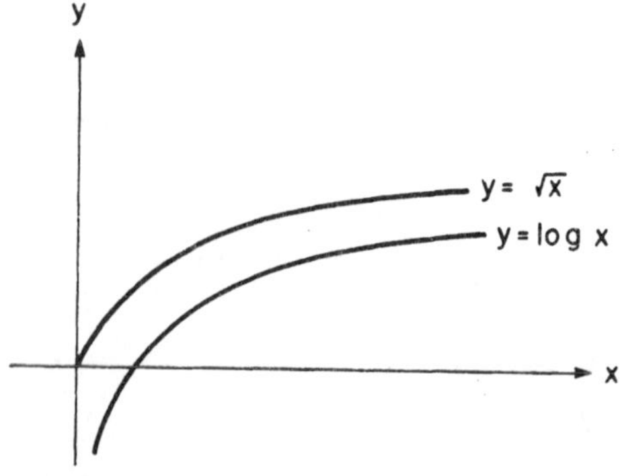

Figure VII.13. $\log x < \sqrt{x}$ if $x > 0$.

$< x < 4$, $f'(x) > 0$; if $x > 4$, $f'(x) < 0$. Hence $f(x)$ attains its maximum value at $x = 4$. $f(4) = -2 + \log 4$.
Since $4 < e^2$, $f(4) < \log e^2 - 2 = 0$. Thus $f(x) < 0$ for all $x > 0$, i.e., $\log x < \sqrt{x}$ for all $x > 0$. The inequality is graphed in Figure VII.13.

14. Let the interval $[2,3]$ be partitioned by n equal increments of length Δx_i by $2 = x_0 < x_1 < x_2 < \cdots < x_n = 3$. Simplify $\lim_{n\to\infty} \sum_{i=1}^{n} (x_i+1)^2 \Delta x_i$.

By the definition of the definite (Riemann) integral,

$$\lim_{n\to\infty} \sum_{i=1}^{n} (x_i+1)^2 \Delta x_i = \int_2^3 (x+1)^2\,dx$$

$$= \frac{1}{3}(x+1)^3\Big|_{x=2}^{x=3} = \frac{1}{3}(4^3 - 3^3) = \boxed{\frac{37}{3}}.$$

(*)15. Let $F(x) = \int_0^x \frac{1}{1+t^2}dt + \int_0^{1/x} \frac{1}{1+t^2}dt$, $x \neq 0$.

(a) Show that $F(x)$ is constant on $(-\infty, 0)$ and constant on $(0,\infty)$.
(b) Evaluate the constant value(s) of $F(x)$.

(a) Let $G(x) = \int_0^x \frac{1}{1+t^2}dt$. Then $F(x) = G(x) + G\left(\frac{1}{x}\right)$. By the Fundamental Theorem of Calculus $G'(x) = \frac{1}{1+x^2}$. Hence

$$F'(x) = G'(x) + G'\left(\frac{1}{x}\right)\frac{d}{dx}\left(\frac{1}{x}\right)$$

$$= G'(x) - \frac{1}{x^2}G'\left(\frac{1}{x}\right) = \frac{1}{1+x^2} - \frac{1}{x^2}\left[\frac{1}{1+\frac{1}{x^2}}\right]$$

$$= \frac{1}{1+x^2} - \frac{1}{1+x^2} = 0 \quad \text{for all} \quad x \neq 0.$$

Hence $F(x) \equiv c_1$, $x > 0$; $F(x) \equiv c_2$, $x < 0$.

(b) It is convenient to evaluate $F(1)$.

$$c_1 = F(1) = 2\int_0^1 \frac{1}{1+t^2}dt = 2(\arctan 1) = \boxed{\frac{\pi}{2}}.$$

If $x < 0$, $F(x) \equiv c_2 = \boxed{-\frac{\pi}{2}}$ since $F(x)$ is an odd function of x.

(*)16. (a) Prove that

$$J = \int_0^a \frac{f(x)}{f(x)+f(a-x)}dx = \frac{a}{2} \qquad (*)$$

(b) Apply this result in evaluating $I=\int_0^2 \frac{x^2}{x^2-2x+2}dx$, and verify your answer by computing I using the Fundamental Theorem of Calculus.

(c) Produce another example where (*) may be applied.

(a) Let $u=a-x \leftrightarrow x=a-u$ in J. Then

$$J=\int_0^a \frac{f(a-u)}{f(a-u)+f(u)}du=\int_0^a \frac{f(a-x)}{f(a-x)+f(x)}dx. \qquad (**)$$

Adding equations (*) and (**), one sees that

$$2J=\int_0^a \frac{f(x)+f(a-x)}{f(x)+f(a-x)}dx=\int_0^a dx=a.\text{ Hence } J=\frac{a}{2}.$$

(b)

$$I=2\int_0^2 \frac{x^2}{2x^2-4x+4}dx=2\int_0^2 \frac{x^2}{x^2+(x-2)^2}dx$$

$$=2\int_0^2 \frac{f(x)}{f(x)+f(2-x)}dx \quad \text{where} \quad f(x)=x^2.$$

Hence by part (a), $I=2\left(\frac{2}{2}\right)=\boxed{2}$. By direct computation,

$$I=\int_0^2 \frac{x^2-2x+2+2x-2}{x^2-2x+2}dx$$

$$=\int_0^2\left[1+\frac{2x-2}{x^2-2x+2}\right]dx=\int_0^2 dx+\int_{x=0}^{x=2} d(\log(x^2-2x+2))$$

$$=2+\log[x^2-2x+2]\Big|_{x=0}^{x=2}=2+\log 2-\log 2=2.$$

(Note that $x^2-2x+2=(x-1)^2+1\neq 0$ for any x.)

(c) Let $f(x)=\sin x$, $a=\frac{\pi}{2}$; $f\left(\frac{\pi}{2}-x\right)=\cos x$. Hence

$$\int_0^{\pi/2} \frac{\sin x}{\sin x+\cos x}dx=\frac{\pi}{4}.$$

17. How many roots does the equation $2x^3-3x^2-12x+1=0$ have? Locate each nonintegral root between two consecutive integers.

Let $f(x)=2x^3-3x^2-12x+1$; $f'(x)=6x^2-6x-12=6(x^2-x-2)=6(x-2)(x+1)$. $f'(x)=0$ when $x=-1,2$. $f(-\infty)=-\infty$, $f(-1)=8$, $f(2)=-19$, $f(+\infty)=+\infty$. Hence $f(x)$ has 3 roots r_1,r_2,r_3: $r_1<-1$, $-1<r_2<2$, $r_3>2$. $f(x)$ cannot have more than 3 roots since $f(x)$ is a cubic. $f(-2)=-3$. Hence $\boxed{-2<r_1<-1}$. $f(0)=1$, $f(1)=-12$. Hence $\boxed{0<r_2<1}$. $f(3)=-8$, $f(4)=33$. Hence $\boxed{3<r_3<4}$.

(*)**18.** Let $\bar{x}$ be a simple root of $f(x)=0$. Show that Newton's method $x_{n+1}=x_n-\dfrac{f(x_n)}{f'(x_n)}$ has order of convergence 2 to $\bar{x}$, i.e. $|x_{n+1}-\bar{x}|\le c|x_n-\bar{x}|^2$, for some constant c.

Assume that $f''(x)$ exists near $\bar{x}$ and that near $\bar{x}$, $|f''(x)|\le M$ for some constant M. Since $\bar{x}$ is a simple root of $f(x)$, $f'(\bar{x})\ne 0$ and hence near $\bar{x}$, $|f'(x)|\ge L$ for some constant $L>0$. Now assume that x_n is close enough to $\bar{x}$ so that $|f''(x)|\le M$ and $|f'(x)|\ge L$ if $|x-\bar{x}|\le|x_n-\bar{x}|$. Using Taylor's formula about x_n,

$$0=f(\bar{x})=f(x_n)+f'(x_n)(\bar{x}-x_n)+\frac{f''(c_n)}{2}(\bar{x}-x_n)^2$$

for some c_n between x_n and $\bar{x}$, i.e. $|c_n-\bar{x}|<|x_n-\bar{x}|$. Hence

$$x_{n+1}=x_n-\frac{f(x_n)}{f'(x_n)}=\bar{x}+\frac{f''(c_n)}{2f'(x_n)}(\bar{x}-x_n)^2$$

$$\leftrightarrow x_{n+1}-\bar{x}=\frac{f''(c_n)}{2f'(x_n)}(\bar{x}-x_n)^2.$$

Hence

$$|x_{n+1}-\bar{x}|\le c|x_n-\bar{x}|^2 \text{ where } c=\frac{M}{2L}.$$

19. Show that $\lim\limits_{x\to 0}\dfrac{(ab)^x-1}{x}=\lim\limits_{x\to 0}\dfrac{a^x-1}{x}+\lim\limits_{x\to 0}\dfrac{b^x-1}{x}$ for $a>0, b>0$.

$(ab)^x=e^{x\log(ab)}=e^{cx}$ where $c=\log(ab)$; $\lim\limits_{x\to 0}\dfrac{e^{cx}-1}{x}=\lim\limits_{x\to 0}\dfrac{f(x)-f(0)}{x}$ $=f'(0)$ where $f(x)=e^{cx}$. $f'(x)=ce^{cx}\Rightarrow f'(0)=c$. Thus $\lim\limits_{x\to 0}\dfrac{(ab)^x-1}{x}$ $=c=\log(ab)$. Similarly, $\lim\limits_{x\to 0}\dfrac{a^x-1}{x}=\log a$, $\lim\limits_{x\to 0}\dfrac{b^x-1}{x}=\log b$. But $\log(ab)=\log a+\log b$. Thus the result is shown.

(*)**20.** Two sequences $\{a_n\}$ and $\{b_n\}$ are said to be *asymptotically equal* if $\lim\limits_{n\to\infty}\dfrac{a_n}{b_n}=1$.

(a) Prove the *Limit Comparison Test*. Assume that for each $n\ge 1$, $a_n>0$, $b_n>0$ and that $\{a_n\}$ and $\{b_n\}$ are asymptotically equal. Then $\sum\limits_{n=1}^{\infty}a_n$ converges if and only if $\sum\limits_{n=1}^{\infty}b_n$ converges.

(b) Show that $\sum\limits_{n=1}^{\infty}\sin\dfrac{1}{n}$ diverges.

(a) Let $c_n = \frac{a_n}{b_n}$. Then $\lim_{n\to\infty} c_n = 1$. Given $\varepsilon > 0$, there is some integer $N(\varepsilon)$ such that $|c_n - 1| < \varepsilon$ if $n \geq N(\varepsilon)$. Let $\varepsilon = \frac{1}{2}$ and choose $N_0 = N\left(\frac{1}{2}\right)$ so that $|c_n - 1| < \frac{1}{2}$ if $n \geq N_0$. Thus $\frac{1}{2} < \frac{a_n}{b_n} < \frac{3}{2}$ for any $n \geq N_0$ $\leftrightarrow \frac{1}{2}b_n < a_n < \frac{3}{2}b_n$ for any $n \geq N_0$. Hence if $\sum_{n=1}^{\infty} b_n$ converges then $\sum_{n=1}^{\infty} a_n$ converges (since $0 < a_n < \frac{3}{2}b_n$) and if $\sum_{n=1}^{\infty} a_n$ converges then $\sum_{n=1}^{\infty} b_n$ converges (since $0 < \frac{1}{2}b_n < a_n$).

(b) Let $a_n = \frac{1}{n}$, $b_n = \sin\frac{1}{n}$. $\lim_{n\to\infty} \frac{a_n}{b_n} = \lim_{x\to 0^+} \frac{x}{\sin x} = 1$ where $x = \frac{1}{n}$. Hence the sequences $\left\{\frac{1}{n}\right\}$ and $\left\{\sin\frac{1}{n}\right\}$ are asymptotically equal. $a_n > 0$, $b_n > 0$ for each $n \geq 1$. $\sum_{n=1}^{\infty} \frac{1}{n}$ diverges. Hence $\sum_{n=1}^{\infty} \sin\frac{1}{n}$ also diverges, by the Limit Comparison Test.

(*)21. Let X be a *fixed* positive number. Find an integer $N(X)$ such that $\frac{X^N}{N!} < \frac{1}{2}$. You are required to find an N that works, *not* necessarily the smallest N.

For each integer $n \geq 1$,

$$\begin{aligned}\log n! &= \log 1 + \log 2 + \cdots + \log n \\ &\geq \int_1^n \log x\,dx = [x\log x - x]\Big|_{x=1}^{x=n} \\ &= n\log n - (n-1). \text{ Thus } n! \geq \frac{n^n}{e^{n-1}}.\end{aligned}$$

Hence $\frac{X^N}{N!} \leq \frac{X^N e^{N-1}}{N^N} = \frac{1}{e}\left(\frac{Xe}{N}\right)^N < \frac{1}{2}\left(\frac{Xe}{N}\right)^N$. Thus $\frac{X^N}{N!} < \frac{1}{2}$ if $\frac{Xe}{N} < 1$, i.e. $\boxed{N > Xe}$.

(*)22. $\hat{x}$ is a *zero* of $f(x)$ if $f(\hat{x}) = 0$. Suppose $J_0(x)$ and $J_1(x)$ are differentiable functions such that

$$\frac{d}{dx}J_0(x) = -J_1(x),$$

$$\frac{d}{dx}[xJ_1(x)] = xJ_0(x).$$

Prove that the zeroes of $J_0(x)$ and $J_1(x)$ interlace, i.e.

(a) between two consecutive zeroes of $J_1(x)$, $J_0(x)$ has a zero and

(b) between two consecutive zeroes of $J_0(x)$, $J_1(x)$ has a zero. (You may assume that $J_0(x)$ and $J_1(x)$ have infinitely many zeroes.)

Case I. Say r_1 and r_2, $r_1 < r_2$, are consecutive zeroes of $J_1(x)$, i.e. $J_1(r_1) = 0$, $J_1(r_2) = 0$. Then $r_1 J_1(r_1) = 0 = r_2 J_1(r_2)$. Hence for some c, $r_1 < c < r_2$, $\frac{d}{dx}[xJ_1(x)] = 0$ at $x = c$ by Rolle's Theorem. $\frac{d}{dx}[xJ_1(x)] = J_1(x) + xJ_1'(x)$ $= J_1(c) + cJ_1'(c)$ at $x = c$. Note that $c \neq 0$ since $J_1(c) \neq 0$. But $\frac{d}{dx}[xJ_1(x)] = xJ_0(x)$. Thus $cJ_0(c) = 0$. Hence $J_0(c) = 0$.

Case II. Say $J_0(R_1) = 0$, $J_0(R_2) = 0$, $R_1 < R_2$, consecutive zeroes of $J_0(x)$. Then $J_0'(C) = 0$ for some C, $R_1 < C < R_2$. But $\frac{d}{dx}J_0(x) = -J_1(x)$. Hence $J_1(C) = 0$.

(*)23. (a) Show that $y = f(x)$ satisfies $\frac{d^2y}{dx^2} + y = 0$ (*) if and only if $y^2 + (y')^2 = \text{constant} = c^2$, say.

(b) Find the general solution of the differential equation $y^2 + (y')^2 = c^2$. Hence show that the general solution of the differential equation (*) is $y = A\sin x + B\cos x$ where A and B are arbitrary constants.

(a) Multiply equation (*) by $2y'$. Then $2y'y'' + 2y'y = 0$

$$\leftrightarrow \frac{d}{dx}\left[(y')^2 + y^2\right] = 0 \leftrightarrow (y')^2 + y^2 = \text{constant} = c^2, \text{ say.}$$

(b) $(y')^2 = c^2 - y^2 \leftrightarrow y' = \pm\sqrt{c^2 - y^2}$. After separation of variables, one has: $\int \frac{dy}{\sqrt{c^2 - y^2}} = \pm \int dx$. Hence $\sin^{-1}\frac{y}{c} = \pm x + \alpha$ where α is an arbitrary constant. Hence $\frac{y}{c} = \sin(\pm x + \alpha) \leftrightarrow y = c\sin(\pm x + \alpha)$. But $\sin(-x + \alpha) = -\sin(x - \alpha)$. Thus the general solution of (*) is $y = a\sin(x + \varphi)$ where a and φ are arbitrary constants. This solution can be expressed as $y = A\sin x + B\cos x$ where A and B are arbitrary constants; the equations $a\cos\varphi = A$, $a\sin\varphi = B$ relate these two representations of the general solution of the differential equation (*).

Supplementary Problems

1. If f and g are strictly increasing on the interval $[a, b]$, prove that the function $(f + g)$ is also strictly increasing on $[a, b]$.

2. (a) Explain the following sentence: "A function $y = f(x)$ is continuous at a point $x = a$".

(b) Use the definition of (a) to check whether the function

$$f(x)=\begin{cases}2x & \text{if} \quad x<1,\\ 0 & \text{if} \quad x=1,\\ 2 & \text{if} \quad x>1,\end{cases}$$

is continuous at $x=1$. If not, can you redefine $f(x)$ at $x=1$ so that it becomes continuous at $x=1$?

(*)3. Consider the function

$$f(x)=\begin{cases}\dfrac{x-1}{|x|-1} & \text{if} \quad x\neq 1,\\ 0 & \text{if} \quad x=1.\end{cases}$$

For each of the following three values of x

(i) $x=-1$; (ii) $x=0$; (iii) $x=1$,

(a) determine whether f is continuous at x. Fully justify your answer.
(b) If x is a point of discontinuity, state whether it is removable or essential.

(*)4. Let

$$f(x)=\begin{cases}1 & \text{if} \quad 1\geq x\geq \frac{1}{2} \text{ or } \frac{1}{3}\geq x\geq \frac{1}{4} \text{ or } \frac{1}{5}\geq x\geq \frac{1}{6}, \text{ etc.,}\\ 2 & \text{if} \quad \frac{1}{2}>x>\frac{1}{3} \text{ or } \frac{1}{4}>x>\frac{1}{5}, \text{ etc.,}\end{cases}$$

as in Fig. S.4. It is clear from the graph that one can make $f(x)$ arbitrarily close to 1 by choosing a suitable positive x sufficiently close to zero. Explain why "$\lim_{x\to 0^+} f(x)=1$" is false.

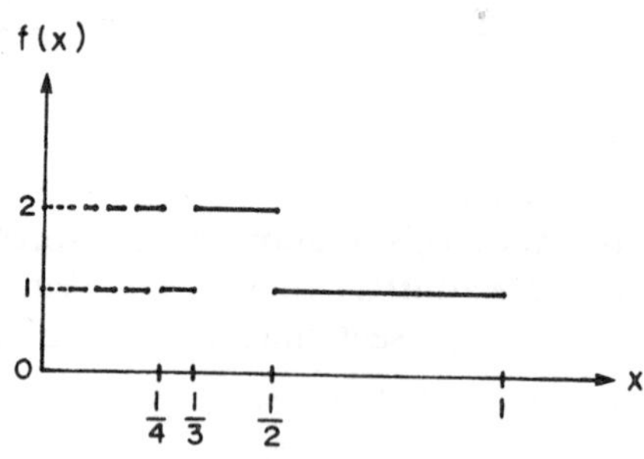

Figure S.4

5. If $\cos x \leq g(x)\leq (x+1)^2$ for $x\in[0,0.1]$, what is $\lim_{x\to 0^+} g(x)$? Explain.

6. If $f(x)=\dfrac{1}{2}x^2+1$, what is the value of $\lim_{h\to 0}\dfrac{f(2+h)-f(2)}{h}$?

7. Given

$$f(x)=\begin{cases}2x-2 & \text{if} \quad x<-1,\\ Ax^3+Bx^2+Cx+D & \text{if} \quad -1\leq x\leq 1,\\ 5x+7 & \text{if} \quad x>1,\end{cases}$$

determine the constants A, B, C, D so that $f(x)$ is differentiable for all real values of x.

8. Is the function $f(x)=|x-2|$
(a) continuous at $x=2$?
(b) differentiable at $x=2$?
Justify your answers.

9. Discuss the continuity and differentiability of the function

$$f(x)=\begin{cases} |x-3| & \text{if } x\geq 1, \\ \dfrac{x^2}{4}-\dfrac{3}{2}x+\dfrac{13}{4} & \text{if } x<1. \end{cases}$$

Give reasons for your answers.

10. If a function has a derivative at $x=5$, is it continuous at $x=5$?

11. Show that $\dfrac{d}{du}|u|=\dfrac{|u|}{u}$, $u\neq 0$. Draw a rough sketch of the function $f(x)=|x^3-x|$ and use the chain rule to find $f'(x)$ for $x\neq 0, 1, -1$. Explain why the chain rule does not apply when $x=0, 1, -1$.

12. Why do the functions $f(x)=\dfrac{x^2+\sin^2 x}{1+x^2}$ and $g(x)=\dfrac{-\cos^2 x}{1+x^2}$ have the same derivative for all x?

13. Give an example of a function continuous on the open interval $(0,1)$ but having no maximum there.

14. Give an example of a function $y=f(x)$ and a number $x=a$ such that $f(a)$ is a local minimum and $f''(a)$ fails to exist.

(*)15. Suppose that $h(x)=xg(x)$ for some function g which is continuous at $x=0$. Prove that the function $h(x)$ has a derivative at $x=0$ and find $h'(0)$ in terms of g.

(*)16. Assume $f(x)$ is defined for all x such that $|x|\leq 1$ and satisfies $x\leq f(x)\leq x+x^2$ (all $|x|\leq 1$). Prove that $f'(0)$ exists and has the value 1.

(*)17. Suppose $f(x)$ is a differentiable function and $\lim\limits_{x\to\infty} f'(x)=0$. Let $g(x)=f(x+1)-f(x)$. Show that $\lim\limits_{x\to\infty} g(x)=0$.

18. State whether the Mean Value Theorem applies to the function $f(x)=x^{2/3}$, on the interval $-2\leq x\leq 2$, and explain your reasoning.

(*)19. $f(x)$ is said to satisfy a *Lipschitz condition* on the closed interval $[a,b]$ if there is a constant $K>0$ such that for all $x_1, x_2\in[a,b]$, $|f(x_2)-f(x_1)|\leq K|x_2-x_1|$.
(a) Prove that if $f'(x)$ is continuous on $[a,b]$ then $f(x)$ satisfies a Lipschitz condition on $[a,b]$.
(b) Prove or disprove the converse of (a).

(c) Give an example of a continuous function defined on some interval $[a, b]$ which does not satisfy a Lipschitz condition on $[a, b]$.

(*)20. Suppose F is an antiderivative of a continuous function f which has an inverse g. Find an antiderivative of g, expressed in terms of F and g. (Assume for simplicity that $F(x)$, $f(x)$, $g(x)$ are defined for all real x.)

(*)21. The following functions all become large for large positive values of x but at different rates. Write a set of inequalities in the form $a > b > c > d$ to denote the relative magnitudes of these functions for large positive x: x^x, $\log x$, e^x, x.

(*)22. Arrange the following functions in order of increasing size as $x \to \infty$: 10^x, e^x, $\log x^5$, e^{x^2}, x^3, e^{20x}.

(*)23. Is it possible for a line to be tangent to a cubic graph $y = x^3 + ax^2 + bx + c$ (a, b, c constants) at two distinct points?

(*)24. Say $f(x)$ and $g(x)$ are functions with continuous first derivatives for all real x such that $fg' - gf'$ is never zero. Prove that between two consecutive roots of $f(x) = 0$ there is exactly one root of $g(x) = 0$. (As an example consider $f(x) = \cos x$, $g(x) = \sin x$.)

25. (a) Give a precise statement of the Intermediate Value Theorem.

(b) Prove that the polynomial $p(x) = x^3 - 3x + 1$ has two roots in the interval $[0, 2]$.

(c) State the Mean Value Theorem and use it to prove that, if $f(x)$ is differentiable on an interval $[a, b]$ with $f'(x) > 0$ for $a < x < b$, then $f(x)$ is increasing on $[a, b]$.

(*)26. Use the relation between Riemann sums and definite integrals to evaluate the limit: $\lim\limits_{n \to \infty} \sum\limits_{i=1}^{n} \dfrac{1}{i+n}$.

27. (a) Given $y = f(x)$, how can one detect the places at which the curve has a horizontal tangent?

(b) How, by examining y'', can one tell where the curve is concave upward and where it is concave downward?

(c) Suppose, for $y = f(x)$, one has a point $x = a$ at which $y' = 0$. How can one tell whether $x = a$ is a relative maximum, a relative minimum or neither?

(d) Given that $\dfrac{d}{dx}H(x) = f(x)$, how can one assign a value to $\int_a^b f(x)\,dx$?

28. Show that $\displaystyle\int_x^1 \frac{1}{1+t^2}\,dt = \int_1^{1/x} \frac{1}{1+v^2}\,dv.$

(*)29. Say $f(x)$ is such that $f(2-x) = -f(x)$ and $f(x)$ is integrable for all real x. Find $a \neq 2$ such that $\int_a^2 f(x)\,dx = 0$.

30. The length L of that portion of a curve $y = y(x)$ between $x = a$ and

$x = b$ is given by the definite integral

$$L = \int_a^b \sqrt{1 + \left(\frac{dy}{dx}\right)^2}\, dx.$$

Use this formula to prove that the length of the circumference of a circle is 2π multiplied by its radius.

31. Prove that if $0 \le x \le 1$ then $\log(1 + x) \le \arctan x$.

Hint: One way to do this is to express each function as a definite integral over the interval $[0, x]$.

(*)32. (a) Say f and g are continuous functions in $[a, b]$ and g does not vanish at any point in $[a, b]$. Prove, by considering the functions

$$F(x) = \int_a^x f(t)g(t)\, dt \text{ and } G(x) = \int_a^x g(t)\, dt, \text{ that}$$

$$\int_a^b f(t)g(t)\, dt = f(c)\int_a^b g(t)\, dt,$$

for some point c in $[a, b]$.

(Hint: Cauchy Mean Value Theorem may be useful here.)

(b) Deduce that $\dfrac{\pi^2}{64} < \displaystyle\int_0^{\pi/4} \frac{x\, dx}{1 + \tan^2 x} < \frac{\pi^2}{32}$.

(*)33. *Prove directly* that if $f(x)$ is a cubic function, i.e. $f(x) = Ax^3 + Bx^2 + Cx + D$ for some constants A, B, C, D, then Simpson's rule for 2 equal subintervals ($n = 2$) gives the *exact* value for $\int_a^b f(x)\, dx$. (Remember that Simpson's rule corresponds to an area approximation under parabolic arches!)

(*)34. Let S_n be a sequence of real numbers such that $S_{n+1} = S_n - S_n^2$.

(a) Show that $\lim\limits_{n \to \infty} S_n = 0$ or $-\infty$.

(b) For which values of S_1 does $\lim\limits_{n \to \infty} S_n = 0$?

35. The hyperbolic sine and cosine functions are defined respectively by

$$\sinh x = \frac{e^x - e^{-x}}{2}, \qquad \cosh x = \frac{e^x + e^{-x}}{2}.$$

(a) Show that $(\cosh x)^2 - (\sinh x)^2 = 1$.

(b) Verify that $\cosh(2x)$ satisfies the differential equation

$$\frac{d^2y}{dx^2} - 4y = 0.$$

(c) Sketch a graph of $y = \sinh x$, and illustrate geometrically that $\sinh x$ has an inverse function, denoted by $\sinh^{-1} x$.

(d) Solve the equation

$$y = \frac{e^x - e^{-x}}{2}$$

for x in terms of y, and hence obtain an expression for $\sinh^{-1}x$.

(*)36. Suppose

$$f(x) = \begin{cases} a_1 + a_2 \log x & \text{if} \quad x > 0, \\ b & \text{if} \quad x = 0, \\ c_1 e^{2x} + \dfrac{c_2}{x} & \text{if} \quad x < 0, \end{cases}$$

where a_1, a_2, b, c_1, c_2 are real constants. For each of the following statements, find equivalent condition(s) about the real constants a_1, a_2, b, c_1, c_2.

(a) $\lim_{x \to 0} f(x) = +\infty$.

(b) $\lim_{x \to 0} f(x) = 4$.

(c) f is continuous at $x = 0$.

(*)37. Find an *even* function f: $\mathbb{R} \to \mathbb{R}$ that is *discontinuous* at *exactly three* points.

(*)38. Find a function f: $\mathbb{R} \to \mathbb{R}$ that is continuous at infinitely many points as well as discontinuous at infinitely many other points.

(*)39. One is interested in the way the largest root of the polynomial $x^3 - x + \alpha$ depends on the parameter α.

(a) Let r denote this largest root. Then r is a function of α. Give a qualitative description of this function and draw a rough sketch of its graph.

(b) Give some *quantitative* information about r. Calculate $\dfrac{dr}{d\alpha}$.

Chapter VIII
Techniques

The problems in this chapter are classified under the following headings.

1. Precalculus	absolute value, inequalities, analytic geometry, trigonometry, logarithms, functions, inverse functions.
2. Differentiation	chain rule, implicit differentiation, differentiation of integrals.
3. Applications of differentiation	tangent lines, normal lines, maxima and minima.
4. Methods of integration	indefinite integrals.
5. Methods of integration	definite integrals.
6. Improper integrals.	
7. Applications of definite integrals	area, volumes, arc length, area of a surface of revolution.
8. Parametric curves.	
9. Polar coordinates.	
10. Sequences and infinite series, Taylor's formula, Taylor series.	
11. Indeterminate forms.	
12. Differential equations.	
13. Miscellaneous.	

Solved Problems

1.1. If $y = 2x + |2 - x|$, express x in terms of y.

$$|2-x| = \begin{cases} 2-x & \text{if} \quad 2-x>0 \leftrightarrow x<2, \\ 0 & \text{if} \quad x=2, \\ x-2 & \text{if} \quad 2-x<0 \leftrightarrow x>2. \end{cases}$$

Hence $y = \begin{cases} 2+x & \text{if} \quad x \le 2, \\ 3x-2 & \text{if} \quad x > 2. \end{cases}$

$$y = 2 + x \leftrightarrow x = y - 2,$$

$$y = 3x - 2 \leftrightarrow x = \frac{y+2}{3}.$$

From the graph (Fig. VIII.1.1) one sees that

$$\boxed{x = \begin{cases} y-2 & \text{if} \quad y \le 4, \\ \dfrac{y+2}{3} & \text{if} \quad y > 4. \end{cases}}$$

1.2. Solve the inequality $\left|\frac{3x-8}{2}\right| \ge 4$.

$$\left|\frac{3x-8}{2}\right| \ge 4 \leftrightarrow |3x-8| \ge 8 \leftrightarrow \left|x - \frac{8}{3}\right| \ge \frac{8}{3},$$

i.e., the distance from x to $\frac{8}{3}$ is at least $\frac{8}{3}$. Hence the solution is $\boxed{x \ge \frac{16}{3} \text{ or } x \le 0}$.

1.3. Solve the inequality $|3x| \ge |6 - 3x|$.

$$|3x| \ge |6 - 3x| \leftrightarrow |x| \ge |x - 2|,$$

i.e., the distance from x to 0 is at least the distance from x to 2. $x = 1$ is equidistant from 0 and 2. Hence the solution is $\boxed{x \ge 1}$.

1.4. Find all x such that $\sqrt{1+x} < 1 + \frac{1}{2}x$.

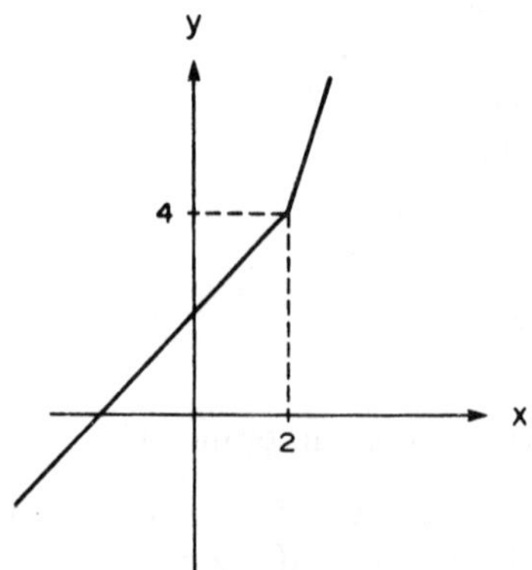

Figure VIII.1.1. Graph of $y = 2x + |2 - x|$.

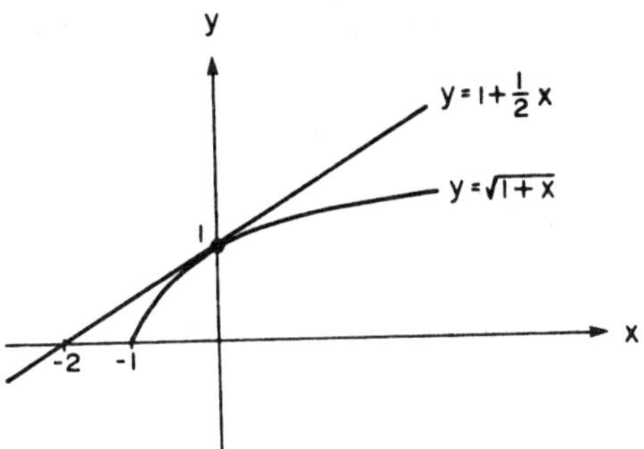

Figure VIII.1.4. Graph of $\sqrt{1+x} < 1 + \frac{1}{2}x$.

The curve $y = \sqrt{1+x}$ is a part of the parabola $y^2 = 1 + x$, $y \geq 0$. The curves $y = \sqrt{1+x}$ and $y = 1 + \frac{1}{2}x$ intersect when $\sqrt{1+x} = 1 + \frac{1}{2}x \Rightarrow 1 + x = \left(1 + \frac{1}{2}x\right)^2 \leftrightarrow 1 + x = 1 + x + \frac{x^2}{4} \leftrightarrow x = 0$.

Graphically (Fig. VIII.1.4) one sees that the solution is $\boxed{x \geq -1, x \neq 0}$.

1.5. Find an equation of a line l through the point $(-3, 2)$ making an angle of radian measure $\pi/4$ with the line having an equation $3x - 2y - 7 = 0$.

Let $y = mx + b$ be the equation of the unknown line l. The equation of the given line is $3x - 2y - 7 = 0 \leftrightarrow y = \frac{3}{2}x - \frac{7}{2}$. Let α be the angle which the given line makes with the x-axis (Fig. VIII.1.5). Then $\tan\alpha = \frac{3}{2}$. $\beta = \frac{\pi}{4}$ is the angle which the unknown line l makes with the given line. Then the slope of l is $m = \tan(\alpha + \beta) = \dfrac{\tan\alpha + \tan\beta}{1 - \tan\alpha\tan\beta} = \dfrac{\frac{3}{2} + 1}{1 - \frac{3}{2}} = -5$.

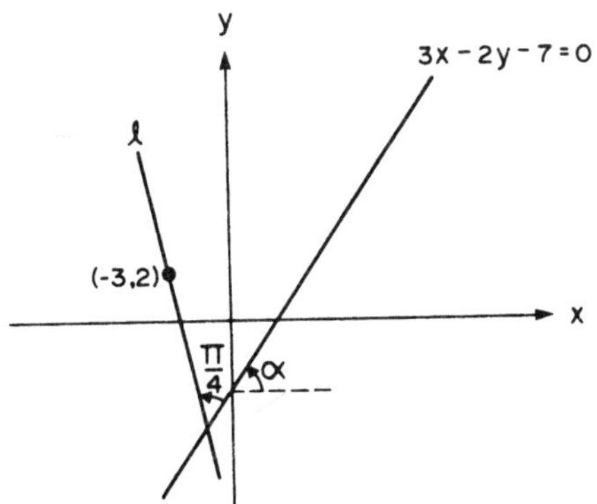

Figure VIII.1.5

l passes through $(-3,2) \Rightarrow 2 = 15 + b \Rightarrow b = -13$. Hence the equation of the unknown line l is $\boxed{y = -5x - 13}$.

1.6. Find the equation of the line which passes through the points of intersection of the circles $x^2 + y^2 - 2x + 4y = 5$, $x^2 + y^2 + 2x + 2y = 11$. Find the coordinates of these points of intersection.

Subtracting the second equation from the first, one finds that the points of intersection lie on the line $-2x + 4y - 2x - 2y = 5 - 11 \leftrightarrow \boxed{y = 2x - 3}$. Substituting for $y = 2x - 3$ in the first equation $\Rightarrow x^2 + (2x-3)^2 - 2x + 4(2x-3) = 5 \leftrightarrow 5x^2 - 6x - 8 = 0 \leftrightarrow (5x+4)(x-2) = 0 \leftrightarrow x = -\frac{4}{5}, 2$. Hence the intersection points are $\boxed{\left(-\frac{4}{5}, -\frac{23}{5}\right)}$ and $\boxed{(2,1)}$. $x^2 + y^2 - 2x + 4y = 5 \leftrightarrow x^2 - 2x + y^2 + 4y = 5 \leftrightarrow (x-1)^2 + (y+2)^2 = 1 + 4 + 5 = 10 \leftrightarrow$ circle of radius $\sqrt{10}$, centred at $(1,-2)$. $x^2 + y^2 + 2x + 2y = 11 \leftrightarrow (x+1)^2 + (y+1)^2 = 1 + 1 + 11 = 13 \leftrightarrow$ circle of radius $\sqrt{13}$, centred at $(-1,-1)$.

(*)1.7. If you were marooned on a desert island without a calculator or table of trigonometric functions, how would you go about determining which is greater.

$$2\tan^{-1}(2\sqrt{2} - 1) \quad \text{or} \quad 3\tan^{-1}\left(\frac{1}{4}\right) + \tan^{-1}\left(\frac{5}{99}\right)?$$

Method 1

Let $y = 2\tan^{-1}(2\sqrt{2} - 1)$, $x = 3\tan^{-1}\left(\frac{1}{4}\right) + \tan^{-1}\left(\frac{5}{99}\right)$, $\alpha = \tan^{-1}\left(\frac{1}{4}\right)$, $\beta = \tan^{-1}\left(\frac{5}{99}\right)$. Then $x = 3\alpha + \beta$. $\tan x = \tan(3\alpha + \beta) = \dfrac{\tan 3\alpha + \tan\beta}{1 - \tan 3\alpha \tan\beta}$.

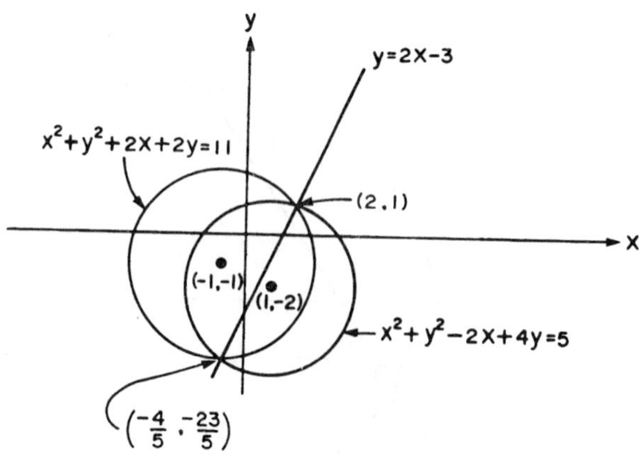

Figure VIII.1.6. Graphical representation of Problem 1.6.

But

$$\tan 3\alpha = \frac{\tan 2\alpha + \tan\alpha}{1 - \tan 2\alpha \tan\alpha} = \frac{\dfrac{2\tan\alpha}{1-\tan^2\alpha} + \tan\alpha}{1 - \dfrac{2\tan^2\alpha}{1-\tan^2\alpha}}$$

$$= \frac{2\tan\alpha + \tan\alpha(1-\tan^2\alpha)}{1 - 3\tan^2\alpha} = \frac{\tan\alpha[3-\tan^2\alpha]}{1-3\tan^2\alpha}$$

$$= \frac{\frac{1}{4}\left[3 - \frac{1}{16}\right]}{1 - \frac{3}{16}} = \frac{47}{52};$$

$$\tan\beta = \frac{5}{99}.$$

Hence $\tan x = \dfrac{\frac{47}{52} + \frac{5}{99}}{1 - \left(\frac{47}{52}\right)\left(\frac{5}{99}\right)} = 1$. Since $2\sqrt{2} - 1 > 1$, $\frac{y}{2} > x$. Hence $y > 2x$,

and $\boxed{y > x}$.

Method 2

If $x > 0$, $\tan^{-1}x < x$. Hence $3\tan^{-1}\left(\frac{1}{4}\right) + \tan^{-1}\left(\frac{5}{99}\right) < (3)\left(\frac{1}{4}\right) + \frac{5}{99} < 1$. Since $2\sqrt{2} - 1 > 1$, $\tan^{-1}(2\sqrt{2} - 1) > \tan^{-1}1 = \frac{\pi}{4}$. Hence $2\tan^{-1}(2\sqrt{2} - 1) > (2)\left(\frac{\pi}{4}\right) = \frac{\pi}{2} > 1$.

1.8. Given that $\log 2 = 0.7$, $\log 5 = 1.6$, evaluate $\log 0.1$.

$$\log 0.1 = \log\frac{1}{10} = \log 1 - \log 10 = -\log 10$$

$$= -\log[(2)(5)] = -[\log 2 + \log 5] = \boxed{-2.3}.$$

1.9. Find the domain and range of the inverse function of the function $f(x) = 2^x$. Find the inverse function.

$y = f(x) = 2^x = e^{x\log 2}$ (Fig. VIII.1.9*a*) is a 1:1 function with domain = $(-\infty, \infty)$ and range = $(0, \infty)$. $\log y = x\log 2$. Hence the inverse function $y = f^{-1}(x)$ (Fig. VIII.1.9*b*) satisfies $\log x = y\log 2 \leftrightarrow f^{-1}(x) = \boxed{\dfrac{\log x}{\log 2}}$ with

$\boxed{\text{domain} = (0, \infty)}$ and $\boxed{\text{range} = (-\infty, \infty)}$.

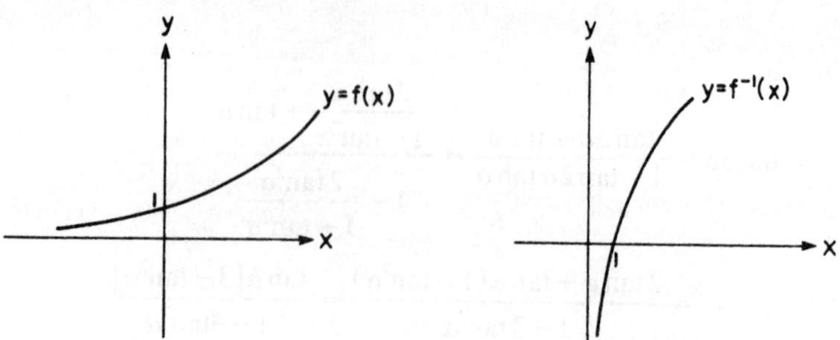

Figure VIII.1.9*a*. Graph of $y = f(x) = 2^x$. *b*. Graph of $y = f^{-1}(x)$.

1.10. (a) Find the inverse of the function $\dfrac{2}{4x-5}$; state the domain and the range of the inverse.

(b) The graph of a function $y = f(x)$ is as shown in Fig. VIII.1.10*a*. Determine the graph of the inverse function $y = f^{-1}(x)$.

(a) $y = f(x) = \dfrac{2}{4x-5}$ is a 1:1 function with domain $= \left\{x|x \neq \dfrac{5}{4}\right\}$ and range $= \{y|y \neq 0\}$. $y = f^{-1}(x)$ satisfies the equation $x = \dfrac{2}{4y-5} \leftrightarrow y = f^{-1}(x) = \boxed{\dfrac{1}{2x} + \dfrac{5}{4}}$. $y = f^{-1}(x)$ has $\boxed{\text{domain} = \{x|x \neq 0\}}$ and $\boxed{\text{range} = \left\{y|y \neq \dfrac{5}{4}\right\}}$.

(b) See the graph in Fig. VIII.1.10*b*.

2.1. Find $\dfrac{dy}{dx}$ if $y = (x^2+1)^{100}(x^3-1)^{10}$.

Let $u = x^2+1$, $v = x^3 - 1$. Then $y = u^{100}v^{10}$.

$$\begin{aligned}\frac{dy}{dx} &= v^{10}\frac{d}{dx}(u^{100}) + u^{100}\frac{d}{dx}(v^{10}) \\ &= 100v^{10}u^{99}\frac{du}{dx} + 10u^{100}v^9\frac{dv}{dx}.\end{aligned}$$

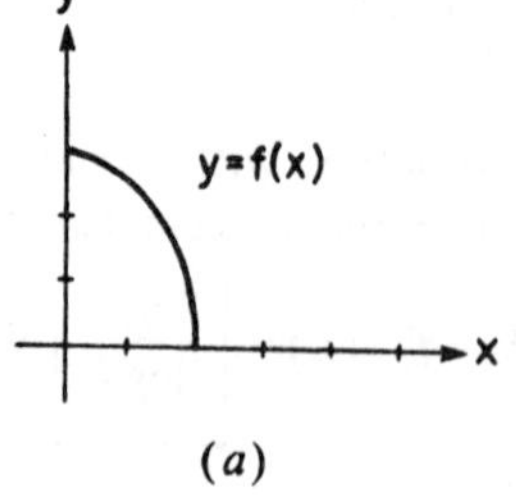

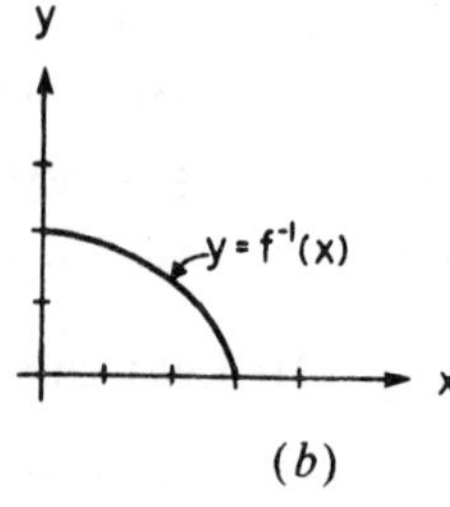

Figures VIII.1.10*a* and *b*.

$$\frac{du}{dx} = 2x, \qquad \frac{dv}{dx} = 3x^2.$$

Hence

$$\frac{dy}{dx} = 200v^{10}u^{99}x + 30u^{100}v^9x^2$$
$$= u^{99}v^9x[200v + 30ux]$$
$$= 10x(x^2+1)^{99}(x^3-1)^9[20x^3 - 20 + 3x^3 + 3x]$$
$$= \boxed{10x(x^2+1)^{99}(x^3-1)^9[23x^3 + 3x - 20]}.$$

2.2. Find $f'(x)$ where $f(x) = \log\left[\sqrt[3]{\dfrac{(x+1)(x+2)(x+3)(x+4)}{x^2}}\right]$.

$$f(x) = \frac{1}{3}\log\left[\frac{(x+1)(x+2)(x+3)(x+4)}{x^2}\right]$$
$$= \frac{1}{3}[\log(x+1) + \log(x+2) + \log(x+3) + \log(x+4) - 2\log x]$$
$$\Rightarrow f'(x) = \boxed{\frac{1}{3}\left[\frac{1}{x+1} + \frac{1}{x+2} + \frac{1}{x+3} + \frac{1}{x+4} - \frac{2}{x}\right]}.$$

Note that $f'(x)$ is undefined when $x = 0,\ -2 \le x \le -1,\ -4 \le x \le -3$.

2.3. Find $f'(x)$ where $f(x) = |x^2 - 4|$.

$$f(x) = \begin{cases} x^2 - 4 & \text{if} \quad x^2 \ge 4 \leftrightarrow |x| \ge 2 \leftrightarrow x \le -2, x \ge 2, \\ 4 - x^2 & \text{if} \quad -2 < x < 2. \end{cases}$$

Hence

$$f'(x) = \boxed{\begin{cases} 2x & \text{if} \quad x < -2, x > 2, \\ -2x & \text{if} \quad -2 < x < 2, \\ \text{undefined} & \text{if} \quad x = \pm 2. \end{cases}}$$

2.4. Find $\dfrac{d^3y}{dx^3}$ when $y = \dfrac{1+x}{1-x}$.

$y = \dfrac{1+x}{1-x} = \dfrac{-1+x+2}{1-x} = -1 + \dfrac{2}{1-x}$. Hence $\dfrac{dy}{dx} = \dfrac{2}{(1-x)^2}$, $\dfrac{d^2y}{dx^2}$ $= \dfrac{4}{(1-x)^3}$, $\dfrac{d^3y}{dx^3} = \boxed{\dfrac{12}{(1-x)^4}}$.

2.5. Find the values of $\dfrac{dy}{dx}$ and $\dfrac{d^2y}{dx^2}$ (a) at the point (7,0) and (b) at the point (32,5), on the curve $x = y^3 - 4y^2 + 7$, if they exist.

On the given curve y is defined implicitly as a function of x. Taking $\frac{d}{dx}$ of each side of the equation $\Rightarrow 1 = 3y^2\frac{dy}{dx} - 8y\frac{dy}{dx} = \frac{dy}{dx}(3y^2 - 8y)$. Hence

$$\frac{dy}{dx} = \frac{1}{3y^2 - 8y}; \qquad \frac{d^2y}{dx^2} = \frac{d\left(\frac{dy}{dx}\right)}{dy}\frac{dy}{dx}$$

$$= -\frac{(6y-8)}{(3y^2-8y)^2}\frac{dy}{dx} = -\frac{(6y-8)}{(3y^2-8y)^3}.$$

(a) At (7,0): $\frac{dy}{dx}$ does not exist, so $\frac{d^2y}{dx^2}$ does not exist.

(b) At (32,5): $\frac{dy}{dx} = \boxed{\frac{1}{35}}$, $\frac{d^2y}{dx^2} = \boxed{\frac{-22}{(35)^3}}$.

2.6. Say $f(x) = e^{-x^2}$, $x \geq 0$.
(a) Sketch the graph of $y = f(x)$.
(b) Sketch the graph of the inverse function, $y = f^{-1}(x)$.
(c) Find the domain of $f^{-1}(x)$.
(d) Find the range of $f^{-1}(x)$.
(e) Compute $\frac{d}{dx}f^{-1}(x)$ at $x = \frac{1}{2}$.

(a) The graph is shown as Fig. VIII.2.6*a*.
(b) The graph is shown as Fig. VIII.2.6*b*.
(c) Domain of $f^{-1}(x)$: $\boxed{(0,1]}$.

(d) Range of $f^{-1}(x)$: $\boxed{[0,\infty)}$.

(e) $y = f^{-1}(x)$ implicitly satisfies $x = e^{-y^2}$. Taking $\frac{d}{dx}$ of each side of this equation $\Rightarrow 1 = -2ye^{-y^2}\frac{dy}{dx} \leftrightarrow \frac{dy}{dx} = -\frac{e^{y^2}}{2y}$. At $x = \frac{1}{2}$, $e^{-y^2} = \frac{1}{2} \leftrightarrow e^{y^2}$

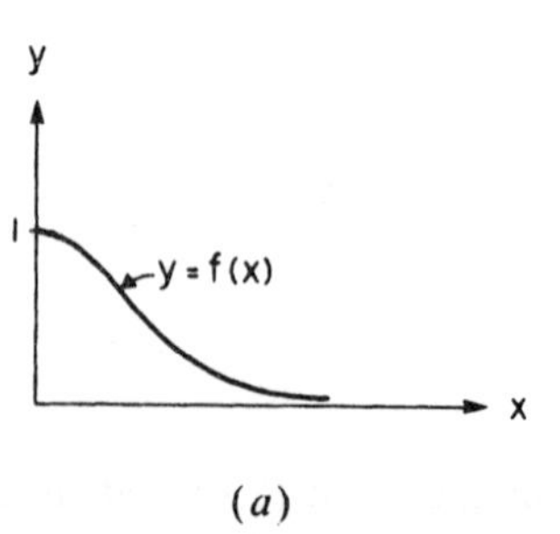

(*a*)

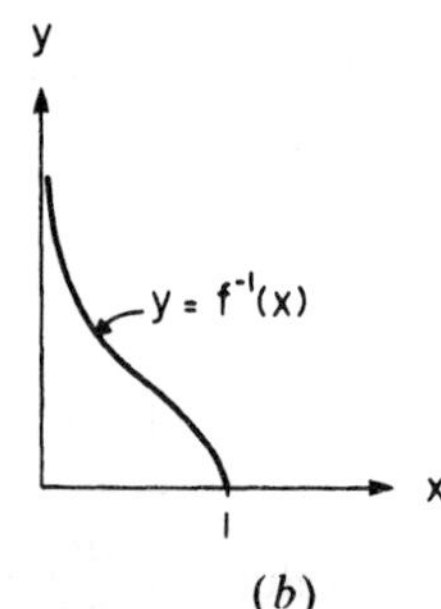

(*b*)

Figures VIII.2.6*a* and *b*.

$= 2 \leftrightarrow y^2 = \log 2$, so $y = \sqrt{\log 2}$ since $y \ge 0$. Hence $\left.\frac{d}{dx} f^{-1}(x)\right|_{x=\frac{1}{2}}$ $= \boxed{-\frac{1}{\sqrt{\log 2}}}$.

2.7. Find $\frac{dy}{dx}$ where $y = \int_0^{x^2} f(t)\,dt$.

Let $u = x^2$. Then $y = F(u) = \int_0^u f(t)\,dt$. $\frac{dy}{dx} = \frac{dF}{du}\frac{du}{dx} = 2x\frac{dF}{du}$. By the Fundamental Theorem of Calculus, $\frac{dF}{du} = f(u)$. Hence $\frac{dy}{dx} = 2xf(u)$ $= \boxed{2xf(x^2)}$.

(*)2.8. Obtain $\frac{dy}{dx}$ for $y = \int_x^{x^2} \sin\sqrt{t}\,dt$.

Let $u = x^2$, $F(x) = \int_0^x \sin\sqrt{t}\,dt$. Then $y = F(u) - F(x)$; $\frac{dy}{dx} = F'(u)\frac{du}{dx} - F'(x)$. By the Fundamental Theorem of Calculus, $F'(x) = \sin\sqrt{x}$. Hence $\frac{dy}{dx} = 2x\sin\sqrt{u} - \sin\sqrt{x} = \boxed{2x\sin x - \sin\sqrt{x}}$, $x > 0$.

(*)2.9. Let $f(x) = \int_{\sin x}^{\tan x} (1 + xt^2)\,dt$; find $\frac{df}{dx}$.

$$f(x) = \int_{\sin x}^{\tan x} dt + x\int_{\sin x}^{\tan x} t^2\,dt.$$

Let

$$G(x) = \int_{a(x)}^{b(x)} g(t)\,dt$$
$$= \int_{a(x)}^{\alpha} g(t)\,dt + \int_{\alpha}^{b(x)} g(t)\,dt$$

for some constant α. Hence $G(x) = \int_{\alpha}^{b(x)} g(t)\,dt - \int_{\alpha}^{a(x)} g(t)\,dt$. By the Fundamental Theorem of Calculus and the chain rule, $G'(x) = b'(x)g(b(x)) - a'(x)g(a(x))$.

$$f'(x) = \int_{\sin x}^{\tan x} t^2\,dt + \frac{d}{dx}\left[\int_{\sin x}^{\tan x} dt\right] + x\frac{d}{dx}\left[\int_{\sin x}^{\tan x} t^2\,dt\right]$$

$$= \boxed{\frac{\tan^3 x - \sin^3 x}{3} + \sec^2 x[1 + x\tan^2 x] - \cos x[1 + x\sin^2 x]}.$$

3.1. Show that the curve $y = x^5 + 15x^3 - 15x^2 + 10x - 4$ is everywhere increasing.

$y' = 5x^4 + 45x^2 - 30x + 10 = 5[x^4 + (9x^2 - 6x + 1) + 1] = 5[x^4 + (3x-1)^2 + 1] > 0$ for all x. Hence the curve is everywhere increasing.

3.2. What conditions must the constants a, b, and c satisfy if the cubic polynomial $y = x^3 + ax^2 + bx + c$ is to have a point of inflection with a horizontal tangent line?

$y' = 3x^2 + 2ax + b$, $y'' = 6x + 2a$. At an inflection point, $y'' = 0$. Hence $x = -\frac{a}{3}$ is the inflection point of such a cubic polynomial. The cubic polynomial has a horizontal tangent line at $x = -\frac{a}{3}$ if $y'\left(-\frac{a}{3}\right) = 0$ $\leftrightarrow 3\left(-\frac{a}{3}\right)^2 + 2a\left(-\frac{a}{3}\right) + b = 0$. Hence $\boxed{b = \frac{a^2}{3},\ c \text{ arbitrary}}$.

3.3. Find the equation of the line tangent to the curve whose equation is $x^3 - 4xy + y^3 = 0$, at the point $(2,2)$.

The given equation defines y implicitly as a function of x. First one must find y' at $(2,2)$. Taking $\frac{d}{dx}$ of the equation $\Rightarrow 3x^2 - 4y - 4xy' + 3y^2y' = 0$. Then at $(2,2)$: $12 - 8 + (-8 + 12)y' = 0$. Hence $y' = -1$ at $(2,2)$. Thus the tangent line to the given curve at $(2,2)$ is $y - 2 = -(x-2) \leftrightarrow \boxed{x + y = 4}$.

(*)3.4. Let l be the line tangent to the astroid $x^{2/3} + y^{2/3} = 4$ at the point $(2\sqrt{2}, 2\sqrt{2})$. Find the area of the triangle formed by l and the coordinate axes (Fig. VIII.3.4a).

First one must find the slope of l. Implicit differentiation of the given equation $\Rightarrow \frac{2}{3}x^{-1/3} + \frac{2}{3}y^{-1/3}\frac{dy}{dx} = 0$. Hence at $(2\sqrt{2}, 2\sqrt{2})$: $\frac{dy}{dx} = -1$. The tangent line l at $(2\sqrt{2}, 2\sqrt{2})$ is $y - 2\sqrt{2} = -(x - 2\sqrt{2}) \leftrightarrow x + y = 4\sqrt{2}$. The $x-$intercept of this line is $b = 4\sqrt{2}$. The $y-$intercept of this line is $h = 4\sqrt{2}$. The area A of the triangle formed by l (Fig. VIII.3.4b) and the coordinate axes is $A = \frac{1}{2}bh = \boxed{16}$.

(*)3.5. Find all points on the curve $y = 3x^3 + 14x^2 + 3x + 8$ where the tangent at that point passes through the origin.

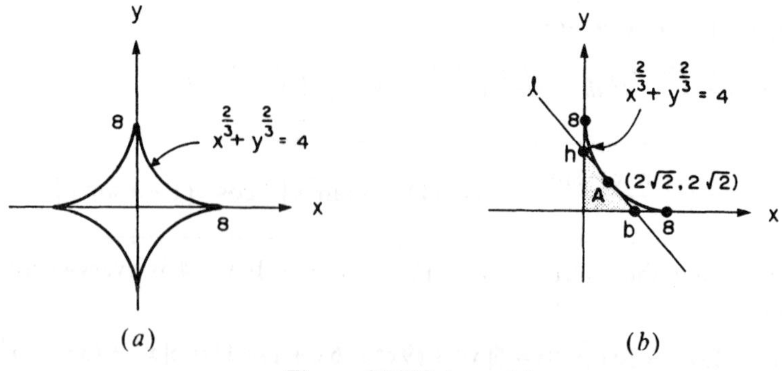

Figures VIII.3.4a and b.

The given curve γ is $y = f(x) = 3x^3 + 14x^2 + 3x + 8$. The tangent line l through $(x_0, y_0) \in \gamma$ is $y - y_0 = f'(x_0)(x - x_0)$. If l passes through the origin $\leftrightarrow (0,0)$, then (x_0, y_0) must satisfy the equation $-y_0 = -x_0 f'(x_0)$. Thus the points $(x_0, y_0) \in \gamma$ whose tangent lines l pass through the origin must simultaneously satisfy the equations $y_0 = f(x_0)$ and $y_0 = x_0 f'(x_0)$. Equivalently,

$$f(x_0) = x_0 f'(x_0) \leftrightarrow 3x_0^3 + 14x_0^2 + 3x_0 + 8 = x_0\left[9x_0^2 + 28x_0 + 3\right]$$
$$\leftrightarrow 3x_0^3 + 7x_0^2 - 4 = 0 \leftrightarrow (x_0 + 1)\left(3x_0^2 + 4x_0 - 4\right) = 0$$
$$\leftrightarrow (x_0 + 1)(3x_0 - 2)(x_0 + 2) = 0.$$

Thus $x_0 = -1, \frac{2}{3}, -2$. Hence there are three such points, namely:

$$\boxed{(-1,16), \left(\frac{2}{3}, \frac{154}{9}\right), (-2,34)}.$$

3.6. At what points on the curve γ given by the equation $y = x^3 - 3x^2 + 2$ is the tangent line l parallel to the line $y = 9x + 4$?

For the given curve γ, $y' = 3x^2 - 6x$. The given line has slope 9. For the tangent line l to the given curve γ at (x_0, y_0) to have slope 9, one must have $3x_0^2 - 6x_0 = 9 \leftrightarrow 3(x_0^2 - 2x_0 - 3) = 3(x_0 - 3)(x_0 + 1) = 0 \Rightarrow x_0 = 3, -1$. But $y_0 = x_0^3 - 3x_0^2 + 2$. Hence the points are $\boxed{(3,2) \text{ and } (-1,-2)}$.

(*)3.7. Find the equation of the common tangent l to the curves $y = x^2$ and $y = \frac{1}{x}$ (Fig. VIII.3.7).

Let $f(x) = x^2$, $g(x) = \frac{1}{x}$; $f'(x) = 2x$, $g'(x) = -\frac{1}{x^2}$. The tangent line

$$l_1 \text{ to } f(x) \text{ at } (a, f(a)) \quad \text{is} \quad y - f(a) = f'(a)(x - a). \tag{1}$$

The tangent line l_2 to $g(x)$ at $(b, g(b))$ is

$$y - g(b) = g'(b)(x - b). \tag{2}$$

But

$$l_1 = l_2 = l.$$

Hence

$$f'(a) = g'(b) \leftrightarrow 2a = -\frac{1}{b^2}. \tag{3}$$

After substituting equation (3) into equation (2) and then subtracting equation (2) from equation (1), one finds that

$$g(b) - f(a) = f'(a)(b - a)$$
$$\leftrightarrow \frac{1}{b} - a^2 = 2a(b - a). \tag{4}$$

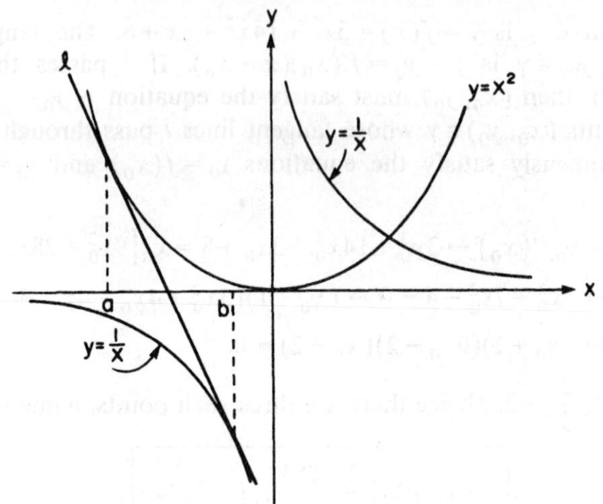

Figure VIII.3.7

Substituting equation (3) into equation (4), one finds that $\frac{1}{b}+\left(\frac{1}{2b^2}\right)^2 = -\frac{1}{b} \leftrightarrow b^3 = -\frac{1}{8}$. Hence $b=-\frac{1}{2}$; $a=-\frac{1}{2b^2}=-2$. $f'(-2)=-4$, $f(-2)=4$. Thus l satisfies the equation $y-4=-4(x+2) \leftrightarrow \boxed{y=-4x-4}$.

3.8. Find the equation of the tangent line and of the normal line to $y=\frac{x+1}{x-2}$ at $x=3$.

$$y=f(x)=\frac{x+1}{x-2}=\frac{x-2+3}{x-2}=1+\frac{3}{x-2};\quad f'(x)=\frac{-3}{(x-2)^2}.$$

$f'(3)=-3$; $f(3)=4$. Hence the tangent line at $x=3$ is $y-4=-3(x-3) \leftrightarrow \boxed{y=-3x+13}$. The normal line at $x=3$ has slope $\frac{-1}{f'(3)}=\frac{1}{3}$. Thus the normal line at $x=3$ is $y-4=\frac{1}{3}(x-3) \leftrightarrow \boxed{y=\frac{x}{3}+3}$.

(*)3.9. The shape of Gage hill can be described by the equation $y=-x^2+17x-66$ $(6 \le x \le 11)$. Say a person with a high-powered rifle is located at $P_0=(2,0)$. Whereabouts on Gage hill is a moose 100% safe?

$y=f(x)=-x^2+17x-66=-(x^2-17x+66)=-(x-11)(x-6)$. It is assumed that a moose is safe if it cannot be seen by the person at P_0. Hence a moose is safe if it is located beyond the point of tangency X of the line l connecting P_0 to the hill where l is tangent to the hill (Fig. VIII.3.9).

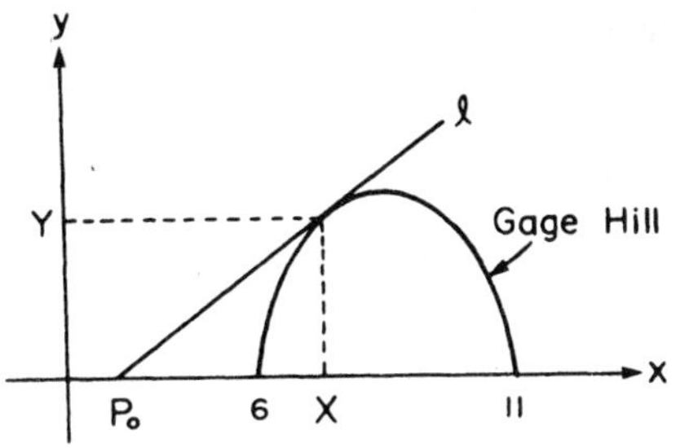

Figure VIII.3.9

$f'(x) = -2x + 17 \Rightarrow f'(X) = -2X + 17$. l has the equation

$$y - Y = (-2X + 17)(x - X) \quad \text{where} \quad Y = -X^2 + 17X - 66.$$

l passes through $(2,0) \Rightarrow (X, Y)$ satisfies the equation $-Y = (-2X+17)(2-X) \leftrightarrow X^2 - 17X + 66 = (2-X)(-2X+17) \leftrightarrow X^2 - 4X - 32 = 0 \leftrightarrow (X-8)(X+4) = 0$. Since $X \geq 6$, one takes $X = 8$. Thus the moose is safe anywhere on Gage hill where $\boxed{x \geq 8}$.

3.10. Determine the maximum value of $y = x^3 - 3x$ on $[-1, 3]$.

The maximum occurs either at a critical point $X \in (-1, 3)$ where $y' = 0$ or at one of the endpoints of the given interval. On the open interval $(-1, 3)$ $y' = 3x^2 - 3 = 0$ when $x = X = 1$. $y(-1) = 2$, $y(1) = -2$, $y(3) = 18$. Hence the maximum value is $\boxed{18}$.

3.11. If $y = 2 + \sin x$ for $0 \leq x \leq 2\pi$, find the absolute maximum and minimum of the function and the points at which they occur.

$y' = \cos x$; $y' = 0 \leftrightarrow \cos x = 0 \leftrightarrow x = \frac{\pi}{2}, \frac{3\pi}{2}$. $y\left(\frac{\pi}{2}\right) = 3$, $y\left(\frac{3\pi}{2}\right) = 1$. At the endpoints: $y(0) = y(2\pi) = 2$. Hence the absolute maximum is $\boxed{3}$ occurring at $\boxed{x = \frac{\pi}{2}}$; the absolute minimum is $\boxed{1}$ at $\boxed{x = \frac{3\pi}{2}}$.

3.12. If $f(x) = \sin x - x\cos x$, show that $0 \leq f(x) \leq \pi$, whenever $0 \leq x \leq \pi$.

$f'(x) = x \sin x \geq 0$ if $0 \leq x \leq \pi$. Hence $f(x)$ is increasing on $[0, \pi]$. $f(0) = 0$, $f(\pi) = \pi$. Thus $0 \leq f(x) \leq \pi$ on $[0, \pi]$.

(*)3.13. Show that the maximum value of $y = a\sin x + b\cos x$ is $\sqrt{a^2 + b^2}$.

$$y' = a\cos x - b\sin x; \qquad y' = 0 \leftrightarrow \sin x = \frac{a}{b}\cos x.$$

Then $\frac{a^2}{b^2}\cos^2 x + \cos^2 x = 1 \leftrightarrow \cos^2 x = \frac{b^2}{a^2 + b^2}$.

Hence at a point where $y' = 0$,

$$\cos x = \frac{\pm b}{\sqrt{a^2+b^2}}, \qquad \sin x = \frac{\pm a}{\sqrt{a^2+b^2}} \text{ and thus}$$

$$y = \frac{\pm(a^2+b^2)}{\sqrt{a^2+b^2}} = \pm\sqrt{a^2+b^2}.$$

Since $f(x) = a\sin x + b\cos x$ is bounded, $\sqrt{a^2+b^2}$ is the maximum value of $f(x)$.

(*)3.14. Let $f(x) = 5x^2 + Ax^{-5}$ for $x > 0$, where A is a positive constant. Find the smallest A such that $f(x) \geq 28$ for all $x > 0$.

$f'(x) = 10x - 5Ax^{-6}$; $f'(x) = 0$ at $x = x^*$, $(x^*)^7 = \frac{A}{2} \leftrightarrow x^* = \left(\frac{A}{2}\right)^{1/7}$. $f(x^*) = 5(x^*)^2 + A(x^*)^{-5} = (x^*)^{-5}[5(x^*)^7 + A] = 7\left(\frac{A}{2}\right)\left(\frac{A}{2}\right)^{-5/7} = 7\left(\frac{A}{2}\right)^{2/7}$. Since $\lim_{x\to 0^+} f(x) = +\infty$ and $\lim_{x\to +\infty} f(x) = +\infty$, $x = x^*$ supplies the minimum value of $f(x)$ for any value of A. Now find the smallest A: $7\left(\frac{A}{2}\right)^{2/7} = 28 \leftrightarrow A = 2(4)^{7/2} = 2^8 = \boxed{256}$.

3.15. What is the absolute minimum and maximum of the function $f(x) = 2x^3 - 3x^2 + 12x$ on the interval $-3 \leq x \leq 3$?

$f'(x) = 6x^2 - 6x + 12 = 6[x^2 - x + 2] = 6\left[\left(x - \frac{1}{2}\right)^2 + \frac{7}{4}\right] > 0$ for all x. Hence the minimum and maximum values of $f(x)$ on $[-3, 3]$ occur at the endpoints. $f(-3) = -117$, $f(3) = 63$, i.e. minimum value is $\boxed{-117}$, maximum value is $\boxed{63}$.

3.16. Determine the absolute maximum and minimum points for the following function or state that they do not exist.

$$f(x) = \frac{x+3}{x}, \qquad -1 \leq x \leq 2.$$

$f(x) = \frac{x+3}{x} = 1 + \frac{3}{x}$. $\lim_{x\to 0^\pm} f(x) = \pm\infty$. Moreover $-1 < 0 < 2$. Hence no absolute maximum or minimum exists for $f(x)$ on the given interval.

4.1. Evaluate $I = \int (7x-1)^{12}\, dx$.

Let $u = 7x - 1$. Then $du = 7\,dx \leftrightarrow dx = \frac{du}{7}$. Hence $I = \frac{1}{7}\int u^{12}\, du = \frac{u^{13}}{(7)(13)} + C = \boxed{\frac{(7x-1)^{13}}{91} + C}$.

4.2. Integrate $I = \int \frac{1}{y\sqrt{y^3-1}}\,dy$.

$I = \int \frac{y^2\,dy}{y^3\sqrt{y^3-1}}$; hence let $u = \sqrt{y^3-1} \Rightarrow u^2 = y^3 - 1 \leftrightarrow y^3 = 1 + u^2$. Thus $2u\,du = 3y^2\,dy \leftrightarrow y^2\,dy = \frac{2}{3}u\,du$. Hence $I = \frac{2}{3}\int \frac{du}{u^2+1} = \frac{2}{3}\arctan u + C$

$= \boxed{\frac{2}{3}\arctan\sqrt{y^3-1} + C}$.

4.3. Integrate $I = \int \frac{dx}{\sqrt{1+e^x}}$.

Let $u = \sqrt{1+e^x} \Rightarrow u^2 = 1 + e^x$. Then $2u\,du = e^x\,dx \leftrightarrow dx = \frac{2u}{u^2-1}du$. Hence $I = 2\int \frac{du}{u^2-1}$. But

$$\frac{1}{u^2-1} = \frac{1}{(u-1)(u+1)} = \frac{\frac{1}{2}}{u-1} - \frac{\frac{1}{2}}{u+1}$$

$$\Rightarrow I = \int \frac{du}{u-1} - \int \frac{du}{u+1} = \log\left|\frac{u-1}{u+1}\right| + C$$

$$= \boxed{\log\left[\frac{\sqrt{1+e^x}-1}{\sqrt{1+e^x}+1}\right] + C}.$$

4.4. Determine $I = \int \frac{x^3+2x^2+x+1}{x^2+2x}\,dx$.

$\frac{x^3+2x^2+x+1}{x^2+2x} = \frac{x^3+2x^2}{x^2+2x} + \frac{x+1}{x^2+2x} = x + \frac{x+1}{x^2+2x}$;

$d(x^2+2x) = 2(x+1)\,dx$. Hence $I = \int x\,dx + \frac{1}{2}\int \frac{d(x^2+2x)}{x^2+2x}$

$= \boxed{\frac{x^2}{2} + \frac{1}{2}\log|x^2+2x| + C}$.

4.5. Evaluate $I = \int \frac{dx}{1+\tan^2 x}$.

$\frac{1}{1+\tan^2 x} = \frac{1}{\sec^2 x} = \cos^2 x = \frac{1}{2}(1+\cos 2x)$. Hence $I = \frac{1}{2}\int [1+\cos 2x]\,dx$

$= \boxed{\frac{x}{2} + \frac{\sin 2x}{4} + C}$.

4.6. Evaluate the integral $I = \int (\sin x)(\sin 3x)\,dx$.

$$\sin A \sin B = \frac{1}{2}[\cos(A-B) - \cos(A+B)]$$

$$\Rightarrow \sin x \sin 3x = \frac{1}{2}[\cos 2x - \cos 4x].$$

Thus $I = \boxed{\frac{1}{4}\sin 2x - \frac{1}{8}\sin 4x + C}$.

4.7. Evaluate the following indefinite integral:

$$I = \int \frac{\sqrt{a^2 - x^2}}{x^4}\,dx.$$

Let $x = a\sin\theta \Rightarrow a^2 - x^2 = a^2\cos^2\theta$, $dx = a\cos\theta\,d\theta$. Hence

$$I = \frac{1}{a^2}\int \frac{\cos^2\theta}{\sin^4\theta}\,d\theta = \frac{1}{a^2}\int \cot^2\theta\csc^2\theta\,d\theta$$

$$= -\frac{1}{a^2}\int \cot^2\theta\,d(\cot\theta) = -\frac{\cot^3\theta}{3a^2} + C$$

$$= -\frac{\cos^3\theta}{3a^2\sin^3\theta} + C = -\frac{1}{3a^2}\left(\frac{a^2 - x^2}{a^2}\right)^{3/2}\frac{a^3}{x^3} + C$$

$$= \boxed{-\frac{(a^2 - x^2)^{3/2}}{3a^2x^3} + C}.$$

4.8. Evaluate: $I = \int \frac{dx}{(x+1)\sqrt{x+6}}$.

Let $u = \sqrt{x+6} \Rightarrow u^2 = x + 6 \Rightarrow 2u\,du = dx$. Hence $I = 2\int \frac{du}{u^2 - 5}$. But

$$\frac{1}{u^2 - 5} = \frac{1}{(u + \sqrt{5})(u - \sqrt{5})}$$

$$= \frac{1}{2\sqrt{5}}\left[\frac{1}{u - \sqrt{5}} - \frac{1}{u + \sqrt{5}}\right] \Rightarrow I = \frac{1}{\sqrt{5}}\log\left|\frac{u - \sqrt{5}}{u + \sqrt{5}}\right| + C$$

$$= \boxed{\frac{1}{\sqrt{5}}\log\left|\frac{\sqrt{x+6} - \sqrt{5}}{\sqrt{x+6} + \sqrt{5}}\right| + C}.$$

(*)**4.9.** Evaluate $I = \int \frac{1}{x^3 - 1}\,dx$.

For all $x \neq 1$, the following relation holds:

$$\frac{1}{x^3-1} = \frac{1}{(x-1)(x^2+x+1)} = \frac{A}{x-1} + \frac{Bx+C}{x^2+x+1} \tag{1}$$

where the constants A, B, C are to be determined. As $x \to 1$, the right hand side (r.h.s.) of equation (1) $\to \frac{A}{x-1}$, the left hand side (l.h.s.) of equation (1) $\to \frac{1}{(1+1+1)(x-1)} = \frac{1}{3(x-1)}$; hence $A = \frac{1}{3}$. As $x \to \infty$, the r.h.s. of equation (1) $\to \frac{A+B}{x}$, the l.h.s. of equation (1) $\to \frac{1}{x^3}$; hence $B + A = 0 \Rightarrow B = -\frac{1}{3}$. At $x = 0$ equation (1) implies $-A + C = -1$; hence $C = A - 1 = -\frac{2}{3}$. Thus $\frac{1}{x^3-1} = \frac{1}{3}\left[\frac{1}{x-1} - \frac{(x+2)}{x^2+x+1}\right]$. Since $x^2+x+1 = \left(x+\frac{1}{2}\right)^2 + \left(\frac{\sqrt{3}}{2}\right)^2$, and $d(x^2+x+1) = (2x+1)\,dx$, it follows that

$$\frac{x+2}{x^2+x+1}dx = \frac{\left\{\frac{1}{2}[2x+1]+\frac{3}{2}\right\}}{x^2+x+1}dx = \frac{1}{2}\frac{d(x^2+x+1)}{x^2+x+1} + \frac{\frac{3}{2}dx}{\left[\left(x+\frac{1}{2}\right)^2+\left(\frac{\sqrt{3}}{2}\right)^2\right]}.$$

Hence

$$I = \frac{1}{3}\int\frac{dx}{x-1} - \frac{1}{6}\int\frac{d(x^2+x+1)}{(x^2+x+1)} - \frac{1}{2}\int\frac{dx}{\left(x+\frac{1}{2}\right)^2+\left(\frac{\sqrt{3}}{2}\right)^2}$$

$$= \boxed{\frac{1}{3}\log|x-1| - \frac{1}{6}\log[x^2+x+1] - \frac{1}{\sqrt{3}}\arctan\left(\frac{2x+1}{\sqrt{3}}\right) + C}.$$

$$\left[\int\frac{du}{u^2+a^2} = \frac{1}{a}\arctan\frac{u}{a} + C\right]$$

4.10. Evaluate

(a) $$I = \int\frac{3x+6x^2}{(x+2)^2(x-1)^2}dx.$$

(b) $$J = \int\frac{x+1}{x^2-2x+10}dx.$$

(a) $$\frac{3x+6x^2}{(x+2)^2(x-1)^2}=\frac{A}{x+2}+\frac{B}{(x+2)^2}+\frac{C}{x-1}+\frac{D}{(x-1)^2}, \qquad (1)$$

for some undetermined constants A, B, C, D. As $x\to-2$, the r.h.s. of equation (1) $\to\dfrac{B}{(x+2)^2}$, the l.h.s. of equation (1) $\to\dfrac{-6+24}{(3)^2(x+2)^2}$ $=\dfrac{2}{(x+2)^2}$. Hence $B=2$. Similarly, $D=1$. As $x\to\infty$, the r.h.s. of equation (1) $\to\dfrac{A+C}{x}$, the l.h.s. of equation (1) $\to\dfrac{6}{x^2}$; hence $A+C=0$. At $x=0$ equation (1) reduces to $\dfrac{A}{2}+\dfrac{B}{4}-C+D=0$; hence $\dfrac{A}{2}-C=-\dfrac{B}{4}-D=$ $-\dfrac{3}{2}$. Thus $\dfrac{3}{2}A=-\dfrac{3}{2}\Rightarrow A=-1\Rightarrow C=1$. Hence

$$I=\int\left[\frac{-1}{x+2}+\frac{2}{(x+2)^2}+\frac{1}{x-1}+\frac{1}{(x-1)^2}\right]dx$$

$$=\boxed{\log\left|\frac{x-1}{x+2}\right|-\frac{2}{x+2}-\frac{1}{x-1}+C}.$$

(b) $x^2-2x+10=(x-1)^2+3^2$; $d(x^2-2x+10)=2(x-1)\,dx$; $\dfrac{x+1}{x^2-2x+10}=\dfrac{1}{2}\dfrac{[2(x-1)]}{x^2-2x+10}+\dfrac{2}{x^2-2x+10}$. Thus

$$J=\int\frac{1}{2}\frac{d(x^2-2x+10)}{x^2-2x+10}+2\int\frac{dx}{(x-1)^2+3^2}$$

$$=\boxed{\log\sqrt{x^2-2x+10}+\frac{2}{3}\arctan\left(\frac{x-1}{3}\right)+C}.$$

4.11. Evaluate $I=\displaystyle\int\frac{3x^2-8x+13}{x^3+x^2-5x+3}dx$.

$$\frac{3x^2-8x+13}{x^3+x^2-5x+3}=\frac{3x^2-8x+13}{(x-1)^2(x+3)}=\frac{A}{x-1}+\frac{B}{(x-1)^2}+\frac{C}{x+3} \qquad (1)$$

for some undetermined constants A, B, C. $B=\lim\limits_{x\to1}\dfrac{3x^2-8x+13}{x+3}=2$. C $=\lim\limits_{x\to-3}\dfrac{3x^2-8x+13}{(x-1)^2}=\dfrac{64}{16}=4$. As $x\to\infty$, the r.h.s. of equation (1) $\to\dfrac{A+C}{x}$, the l.h.s. of equation (1) $\to\dfrac{3}{x}$; hence $A+C=3\Rightarrow A=3-C=$

-1. Thus

$$I=\int\left[\frac{-1}{x-1}+\frac{2}{(x-1)^2}+\frac{4}{x+3}\right]dx$$

$$=\boxed{\log\left[\frac{(x+3)^4}{|x-1|}\right]-\frac{2}{x-1}+C}.$$

(*)4.12. Evaluate $I=\int\frac{e^{4t}}{e^{2t}+3e^t+2}dt$.

Let $u=e^t$; $du=e^t\,dt$. Thus $I=\int\frac{u^3}{u^2+3u+2}du$. $\frac{u^3}{u^2+3u+2}=(u-3)$ $+\frac{7u+6}{u^2+3u+2}$; $\frac{7u+6}{u^2+3u+2}=\frac{7u+6}{(u+1)(u+2)}=\frac{A}{u+1}+\frac{B}{u+2}$ for some undetermined constants A and B. $A=\lim_{u\to-1}\frac{7u+6}{u+2}=-1$; $B=\lim_{u\to-2}\frac{7u+6}{u+1}$ $=8$. Hence

$$I=\int\left[(u-3)-\frac{1}{u+1}+\frac{8}{u+2}\right]du$$

$$=\frac{u^2}{2}-3u-\log|u+1|+8\log|u+2|+C$$

$$=\boxed{\frac{1}{2}e^{2t}-3e^t+8\log(e^t+2)-\log(e^t+1)+C}.$$

(*)4.13. Evaluate $I=\int\frac{dx}{\sin x-\cos x+1}$.

$$\frac{1}{\sin x-\cos x+1}=\frac{\sin x-\cos x-1}{(\sin x-\cos x+1)(\sin x-\cos x-1)}$$

$$=\frac{\sin x-\cos x-1}{\sin^2x+\cos^2x-1-2\sin x\cos x}=\frac{1+\cos x-\sin x}{2\sin x\cos x}$$

$$=\frac{\cos^2x+\sin^2x+\cos x-\sin x}{2\sin x\cos x}=\frac{\cos x(1+\cos x)-\sin x(1-\sin x)}{2\sin x\cos x}$$

$$=\frac{1}{2}\left[\frac{1+\cos x}{\sin x}-\frac{(1-\sin x)}{\cos x}\right]=\frac{1}{2}\left[\frac{(1+\cos x)(1-\cos x)}{\sin x(1-\cos x)}\right.$$

$$\left.-\frac{(1-\sin x)(1+\sin x)}{\cos x(1+\sin x)}\right]=\frac{1}{2}\left[\frac{\sin x}{1-\cos x}-\frac{\cos x}{1+\sin x}\right].$$

Hence

$$I=\frac{1}{2}\int\left[\frac{d(1-\cos x)}{1-\cos x}-\frac{d(1+\sin x)}{1+\sin x}\right]=\boxed{\frac{1}{2}\log\left[\frac{1-\cos x}{1+\sin x}\right]+C}.$$

(An alternative method is to use the substitution $u=\tan\frac{x}{2}$ and thus reduce the integrand to a rational function of u. Then use the method of partial fractions.)

5.1. Evaluate $I=\int_0^{\pi/2}\cos^2 3\theta\,d\theta$.

By comparing areas one sees that $J=\int_0^{\pi/2}\cos^2\theta\,d\theta=\int_0^{\pi/2}\sin^2\theta\,d\theta$ $=\int_{n\pi/2}^{(n+1)\pi/2}\cos^2\theta\,d\theta=\int_{n\pi/2}^{(n+1)\pi/2}\sin^2\theta\,d\theta$ for $n=0,\pm1,\pm2,\ldots$. Since $\cos^2\theta+\sin^2\theta\equiv 1$, $J=\int_0^{\pi/2}(1-\sin^2\theta)\,d\theta=\int_0^{\pi/2}d\theta-J$. Hence $2J=\int_0^{\pi/2}d\theta$; $J=\int_0^{\pi/2}\frac{1}{2}d\theta=\frac{\pi}{4}$. In general $\int_0^{m\pi/2}\cos^2 k\theta\,d\theta=\int_0^{m\pi/2}\sin^2 k\theta\,d\theta=\int_0^{m\pi/2}\frac{1}{2}\,d\theta=\frac{m\pi}{4}$ for any integers k, $m=\pm1,\pm2,\ldots$. Thus $I=\int_0^{\pi/2}\cos^2 3\theta\,d\theta=\int_0^{\pi/2}\sin^2 3\theta\,d\theta=\int_0^{\pi/2}\frac{1}{2}d\theta=\boxed{\frac{\pi}{4}}$.

5.2. Evaluate $\int_{-1}^{1}\sin(x^3)\,dx$.

$f(x)=\sin(x^3)$ is an odd function of x, i.e., $f(-x)=-f(x)$ for all x. Hence $\int_{-1}^{1}f(x)\,dx=\boxed{0}$.

5.3. Evaluate the integral $I=\int_0^{\pi/2}\sin^5\theta\cos^5\theta\,d\theta$.

$\sin\theta\cos\theta=\frac{\sin 2\theta}{2}$. Hence $I=\frac{1}{32}\int_0^{\pi/2}\sin^5 2\theta\,d\theta$ $=\frac{1}{32}\int_0^{\pi/2}(1-\cos^2 2\theta)^2\sin 2\theta\,d\theta$. Let $u=\cos 2\theta$. Then $\theta=0\leftrightarrow u=1$; $\theta=\frac{\pi}{2}$ $\leftrightarrow u=-1$; $2\sin 2\theta\,d\theta=-du$. Thus $I=-\frac{1}{64}\int_1^{-1}(1-u^2)^2\,du=$ $\frac{1}{32}\int_0^1(1-u^2)^2\,du=\frac{1}{32}\int_0^1[1-2u^2+u^4]\,du=\frac{1}{32}\left[u-\frac{2u^3}{3}+\frac{u^5}{5}\right]\Bigg|_{u=0}^{u=1}=$ $\frac{1}{32}\left[1-\frac{2}{3}+\frac{1}{5}\right]=\boxed{\frac{1}{60}}$.

(*)5.4. Find $I=\int_0^{\pi/2}\cos^8\theta\,d\theta$ using the information that $J=\int_0^{\pi/2}\cos^6\theta\,d\theta = \frac{5\pi}{32}$.

$$I=\int_0^{\pi/2}\cos^6\theta(1-\sin^2\theta)\,d\theta = J-K$$

where

$$\begin{aligned}K&=\int_0^{\pi/2}\cos^6\theta\sin^2\theta\,d\theta\\&=\int_0^{\pi/2}\cos^6\theta\sin\theta\sin\theta\,d\theta\\&=-\int_0^{\pi/2}\frac{\sin\theta}{7}\,d\cos^7\theta=-\left.\frac{\sin\theta\cos^7\theta}{7}\right|_{\theta=0}^{\theta=\pi/2}+\frac{I}{7}=\frac{I}{7}.\end{aligned}$$

Hence $I=J-\frac{I}{7}\leftrightarrow\frac{8}{7}I=J;\ I=\left(\frac{7}{8}\right)\left(\frac{5\pi}{32}\right)=\boxed{\frac{35\pi}{256}}$.

5.5. Evaluate $I=\int_{-2}^{3}|1-x^2|\,dx$.

$$|1-x^2|=\begin{cases}1-x^2 & \text{if}\quad x^2<1,\\ x^2-1 & \text{if}\quad x^2\geq 1.\end{cases}$$

Hence

$$\begin{aligned}I&=\int_{-2}^{-1}|1-x^2|\,dx+\int_{-1}^{1}|1-x^2|\,dx+\int_1^3|1-x^2|\,dx\\&=\int_{-2}^{-1}(x^2-1)\,dx+\int_{-1}^{1}(1-x^2)\,dx+\int_1^3(x^2-1)\,dx\\&=\left[\frac{x^3}{3}-x\right]\Bigg|_{x=-2}^{x=-1}+\left[x-\frac{x^3}{3}\right]\Bigg|_{x=-1}^{x=1}+\left[\frac{x^3}{3}-x\right]\Bigg|_{x=1}^{x=3}\\&=\frac{1}{3}(-1+8)-(-1+2)+2\left[1-\frac{1}{3}\right]+\frac{1}{3}(27-1)-2\\&=\boxed{\frac{28}{3}}.\end{aligned}$$

5.6. Compute $I=\int_0^{2\pi}|\cos t|\,dt$.

Graphically one sees that

$$I=4\int_0^{\pi/2}|\cos t|\,dt=4\int_0^{\pi/2}\cos t\,dt=4\sin t\Big|_{t=0}^{t=\pi/2}=\boxed{4}.$$

(*)5.7. Evaluate $I=\int_0^{2\pi}|\sin x-\cos x|\,dx$.

On the interval $[0,2\pi]$,

$$\sin x-\cos x\geq 0\leftrightarrow\begin{cases}\tan x\geq 1 \quad\text{and}\quad \cos x\geq 0\leftrightarrow\dfrac{\pi}{4}\leq x\leq\dfrac{\pi}{2},\\ \tan x\leq 1\quad\text{and}\quad \cos x\leq 0\leftrightarrow\dfrac{\pi}{2}\leq x\leq\dfrac{5\pi}{4}\end{cases}$$

$$\leftrightarrow\frac{\pi}{4}\leq x\leq\frac{5\pi}{4}.$$

But for any a, b,

$$\int_a^b|\sin x-\cos x|\,dx=\int_{a+\pi}^{b+\pi}|\sin x-\cos x|\,dx. \tag{1}$$

Thus $\int_0^{\pi/4}|\sin x-\cos x|\,dx=\int_\pi^{5\pi/4}|\sin x-\cos x|\,dx$ and $\int_{5\pi/4}^{2\pi}|\sin x-\cos x|\,dx=\int_{\pi/4}^{\pi}|\sin x-\cos x|\,dx$. Hence

$$I=2\int_{\pi/4}^{5\pi/4}|\sin x-\cos x|\,dx=2\int_{\pi/4}^{5\pi/4}[\sin x-\cos x]\,dx$$

$$=2[-\cos x-\sin x]\Big|_{x=\pi/4}^{x=5\pi/4}=(2)\left(\frac{4}{\sqrt{2}}\right)=\boxed{4\sqrt{2}}.$$

(To show equation (1), let $x=X-\pi$. Then $\int_a^b|\sin x-\cos x|\,dx$ $=\int_{a+\pi}^{b+\pi}|\sin(X-\pi)-\cos(X-\pi)|\,dX=\int_{a+\pi}^{b+\pi}|-\sin X+\cos X|\,dX=$ $\int_{a+\pi}^{b+\pi}|\sin x-\cos x|\,dx$.)

(*)5.8. Determine $I=\int_2^4\frac{\sqrt{x^2-4}}{x}dx$.

Let $u=\sqrt{x^2-4}\Rightarrow u^2=x^2-4\Rightarrow u\,du=x\,dx$; $x=2\leftrightarrow u=0$; $x=4\leftrightarrow u=2\sqrt{3}$; $\frac{dx}{x}=\frac{u\,du}{x^2}=\frac{u\,du}{u^2+4}$. Hence $I=\int_0^{2\sqrt{3}}\frac{u^2}{u^2+4}du$. $\frac{u^2}{u^2+4}=\frac{u^2+4-4}{u^2+4}$ $=1-\frac{4}{u^2+4}$. Thus $I=\int_0^{2\sqrt{3}}\left[1-\frac{4}{u^2+4}\right]du=2\sqrt{3}-4J$ where J $=\int_0^{2\sqrt{3}}\frac{1}{u^2+4}du$. Let $u=2v$. Then $J=\frac{1}{2}\int_0^{\sqrt{3}}\frac{1}{1+v^2}dv=\frac{\tan^{-1}\sqrt{3}}{2}=$ $\frac{\pi}{6}$. Hence $I=\boxed{2\sqrt{3}-\frac{2\pi}{3}}$. (Alternatively, one could use the trigonometric substitution $x=2\sec\theta$.)

5.9. Evaluate $I = \int_{-2}^{-1}\left(\frac{-3x^2+6x-2}{3x^2-2x}\right)dx.$

$$\frac{-3x^2+6x-2}{3x^2-2x} = \frac{-(3x^2-2x)+4x-2}{3x^2-2x} = -1+\frac{4x-2}{3x^2-2x};$$

$$\frac{4x-2}{3x^2-2x} = \frac{4x-2}{x(3x-2)} = \frac{x+(3x-2)}{x(3x-2)} = \frac{1}{3x-2}+\frac{1}{x}.$$

Hence

$$I = \int_{-2}^{-1}\left[-1+\frac{1}{x}+\frac{\frac{1}{3}}{x-\frac{2}{3}}\right]dx$$

$$= \left[-x+\log|x|+\frac{1}{3}\log\left|x-\frac{2}{3}\right|\right]\Bigg|_{x=-2}^{x=-1}$$

$$= \left(1+\frac{1}{3}\log\frac{5}{3}\right)-\left(2+\log 2+\frac{1}{3}\log\frac{8}{3}\right)$$

$$= \boxed{\frac{1}{3}\log 5-2\log 2-1}.$$

(*)5.10. Say $0 < x_0 < x_1$. Evaluate $I = \int_{x_0}^{x_1}\sqrt{(x-x_0)(x_1-x)}\,dx.$

Let $a = \frac{x_1-x_0}{2} > 0$ and let $u = x - \frac{(x_0+x_1)}{2}$. Note that $\frac{x_0+x_1}{2}$ is the midpoint of the interval of integration and a is the distance from the midpoint to either endpoint of the interval of integration $[x_0, x_1]$.

$x - x_0 = u + a, \quad x - x_1 = u - a; \quad x = x_0 \leftrightarrow u = -a; \quad x = x_1 \leftrightarrow u = a.$

Hence $I = \int_{-a}^{a}\sqrt{(u+a)(a-u)}\,du = \int_{-a}^{a}\sqrt{a^2-u^2}\,du$, the area A under the semi-circle $y = \sqrt{a^2-x^2}$, $-a \le x \le a$ (cf. Fig. VIII.5.10). Thus $I = \frac{\pi a^2}{2}$

$= \boxed{\frac{\pi}{8}(x_1-x_0)^2}$.

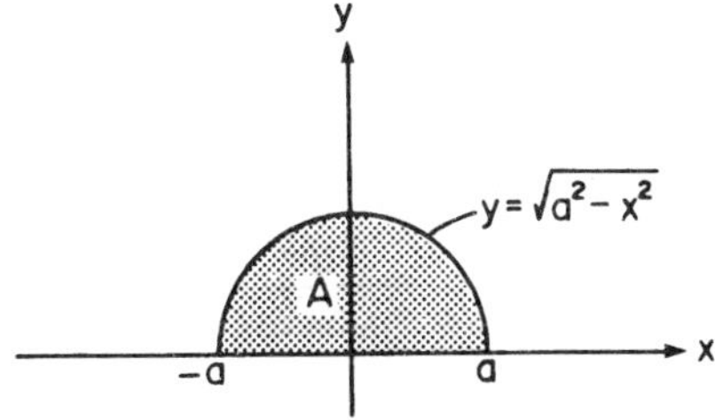

Figure VIII.5.10

(Alternatively, completing the square, $(x-x_0)(x_1-x)=$
$-[x^2-(x_0+x_1)x]-x_0x_1=-\left[x-\left(\frac{x_0+x_1}{2}\right)\right]^2+\left(\frac{x_0+x_1}{2}\right)^2-x_0x_1$
$=\left[\frac{x_0-x_1}{2}\right]^2-\left[x-\left(\frac{x_0+x_1}{2}\right)\right]^2$, etc.)

6.1. Determine whether or not $I=\int_0^{\infty}\frac{dx}{(x+1)^{3/2}}$ converges and find its value if it does.

Let $u=x+1$. Then

$$I=\int_1^{\infty}u^{-3/2}\,du=\lim_{b\to+\infty}\int_1^b u^{-3/2}\,du=\lim_{b\to+\infty}-2u^{-1/2}\Big|_{u=1}^{u=b}$$

$=\lim_{b\to+\infty}\left[2-\frac{2}{\sqrt{b}}\right]=2$. Hence $\boxed{I \text{ converges to the value } 2}$.

6.2. Evaluate $I=\int_{-1}^{1}\frac{dx}{\sqrt{|x|}}$.

I is an improper integral since its integrand $f(x)=\frac{1}{\sqrt{|x|}}$ is singular at $x=0\in[-1,1]$. $f(x)$ is even

$$\Rightarrow I=2\int_0^1\frac{1}{\sqrt{|x|}}dx=2\int_0^1\frac{1}{\sqrt{x}}dx=2\lim_{b\to0^+}\int_b^1\frac{1}{\sqrt{x}}dx$$

$=2\lim_{b\to0^+}2\sqrt{x}\Big|_{x=b}^{x=1}=4\lim_{b\to0^+}[1-\sqrt{b}]=\boxed{4}$, i.e. I converges to 4.

6.3. Determine if the following improper integral converges:

$$I=\int_2^{\infty}\frac{x^2}{\sqrt{x^7+1}}dx.$$

As $x\to+\infty$, $\frac{x^2}{\sqrt{x^7+1}}$ behaves like $\frac{x^2}{x^{7/2}}=\frac{1}{x^{3/2}}$. Hence one expects I to converge. More rigorously, if $x\geq2$, then $\frac{x^2}{\sqrt{x^7+1}}\leq\frac{x^2}{\sqrt{x^7}}=\frac{1}{x^{3/2}}$. Thus $0<I<\int_2^{\infty}\frac{dx}{x^{3/2}}$, a convergent integral $\Rightarrow I$ $\boxed{\text{converges}}$.

(*)6.4. Evaluate the following improper integral:

$$I=\int_2^4\frac{w\,dw}{\sqrt{|w^2-9|}}.$$

$$|w^2-9|=\begin{cases}9-w^2 & \text{if } 2\leq w\leq3,\\ w^2-9 & \text{if } 3\leq w\leq4.\end{cases}$$

Hence $I = J + K$ where $J = \int_2^3 \frac{w\,dw}{\sqrt{9-w^2}}$, $K = \int_3^4 \frac{w\,dw}{\sqrt{w^2-9}}$.

$$J = \lim_{b\to 3^-} \int_2^b \frac{w\,dw}{\sqrt{9-w^2}} = \lim_{b\to 3^-} \left[-(9-w^2)^{1/2}\right]\Big|_{w=2}^{w=b}$$

$$= \lim_{b\to 3^-} (\sqrt{5} - \sqrt{9-b^2}) = \sqrt{5};$$

$$K = \lim_{b\to 3^+} \int_b^4 \frac{w\,dw}{\sqrt{w^2-9}} = \lim_{b\to 3^+} \left[(w^2-9)^{1/2}\right]\Big|_{w=b}^{w=4}$$

$$= \lim_{b\to 3^+} (\sqrt{7} - \sqrt{b^2-9}) = \sqrt{7}.$$

Thus $I = J + K = \boxed{\sqrt{5} + \sqrt{7}}$.

(*)6.5. Test for convergence, giving reasons:

$$I = \int_0^\infty e^{-x} \sin x\,dx.$$

$$|I| = \lim_{b\to+\infty} \left| \int_0^b e^{-x} \sin x\,dx \right| \le \lim_{b\to+\infty} \int_0^b |e^{-x} \sin x|\,dx$$

$$\le \lim_{b\to+\infty} \int_0^b |e^{-x}|\,dx \quad \text{since} \quad |\sin x| \le 1 \quad \text{for each } x.$$

Hence $|I| \le \lim_{b\to+\infty} \int_0^b e^{-x}\,dx = 1 \Rightarrow I$ $\boxed{\text{converges}}$.

(*)6.6. Show whether each of the following integrals converges or diverges. In the case of convergence, give a "good" upper bound for the value of the integral.

(a) $I_1 = \int_0^\infty \frac{1}{\sqrt{x}}\,dx;$

(b) $I_2 = \int_0^\infty \frac{1}{\sqrt{x+x^4}}\,dx.$

(a) I_1 is "doubly improper" since the domain of integration is infinite and its integrand is singular at $x = 0$.

$$I_1 = \lim_{a\to 0^+} \int_a^1 \frac{dx}{\sqrt{x}} + \lim_{b\to+\infty} \int_1^b \frac{dx}{\sqrt{x}}.$$

$\lim_{b\to+\infty} \int_1^b \frac{dx}{\sqrt{x}} = 2 \lim_{b\to+\infty} \sqrt{x}\Big|_{x=1}^{x=b} = +\infty$. Hence I_1 $\boxed{\text{diverges}}$.

(b) For the same reasons as for I_1, I_2 is "doubly improper". As $x \to 0^+$, $f(x) = \frac{1}{\sqrt{x+x^4}} \sim \frac{1}{\sqrt{x}}$; as $x \to +\infty$, $f(x) = \frac{1}{\sqrt{x+x^4}} \sim \frac{1}{\sqrt{x^4}} = \frac{1}{x^2}$. Since $\int_0^1 \frac{1}{\sqrt{x}}$ and $\int_1^\infty \frac{1}{x^2}\,dx$ both converge, one expects I_2 to converge. Clearly

$I_2 > 0$ since $f(x) > 0$ if $x > 0$. If $x \geq 0$: $x + x^4 \geq x$, and $x + x^4 \geq x^4$. Hence $I_2 \leq J + K$ where

$$J = \int_0^1 \frac{1}{\sqrt{x}}\,dx, \qquad K = \int_1^\infty \frac{1}{x^2}\,dx.$$

$$J = \lim_{a \to 0^+} \int_a^1 \frac{1}{\sqrt{x}}\,dx = \lim_{a \to 0^+} 2\sqrt{x}\,\Big|_{x=a}^{x=1} = 2.$$

$$K = \lim_{b \to +\infty} \int_1^b \frac{1}{x^2}\,dx = \lim_{b \to +\infty} -\frac{1}{x}\Big|_{x=1}^{x=b} = 1.$$

Thus $\boxed{0 < I_2 \leq 3}$. (The symbol $\sim$ means *behaves like* or *is asymptotic to*.)

(*)6.7. For a certain real C the improper integral $I = \int_0^\infty \left(\frac{2x}{x^2+1} - \frac{C}{2x+1} \right) dx$ converges. Find C and evaluate the integral. (Note C is a constant.)

$$\begin{aligned} I &= \lim_{b \to +\infty} \int_0^b \left(\frac{2x}{x^2+1} - \frac{C}{2x+1} \right) dx \\ &= \lim_{b \to +\infty} \left[\log(x^2+1) - \frac{C}{2}\log(2x+1) \right] \Big|_{x=0}^{x=b} \\ &= \lim_{b \to +\infty} \log\left[\frac{b^2+1}{(2b+1)^{C/2}} \right] = \log\left[\lim_{b \to +\infty} \frac{b^2+1}{(2b+1)^{C/2}} \right]. \end{aligned}$$

Let

$$\begin{aligned} A &= \lim_{b \to +\infty} \frac{b^2+1}{(2b+1)^{C/2}} = \lim_{b \to +\infty} \left[b^{2-C/2} \frac{(1+b^{-2})}{\left(2+\frac{1}{b}\right)^{C/2}} \right] \\ &= \frac{1}{2^{C/2}} \lim_{b \to +\infty} b^{2-C/2}. \end{aligned}$$

Case I. $2 - \frac{C}{2} > 0 \leftrightarrow C < 4$: $A = +\infty$. Hence I diverges to $+\infty$.

Case II. $C = 4$: $A = \frac{1}{4}$. Hence I $\boxed{\text{converges to } -\log 4}$.

Case III. $C > 4$: $A = 0^+$. Hence I diverges to $-\infty$.

7.1. Find the area A between $y = x^2 - 6x + 8$ and $y = 2x - 7$.

Let $f(x) = x^2 - 6x + 8$, $g(x) = 2x - 7$. The curves $y = f(x)$, $y = g(x)$ (Fig. VIII.7.1) intersect when $f(x) = g(x) \leftrightarrow x^2 - 6x + 8 = 2x - 7 \leftrightarrow x^2 - 8x + 15$

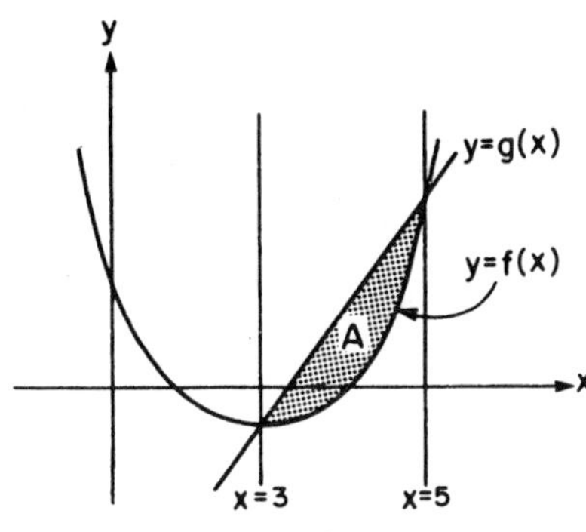

Figure VIII.7.1

$= (x-5)(x-3) = 0 \leftrightarrow x = 3, 5$. Hence $A = \int_3^5 [g(x) - f(x)]dx = -\int_3^5 (x-5)(x-3)\,dx$. Let $u = x - 4$. Then $A = -\int_{-1}^{1} (u^2 - 1)\,du = 2\int_0^1 (1-u^2)\,du = 2\left[u - \frac{u^3}{3}\right]\Big|_{u=0}^{u=1} = \boxed{\frac{4}{3}}$.

(*)7.2. Calculate the area A between the curves $y = x^4 + x^3 + 16x - 4$ and $z = x^4 + 6x^2 + 8x - 4$.

$y - z = x^3 - 6x^2 + 8x = x(x^2 - 6x + 8) = x(x-4)(x-2) = f(x)$. Hence

$$A = \int_0^4 |f(x)|\,dx = A_1 + A_2 \quad \text{where} \quad A_1 = \int_0^2 f(x)\,dx,$$

$A_2 = -\int_2^4 f(x)\,dx$ (Fig. VIII.7.2). Then $A = \int_0^2 (x^3 - 6x^2 + 8x)\,dx$

$$+ \int_2^4 (-x^3 + 6x^2 - 8x)\,dx$$

$$= \left[\frac{x^4}{4} - 2x^3 + 4x^2\right]\Big|_{x=0}^{x=2} + \left[-\frac{x^4}{4} + 2x^3 - 4x^2\right]\Big|_{x=2}^{x=4} = \boxed{8}.$$

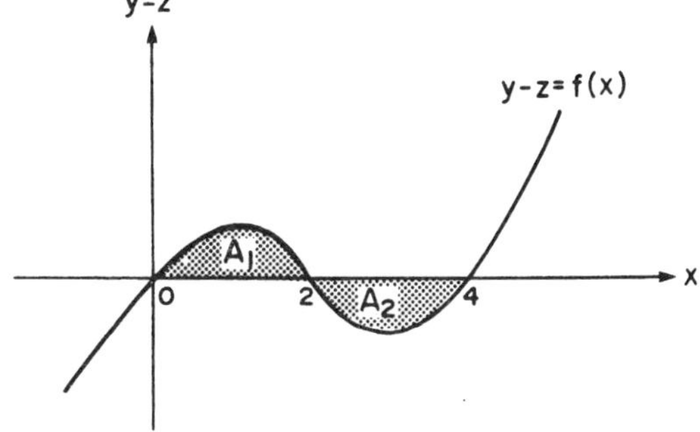

Figure VIII.7.2

7.3. Find the area A bounded by the curves $y^2 = x - 1$ and $y = x - 3$.

$y = x - 3 \leftrightarrow x = f(y) = y + 3$; $y^2 = x - 1 \leftrightarrow x = g(y) = y^2 + 1$ (Fig. VIII.7.3). The curves intersect when $y + 3 = y^2 + 1 \leftrightarrow y^2 - y - 2 = 0 \leftrightarrow (y-2)(y+1) = 0 \leftrightarrow y = -1, 2$. Thus

$$A = \int_{-1}^{2} [f(y) - g(y)]\, dy = \int_{-1}^{2} [y + 2 - y^2]\, dy$$

$$= \left[\frac{y^2}{2} + 2y - \frac{y^3}{3}\right]\Bigg|_{y=-1}^{y=2} = \boxed{\frac{9}{2}}.$$

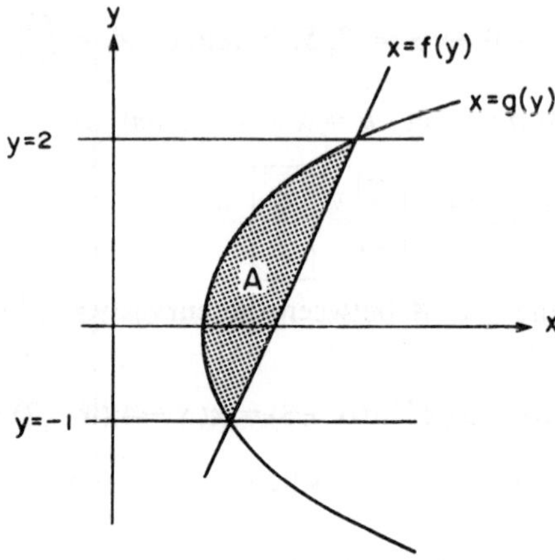

Figure VIII.7.3

7.4. Find the area A of the region bounded by $f(x) = \sin x$, $g(x) = \cos x$ and the lines $x = \frac{\pi}{2}$ and $x = \frac{5\pi}{4}$.

The given curves intersect when $\sin x = \cos x \leftrightarrow \tan x = 1 \leftrightarrow x = \frac{5\pi}{4}$ in the

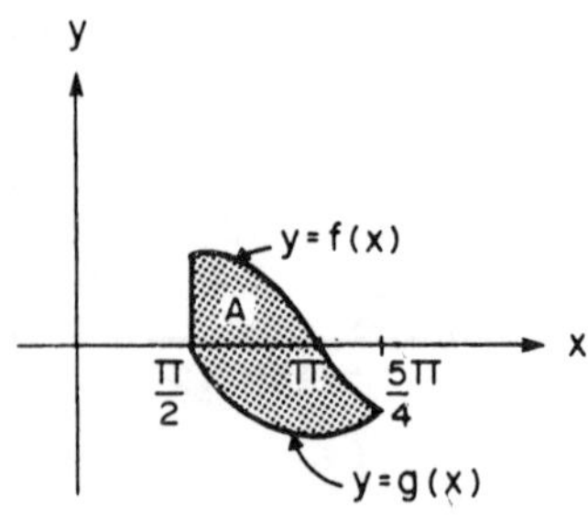

Figure VIII.7.4

interval $\left[\frac{\pi}{2}, \frac{5\pi}{4}\right]$ (Fig. VIII.7.4). Hence

$$A = \int_{\pi/2}^{5\pi/4} [f(x) - g(x)]\,dx = \int_{\pi/2}^{5\pi/4} [\sin x - \cos x]\,dx$$

$$= -[\cos x + \sin x]\Big|_{x=\pi/2}^{x=5\pi/4} = \boxed{1+\sqrt{2}}.$$

7.5. Sketch the plane region consisting of those points whose coordinates (x, y) satisfy the four inequalities

$$x \ge 0, \qquad y \ge 0, \qquad xy \ge 1, \qquad x + y \le \frac{5}{2},$$

and find the area A of the region.

First one should sketch the boundary curves $x = 0$, $y = 0$, $xy = 1\left(y = \frac{1}{x}\right)$, and $x + y = \frac{5}{2}$, in the first quadrant $(x \ge 0, y \ge 0)$ (Fig. VIII.7.5). $y = \frac{1}{x}$ intersects $x + y = \frac{5}{2}$ when $\frac{1}{x} = \frac{5}{2} - x \leftrightarrow x^2 - \frac{5}{2}x + 1 = 0 \leftrightarrow 2x^2 - 5x + 2 = 0 \leftrightarrow (2x-1)(x-2) = 0 \leftrightarrow x = \frac{1}{2}, 2$. Hence

$$A = \int_{1/2}^{2} \left[\frac{5}{2} - x - \frac{1}{x}\right] dx = \left[\frac{5}{2}x - \frac{x^2}{2} - \log x\right]\Bigg|_{x=1/2}^{x=2} = \boxed{\frac{15}{8} - \log 4}.$$

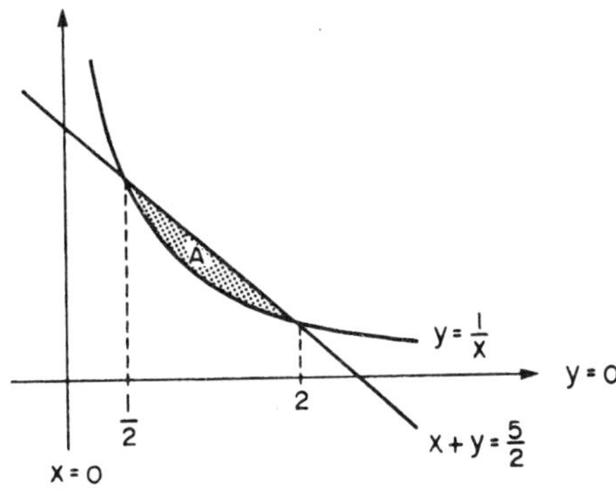

Figure VIII.7.5

7.6. (a) Draw a clear sketch of the graphs of the functions $y = 2^x$ and $y = e^x$, showing their relative positions. Shade the region R bounded by the lines $x = 1$, $x + y = 1$ and the curve $y = 2^x$.

(b) Find the area A of the region R.

(a) $2^x = e^{x\log 2} < e^x$ if $x > 0$ (Fig. VIII.7.6).

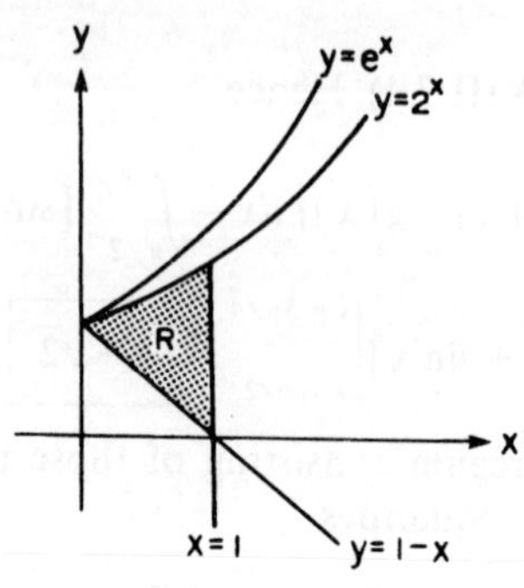

Figure VIII.7.6

(b)
$$A = \int_0^1 [2^x - 1 + x]\,dx = \int_0^1 [e^{x\log 2} - 1 + x]\,dx$$

$$= \left[\frac{2^x}{\log 2} - x + \frac{x^2}{2}\right]\Bigg|_{x=0}^{x=1} = \boxed{\frac{1}{\log 2} - \frac{1}{2}}.$$

(*)7.7. Compute the area A of the shaded region in Fig. VIII.7.7.

$\sin^{-1}x = y \leftrightarrow x = \sin y = f(y)$. Hence

$$A = \int_0^{\pi/2} f(y)\,dy = -\cos y\Big|_{y=0}^{y=\pi/2} = \boxed{1}.$$

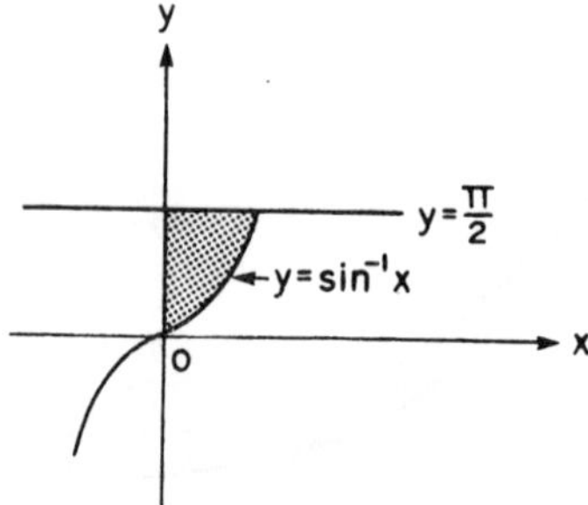

Figure VIII.7.7

7.8. Consider the region bounded by the curves $y = \cos x$, $y = 1$, and $x = \frac{\pi}{2}$.

(a) Write as a definite integral the expression for the volume V of the solid obtained by rotating this region about the x axis

(i) using slices,

(ii) using cylindrical shells.

(b) Calculate the volume using either expression.

Let $g(x) = 1, f(x) = \cos x$; $y = f(x) \leftrightarrow x = f^{-1}(y)$ (Fig. VIII.7.8).

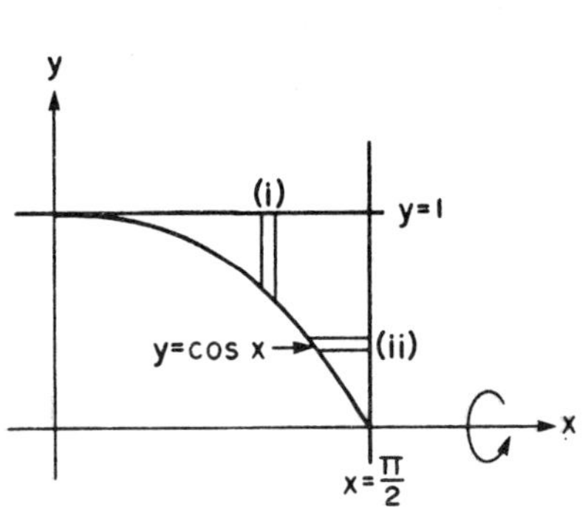

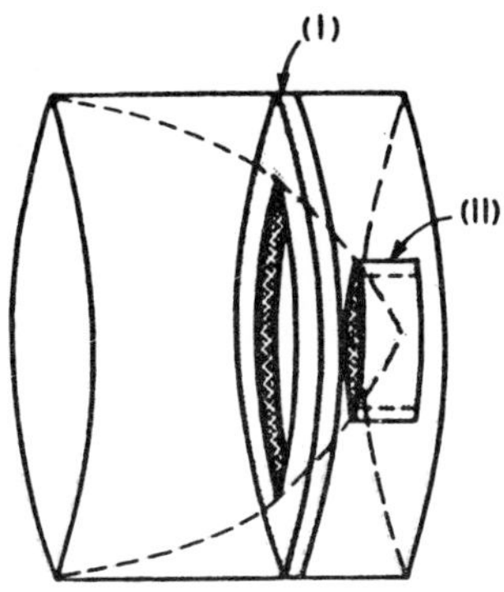

Figure VIII.7.8

(a) (i) The element of volume is $dV = \pi[g^2(x) - f^2(x)]\,dx = \pi[1 - \cos^2 x]\,dx = \pi \sin^2 x\,dx$. Hence

$$V = \pi \int_0^{\pi/2} \sin^2 x\,dx. \tag{1}$$

(ii) $dV = 2\pi y[\pi/2 - x]\,dy = 2\pi y[\pi/2 - f^{-1}(y)]\,dy = 2\pi y \arcsin y\,dy$;

$$V = 2\pi \int_0^1 y \arcsin y\,dy. \tag{2}$$

(b) Using equation (1),

$$V = \pi \int_0^{\pi/2} \frac{1}{2}\,dx = \boxed{\frac{\pi^2}{4}}.$$

Using equation (2),

$$V = 2\pi \int_0^1 \arcsin y\,d\left(\frac{y^2}{2}\right)$$

$$= \pi y^2 \arcsin y \Big|_{y=0}^{y=1} - \pi \int_0^1 \frac{y^2}{\sqrt{1-y^2}}\,dy.$$

Let $y = \sin\theta$. Then $I = \int_0^1 \frac{y^2}{\sqrt{1-y^2}}\,dy = \int_0^{\pi/2} \sin^2\theta\,d\theta = \frac{\pi}{4}$. Thus $V = \pi y^2 \arcsin y \Big|_{y=0}^{y=1} - \pi I = \frac{\pi^2}{2} - \frac{\pi^2}{4} = \boxed{\frac{\pi^2}{4}}$.

7.9. Find the volume V of the solid generated by revolving $y = x^2(2 - x^2)$, $0 \le x \le \sqrt{2}$, about the y-axis.

$$y = f(x) = x^2(2 - x^2) = 2x^2 - x^4.$$

Clearly one wants to use the method of cylindrical shells (Fig. VIII.7.9).

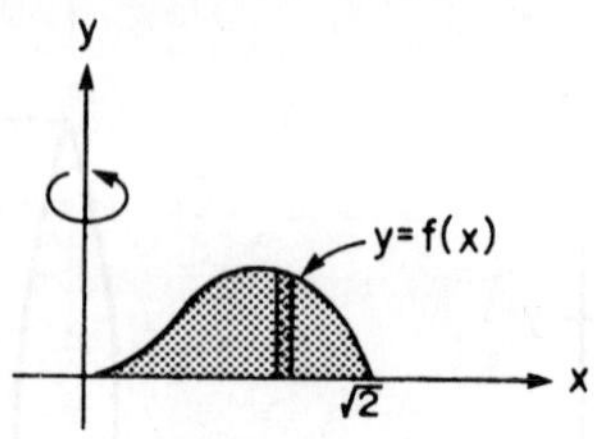

Figure VIII.7.9

Then

$$V = 2\pi \int_0^{\sqrt{2}} x f(x)\, dx = 2\pi \int_0^{\sqrt{2}} [2x^3 - x^5]\, dx$$

$$= 2\pi \left[\frac{x^4}{2} - \frac{x^6}{6} \right] \Bigg|_{x=0}^{x=\sqrt{2}} = \boxed{\frac{4\pi}{3}}.$$

7.10. Let R be the region bounded by the x-axis and $y = \sin x$ for $0 \le x \le \pi$. Find the volume V of the solid generated when R is rotated about
(a) the x-axis;
(b) the y-axis.

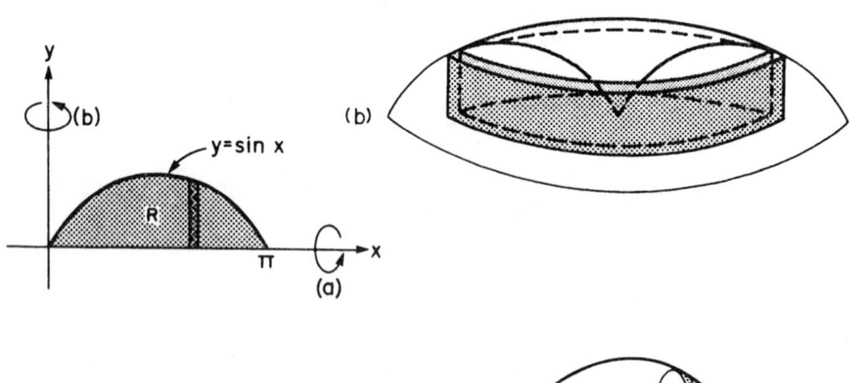

Figure VIII.7.10

(a) The use of slices is convenient (Fig. VIII.7.10).

$$V = \pi \int_0^{\pi} \sin^2 x \, dx = \pi \int_0^{\pi} \frac{1}{2} dx = \boxed{\frac{\pi^2}{2}}.$$

(b) Here the use of cylindrical shells is convenient (Fig. VIII.7.10).

$$V = 2\pi \int_0^{\pi} x \sin x \, dx = -2\pi \int_0^{\pi} x \, d(\cos x)$$

$$= 2\pi[-x\cos x + \sin x]\Big|_{x=0}^{x=\pi} = \boxed{2\pi^2}.$$

7.11. Find the volume V of the solid generated by rotating the triangle bounded by the lines $x = 0$, $y = 0$ and $2x + y = 2$ about the line $x = -1$.

Using cylindrical shells (Fig. VIII.7.11), the volume element is $dV = 2\pi(x - (-1))y\,dx = 2\pi(x+1)y\,dx = 4\pi(1+x)(1-x)\,dx = 4\pi(1-x^2)\,dx$.

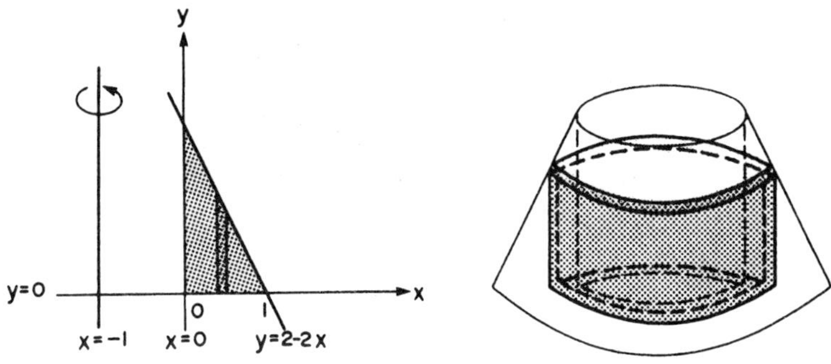

Figure VIII.7.11

Hence $$V = 4\pi \int_0^1 (1-x^2)\,dx = 4\pi\left[1 - \frac{1}{3}\right] = \boxed{\frac{8\pi}{3}}.$$

(*)7.12. Find the volume V of the solid generated by revolving around the line $x = -\frac{\pi}{4}$ the area bounded by $x = -\frac{\pi}{4}$, the two curves $y = \cos x$ and $y = \sin x$, and by $x = \frac{3\pi}{4}$.

Let $f(x) = \cos x$, $g(x) = \sin x$. Using cylindrical shells (Fig. VIII.7.12), the volume element $dV = 2\pi\left(x + \frac{\pi}{4}\right)|f(x) - g(x)|\,dx$;

$$V = 2\pi \int_{-\pi/4}^{3\pi/4} \left(x + \frac{\pi}{4}\right)|f(x) - g(x)|\,dx$$

$$= 2\pi \int_{-\pi/4}^{\pi/4} \left(x + \frac{\pi}{4}\right)(\cos x - \sin x)\,dx - 2\pi \int_{\pi/4}^{3\pi/4} \left(x + \frac{\pi}{4}\right)(\cos x - \sin x)\,dx.$$

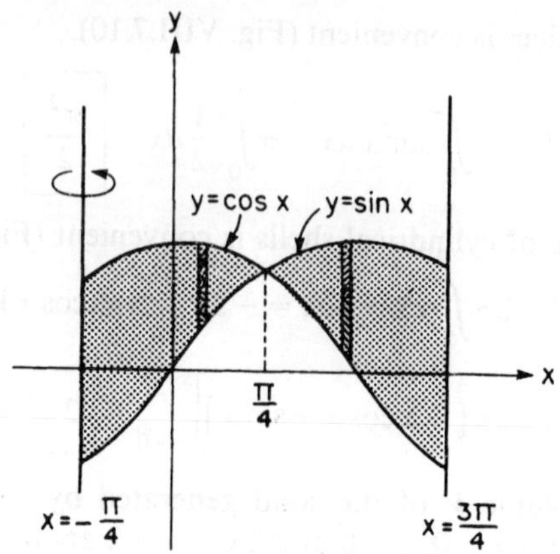

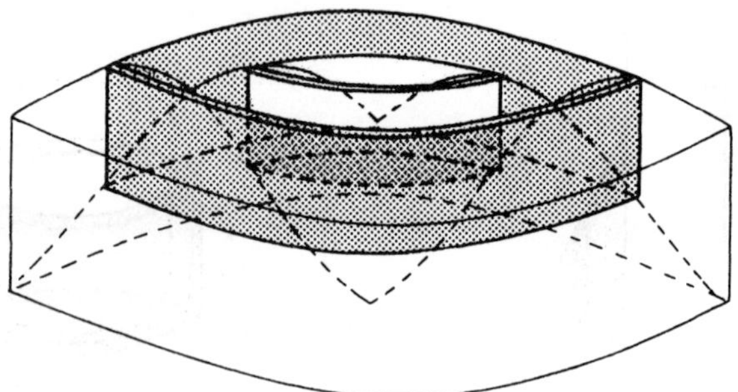

Figure VIII.7.12

In the second integral let $X = x - \frac{\pi}{2}$. Then

$$V = 2\pi \int_{-\pi/4}^{\pi/4} \left[\left(x + \frac{\pi}{4} \right)(\cos x - \sin x) + \left(x + \frac{3\pi}{4} \right)(\cos x + \sin x) \right] dx$$

$$= 2\pi \int_{-\pi/4}^{\pi/4} \pi \cos x \, dx = 2\pi^2 \sin x \Big|_{x=-\pi/4}^{x=\pi/4} = \boxed{2\sqrt{2}\,\pi^2}.$$

(Note that $\int_{-\pi/4}^{\pi/4} F(x)\,dx = 0$ if $F(x)$ is an odd function of x, i.e., if $F(-x) = -F(x)$. In the computation of V, $x \cos x$ and $\sin x$ are odd functions of x, obtained after terms are collected in the integrand.)

7.13. Find the value of $a > 0$ such that when the area bounded by the curve $y = 1 + \sqrt{x}\, e^{x^2} = f(x)$, the line $y = 1$ and the line $x = a$ is rotated about the line $y = 1$, a volume of 2π is generated.

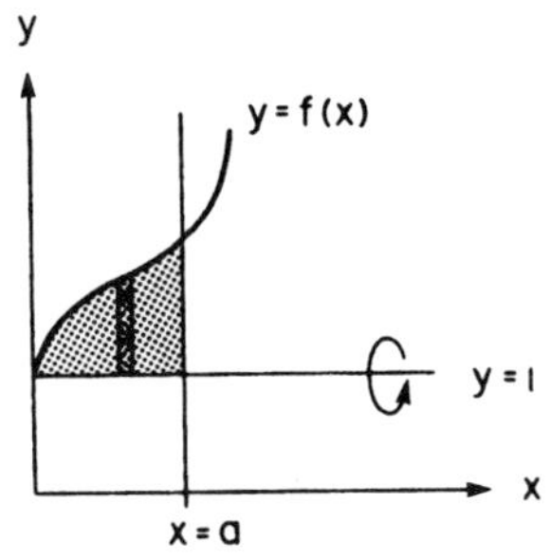

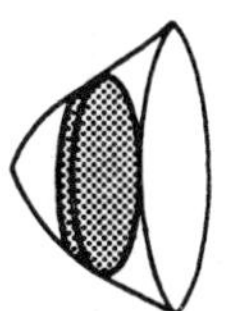

Figure VIII.7.13

Using slices (Fig. VIII.7.13), the volume of

$$2\pi = \pi\int_0^a [f(x)-1]^2\,dx = \pi\int_0^a xe^{2x^2}\,dx.$$

Let

$$u = 2x^2. \text{ Then } 2 = \frac{1}{4}\int_0^{2a^2} e^u\,du \leftrightarrow 8 = e^u\Big|_{u=0}^{u=2a^2} = e^{2a^2}-1 \leftrightarrow e^{2a^2} = 9$$

$$\leftrightarrow 2a^2 = \log 9 \leftrightarrow a = \sqrt{\frac{\log 9}{2}} = \boxed{\sqrt{\log 3}}.$$

7.14. Find the volume V of the region obtained by rotating about the x-axis the graph of $y = \dfrac{1}{x}$ between $x = 1$ and $x = \infty$.

Using slices (Fig. VIII.7.14),

$$V = \pi\int_1^{\infty}\frac{dx}{x^2} = \pi\lim_{b\to\infty}\left[-\frac{1}{x}\right]\Bigg|_{x=1}^{x=b}$$

$$= \boxed{\pi}.$$

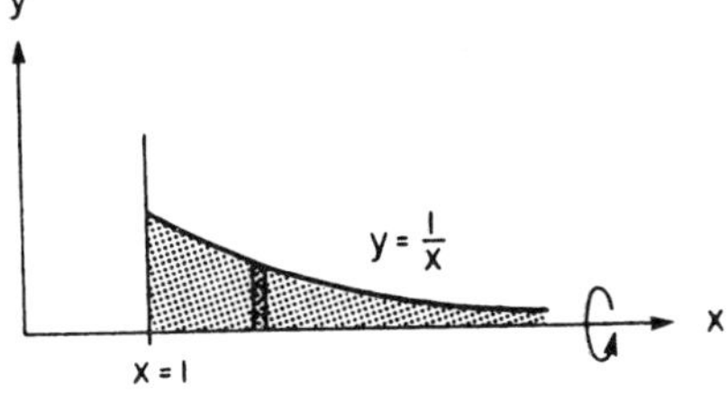

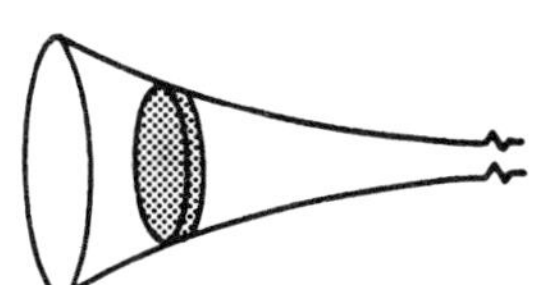

Figure VIII.7.14

7.15. The shaded region $OABCD$ has boundaries $g(x) = \frac{1}{4}x^{3/2}$, $f(x) = \frac{3}{4}x - 1$, $y = 0$ and $x = 8$ (Fig. VIII.7.15). Set up the necessary integrals and

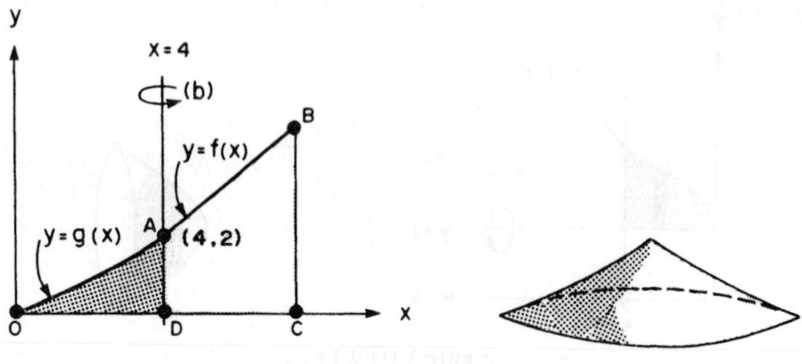

Figure VIII.7.15

expressions in terms of a single variable which give the following values:

(a) The area S of the shaded region.

(b) The volume V generated when the area OAD is revolved about the line $x = 4$.

(c) The perimeter P of the shaded region.

(a) In Fig. VIII.7.15: $D \leftrightarrow (4,0)$, $C \leftrightarrow (8,0)$, $B \leftrightarrow (8,5)$. If $0 \le x \le 4$ then the curve from O to $A \leftrightarrow y = g(x)$; if $4 \le x \le 8$ then the curve from A to $B \leftrightarrow y = f(x)$.

$$S = \int_0^4 g(x)\,dx + \int_4^8 f(x)\,dx = \boxed{\int_0^4 \frac{1}{4}x^{3/2}\,dx + \int_4^8 \left(\frac{3}{4}x - 1\right)dx}.$$

(b) Using cylindrical shells (Fig. VIII.7.15),

$$V = 2\pi \int_0^4 |x-4| g(x)\,dx$$

$$= \boxed{\frac{\pi}{2}\int_0^4 (4-x)x^{3/2}\,dx}.$$

(c) $P = l_1 + l_2 + l_3 + l_4$ where:

l_1 is the length of $OC = 8$;

l_2 is the length of $BC = 5$;

l_3 is the length of $AB = \sqrt{4^2 + 3^2} = 5$;

l_4 is the length of the curve from O to $A = \int_0^4 \sqrt{1 + [g'(x)]^2}\,dx$.

$$g'(x) = \frac{3}{8}x^{1/2}. \text{ Hence } P = \boxed{18 + \int_0^4 \sqrt{1 + \frac{9}{64}x}\,dx}.$$

(*)7.16. Consider a sphere of radius a centred at the origin. Find the surface area S of that portion of the sphere which lies between two parallel planes at distances d_1 and d_2 from the centre where $0 \le d_1 \le d_2 < a$.

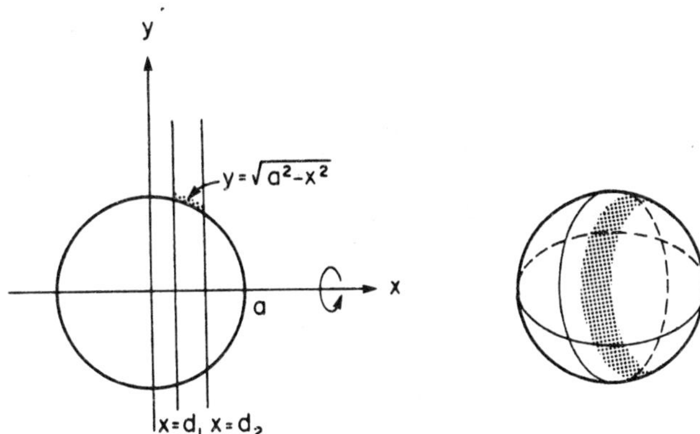

Figure VIII.7.16

S can be generated by rotating $y = \sqrt{a^2 - x^2}$, $d_1 \le x \le d_2$, about the x-axis (Fig. VIII.7.16). The surface area element is $dS = 2\pi y dl$ where the length element of the curve to be rotated is $dl = \sqrt{1 + \left(\frac{dy}{dx}\right)^2} dx$; $\left|\frac{dy}{dx}\right| = \frac{x}{\sqrt{a^2 - x^2}}$ $\Rightarrow dl = \frac{a}{\sqrt{a^2 - x^2}} dx$. Hence $S = 2\pi a \int_{d_1}^{d_2} dx = \boxed{2\pi a(d_2 - d_1)}$.

8.1. Obtain a parametric representation for $2x + 3y = 5$.

Let x be the parameter t, $-\infty < t < \infty$. Then $y = \frac{1}{3}(5 - 2x) = \frac{1}{3}(5 - 2t)$ (Fig. VIII.8.1). Hence a parametric representation of the given line is

$$\left\{ \begin{array}{l} x = t \\ y = \frac{1}{3}(5 - 2t) \end{array} \right\}, \qquad -\infty < t < \infty.$$

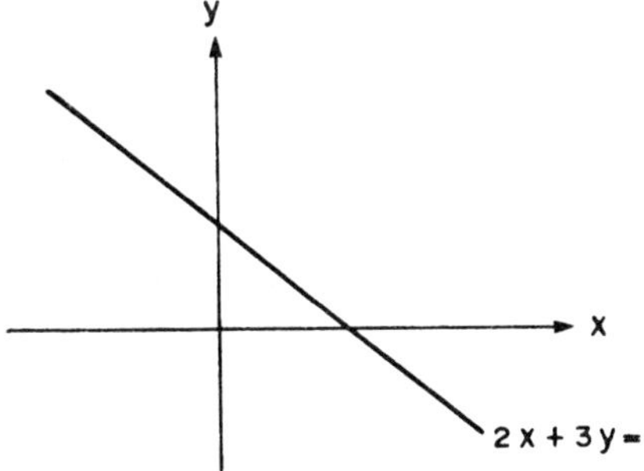

Figure VIII.8.1

8.2. Parametrize the circle Γ given by the equation $x^2 + y^2 - ax = 0$.

$x^2 + y^2 - ax = \left(x - \frac{a}{2}\right)^2 + y^2 - \frac{a^2}{4}$. Hence $\Gamma \leftrightarrow \left(x - \frac{a}{2}\right)^2 + y^2 = \frac{a^2}{4}$, a

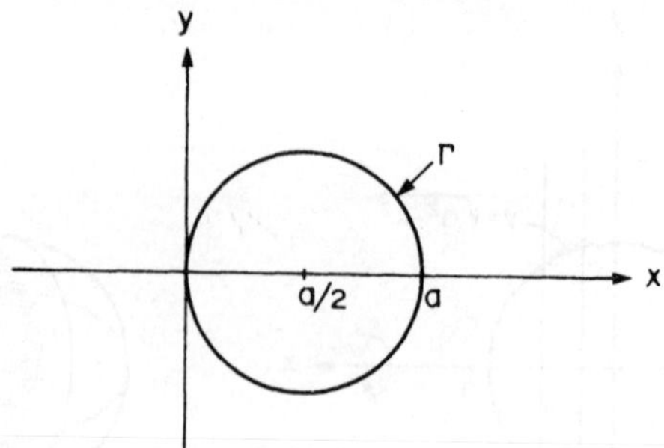

Figure VIII.8.2

circle of radius $\frac{a}{2}$, centred at $\left(\frac{a}{2},0\right)$ (Fig. VIII.8.2).

Let $x-\frac{a}{2}=\frac{a}{2}\cos t$, $y=\frac{a}{2}\sin t$, $0\le t<2\pi$. Then Γ is parametrized by

$$\left\{\begin{array}{l} x=\frac{a}{2}(1+\cos t) \\ y=\frac{a}{2}\sin t \end{array}\right\},\qquad 0\le t<2\pi.$$

8.3. Describe and sketch the graph of the curve C whose parametric equations are

$$x=2t,\qquad y=1-t^2\quad\text{where}\quad -1\le t\le 1.$$

Since $t\in[-1,1]$, $-2\le x\le 2$; $t=\frac{x}{2}$, and thus $y=1-\frac{x^2}{4}$. Hence C is the portion of the parabola $y-1=-\frac{x^2}{4}$ where $-2\le x\le 2$. The arrow in Fig. VIII.8.3 points in the direction of increasing t.

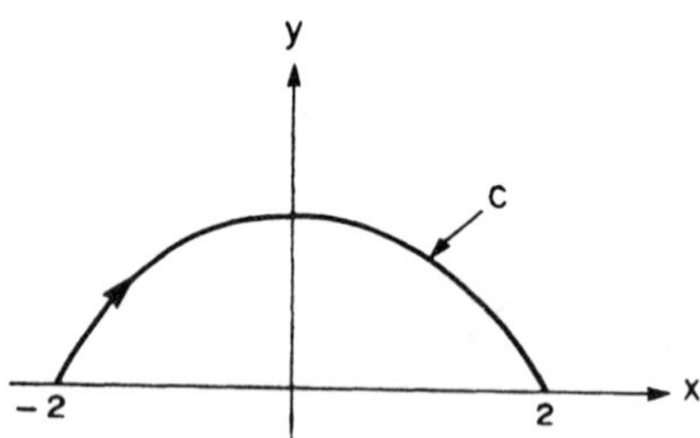

Figure VIII.8.3

8.4. The equations of a certain curve are given parametrically as: $x=2t-1$, $y=4t^2-2t$. Write the equation of this same curve in rectangular coordinates; that is, eliminate the parameter t and write the equation in the form $y=f(x)$. Use this equation to compute $\frac{dy}{dx}$ at $(1,2)$ and hence find the equation of the tangent to the curve at the point $(1,2)$.

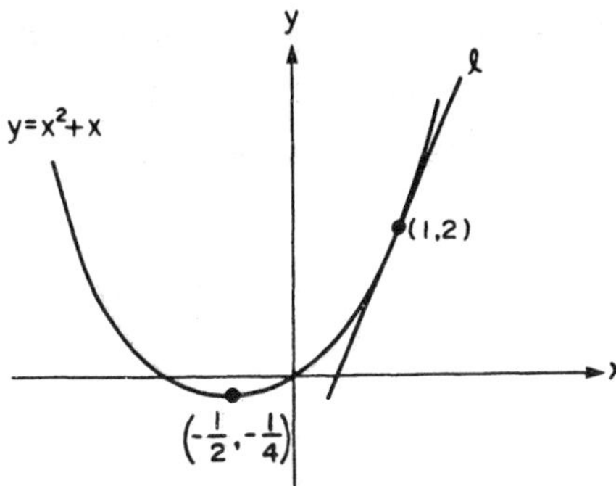

Figure VIII.8.4

It is understood that $-\infty < t < \infty$. $x = 2t - 1 \leftrightarrow 2t = x + 1 \leftrightarrow t = \dfrac{x+1}{2}$; $y = 4t^2 - 2t = (x+1)^2 - (x+1) = x^2 + x = \left(x + \dfrac{1}{2}\right)^2 - \dfrac{1}{4}$. The curve is a parabola with vertex at $\left(-\dfrac{1}{2}, -\dfrac{1}{4}\right)$ (Fig. VIII.8.4).

$$y = f(x) = x^2 + x;\ \frac{dy}{dx} = \boxed{2x+1}.\ \text{At } (1,2)\text{: } \frac{dy}{dx} = \boxed{3}.$$

The tangent to the curve at (1,2) is the line l with the equation $\dfrac{y-2}{x-1} = 3$ $\leftrightarrow \boxed{y = 3x - 1}$.

8.5. Find the slope of the cycloid $x = 2(t - \sin t)$, $y = 2(1 - \cos t)$ at the point where $t = \dfrac{\pi}{2}$.

$$t = \frac{\pi}{2} \leftrightarrow (x, y) = (\pi - 2, 2).$$

The slope of the cycloid (Fig. VIII.8.5) at any point is

$$\frac{dy}{dx} = \frac{\dfrac{dy}{dt}}{\dfrac{dx}{dt}} = \frac{2\sin t}{2(1-\cos t)} = \frac{\sin t}{1 - \cos t}.$$

At $t = \dfrac{\pi}{2}$: $\dfrac{dy}{dx} = \boxed{1}$.

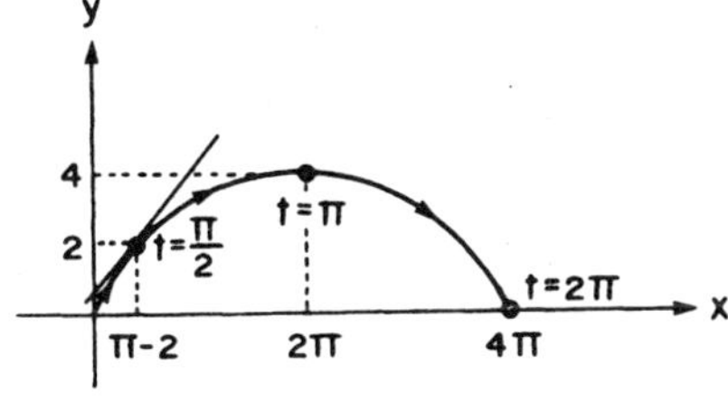

Figure VIII.8.5. Graph of cycloid for $0 \le t \le 2\pi$.

8.6. (a) Find the coordinates of the point P on the path C_1:

$$x = 4\cos t, \qquad y = 4\sin t$$

which corresponds to $t = \pi/6$ and show that P also lies on the path C_2:

$$x = \sqrt{6}\,u, \qquad y = u^2.$$

(b) Find the tangent vectors to each of the two paths at the point P and draw a figure showing the two paths, the point P and arrows representing the tangent vectors.

(a) At $t = \frac{\pi}{6}$: $x = 4\cos t = 4\cos\frac{\pi}{6} = \boxed{2\sqrt{3}}$, $y = 4\sin\frac{\pi}{6} = \boxed{2}$. Thus at $t = \frac{\pi}{6}$: $y = \frac{x^2}{6}$. If a point lies on the path $x = \sqrt{6}\,u$, $y = u^2$, then $y = \left(\frac{x}{\sqrt{6}}\right)^2 = \frac{x^2}{6}$. Hence P lies on both paths (Fig. VIII.8.6).

(b) For the first path C_1, $\frac{dy}{dx} = \frac{\frac{dy}{dt}}{\frac{dx}{dt}} = \frac{4\cos t}{-4\sin t} = -\cot t$; $\frac{dy}{dx} = -\sqrt{3}$ at $t = \frac{\pi}{6}$. For the path C_2, $\frac{dy}{dx} = \frac{x}{3}$; $t = \frac{\pi}{6} \leftrightarrow x = 2\sqrt{3}$ for C_1. Hence at $x = 2\sqrt{3}$: $\frac{dy}{dx} = \frac{2}{\sqrt{3}}$ for C_2. $C_1 \leftrightarrow$ circle $x^2 + y^2 = 16$; $C_2 \leftrightarrow$ parabola $y = \frac{x^2}{6}$.

In Fig. VIII.8.6, $\vec{v}_1 \leftrightarrow$ tangent vector to C_1 at P; $\vec{v}_2 \leftrightarrow$ tangent vector to C_2 at P.

$$\boxed{\vec{v}_1 = \mathbf{i} - \sqrt{3}\,\mathbf{j}}, \quad \boxed{\vec{v}_2 = \mathbf{i} + \frac{2}{\sqrt{3}}\mathbf{j}}.$$

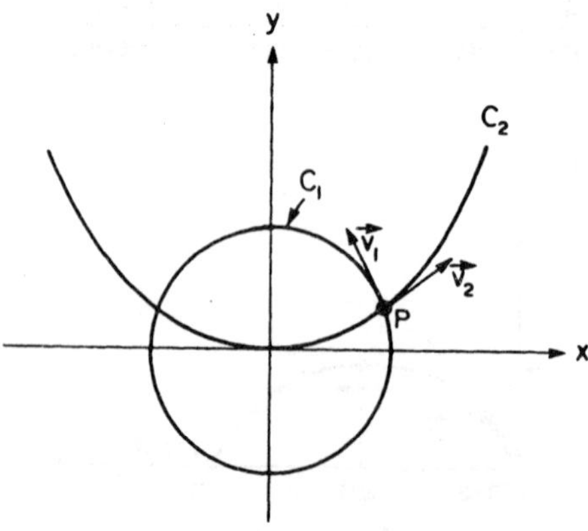

Figure VIII.8.6

8.7. Find $\dfrac{d^2y}{dx^2}$ if $\left\{\begin{matrix} x = t^4 + t \\ y = t^3 - t \end{matrix}\right\}$.

$$\frac{dy}{dx} = \frac{\dfrac{dy}{dt}}{\dfrac{dx}{dt}}; \qquad \frac{d^2y}{dx^2} = \frac{\dfrac{d}{dt}\left(\dfrac{dy}{dx}\right)}{\dfrac{dx}{dt}} = \left(\frac{dx}{dt}\right)^{-1}\left[\frac{\dfrac{d^2y}{dt^2}}{\dfrac{dx}{dt}} - \frac{\dfrac{dy}{dt}\dfrac{d^2x}{dt^2}}{\left(\dfrac{dx}{dt}\right)^2}\right].$$

Let $\dot{x} = \dfrac{dx}{dt}$, $\dot{y} = \dfrac{dy}{dt}$, $\ddot{x} = \dfrac{d^2x}{dt^2}$, $\ddot{y} = \dfrac{d^2y}{dt^2}$. Then $\dfrac{d^2y}{dx^2} = \dfrac{\dot{x}\ddot{y} - \dot{y}\ddot{x}}{(\dot{x})^3}$. For this problem: $\dot{x} = 4t^3 + 1$, $\dot{y} = 3t^2 - 1$; $\ddot{x} = 12t^2$, $\ddot{y} = 6t$. Hence

$$\frac{d^2y}{dx^2} = \frac{(4t^3+1)(6t) - (3t^2-1)(12t^2)}{(4t^3+1)^3}$$

$$= \boxed{\frac{6[-2t^4 + 2t^2 + t]}{(4t^3+1)^3}}.$$

8.8. Find the area A bounded by the x-axis, the line $x = 1$, and the curve having parametric equations

$$x = \log t, \qquad y = \frac{t - t^{-1}}{2}.$$

Clearly $t > 0$. $t = e^x$, $y = \dfrac{e^x - e^{-x}}{2} = \sinh x$, $-\infty < x < \infty$.

$$A = \int_0^1 y\,dx \text{ (Fig. VIII.8.8).}$$

Method 1 (Parametrization method).

$$x = 0 \leftrightarrow t = 1, \qquad x = 1 \leftrightarrow t = e.$$

$$A = \int_0^1 y\,dx = \int_1^e y(t)\frac{dx}{dt}\,dt = \int_1^e \left(\frac{t - t^{-1}}{2}\right)\frac{1}{t}\,dt$$

$$= \frac{1}{2}\int_1^e [1 - t^{-2}]\,dt = \frac{1}{2}[t + t^{-1}]\Big|_{t=1}^{t=e}$$

$$= \frac{e + e^{-1}}{2} - 1 = \boxed{\cosh 1 - 1}.$$

Method 2 (Substitution method where $y = f(x) = \sinh x$).

$$A = \int_0^1 \sinh x\,dx = \cosh x\Big|_{x=0}^{x=1} = \boxed{\cosh 1 - 1}.$$

(Note that usually one is not able to find $y = f(x)$ explicitly, as in this problem.)

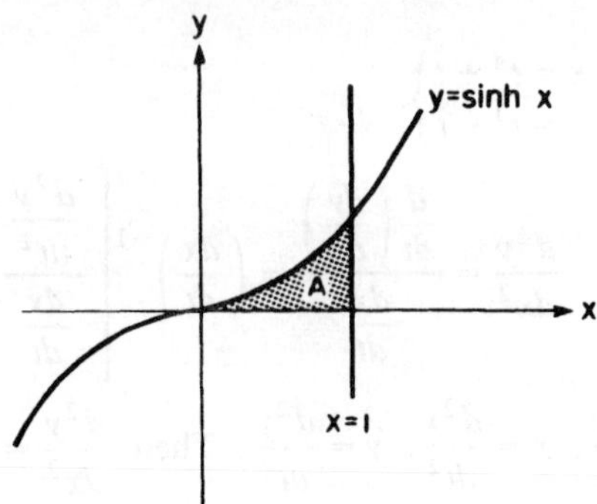

Figure VIII.8.8

8.9. Find the length l of the curve given parametrically by

$$\begin{cases} x = t^2 \\ y = t^3 \end{cases}, \quad 0 \le t \le 1.$$

$$l = \int_0^1 \sqrt{\left(\frac{dx}{dt}\right)^2 + \left(\frac{dy}{dt}\right)^2}\, dt; \qquad \frac{dx}{dt} = 2t, \qquad \frac{dy}{dt} = 3t^2.$$

Hence $l = \int_0^1 \sqrt{4t^2 + 9t^4}\, dt = \int_0^1 t\sqrt{4 + 9t^2}\, dt$. Let $u = \sqrt{4 + 9t^2}$. Then $u^2 = 4 + 9t^2 \Rightarrow t\, dt = \frac{udu}{9}$. $t = 0 \leftrightarrow u = 2$, $t = 1 \leftrightarrow u = \sqrt{13}$. Thus

$$l = \frac{1}{9}\int_2^{\sqrt{13}} u^2\, du$$

$$= \frac{u^3}{27}\Bigg|_{u=2}^{u=\sqrt{13}} = \boxed{\frac{13\sqrt{13} - 8}{27}}.$$

8.10. Sketch the arc of the curve

$$\begin{cases} x = a\cos^3 t \\ y = a\sin^3 t \end{cases} \quad \text{(where } a \text{ is a positive number)}$$

that lies in the interval $0 \le t \le \frac{\pi}{2}$, and calculate the length l of this arc.

$x^{2/3} + y^{2/3} = a^{2/3}$. As t runs from 0 to $\frac{\pi}{2}$, y runs from 0 to a and x runs from a to 0 (Fig. VIII.8.10.). $\frac{dx}{dt} = -3a\cos^2 t\sin t$, $\frac{dy}{dt} = 3a\sin^2 t\cos t$. Hence

$$l = \int_0^{\pi/2} \sqrt{\left(\frac{dx}{dt}\right)^2 + \left(\frac{dy}{dt}\right)^2}\, dt = 3a\int_0^{\pi/2} \sqrt{\cos^4 t\sin^2 t + \sin^4 t\cos^2 t}\, dt$$

$$= 3a\int_0^{\pi/2} \cos t \sin t\, dt = 3a\int_{t=0}^{t=\pi/2} \sin t\, d(\sin t)$$

$$= \frac{3a}{2}\sin^2 t\Bigg|_{t=0}^{t=\pi/2} = \boxed{\frac{3a}{2}}.$$

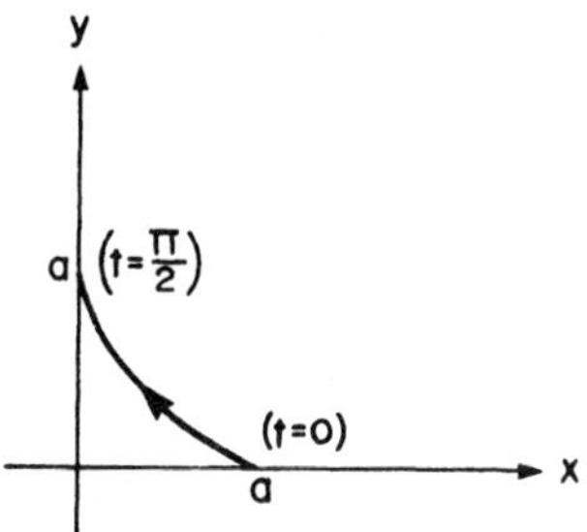

Figure VIII.8.10

8.11. (a) Set up an integral that gives the perimeter l of the ellipse given by the parametric equations

$$\begin{cases} x = 2\cos t, \\ y = 3\sin t. \end{cases}$$

(b) Prove that this length is less than 6π.

(a) Since t runs from 0 to 2π,

$$l = \int_0^{2\pi} \sqrt{\left(\frac{dx}{dt}\right)^2 + \left(\frac{dy}{dt}\right)^2}\, dt = \boxed{\int_0^{2\pi} \sqrt{4\sin^2 t + 9\cos^2 t}\, dt}$$

(an elliptic integral).

(b) $4\sin^2 t \le 9\sin^2 t$. Hence $l \le \int_0^{2\pi} \sqrt{9(\sin^2 t + \cos^2 t)}\, dt = 3\int_0^{2\pi} dt = 6\pi$. (Note that 6π is the circumference of a circle of radius 3.)

9.1. Find the polar coordinate equation corresponding to the equation

$$3x - 5y = 7.$$

In polar coordinates: $x = r\cos\theta$, $y = r\sin\theta$. Thus $3x - 5y = 7 \leftrightarrow 3r\cos\theta - 5r\sin\theta = 7 \leftrightarrow \boxed{r = \frac{7}{3\cos\theta - 5\sin\theta}}$.

9.2. Express the equation $r = \cos\theta + \sin\theta$ in cartesian coordinates and identify the locus.

$$r = \cos\theta + \sin\theta \leftrightarrow r^2 = r\cos\theta + r\sin\theta \leftrightarrow x^2 + y^2 = x + y$$

$$\leftrightarrow \boxed{\left(x - \frac{1}{2}\right)^2 + \left(y - \frac{1}{2}\right)^2 = \left(\frac{1}{\sqrt{2}}\right)^2}$$

$$\leftrightarrow \text{circle of radius } \boxed{\frac{1}{\sqrt{2}}}, \text{ centred at } \boxed{\left(\frac{1}{2}, \frac{1}{2}\right)}.$$

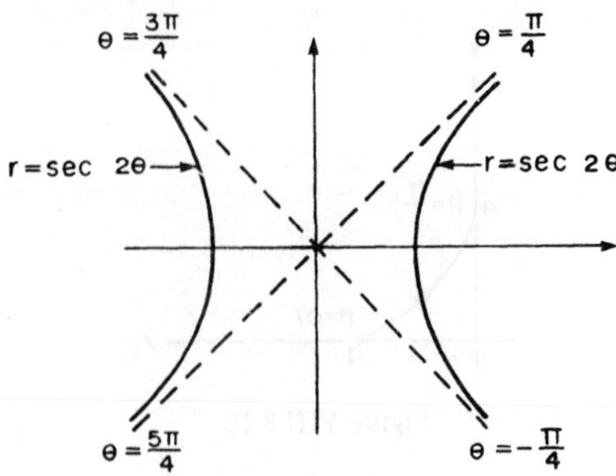

Figure VIII.9.3

9.3. If r and θ are polar coordinates, sketch a graph of the curve $r = \sec 2\theta$.

$$r = f(\theta) = \sec 2\theta; \qquad f(\theta + \pi) = f(\theta), \qquad f(-\theta) = f(\theta).$$

θ	0	$\frac{\pi}{8}$	$\frac{\pi}{4}$
r	1	$\sqrt{2}$	$+\infty$

$r \geq 0 \leftrightarrow \cos 2\theta \geq 0 \leftrightarrow -\frac{\pi}{2} \leq 2\theta \leq \frac{\pi}{2}, \ \frac{3\pi}{2} \leq 2\theta \leq \frac{5\pi}{2} \leftrightarrow -\frac{\pi}{4} \leq \theta \leq \frac{\pi}{4}$, $\frac{3\pi}{4} \leq \theta \leq \frac{5\pi}{4}$. $\frac{dr}{d\theta} = 2\sec 2\theta \tan 2\theta > 0$ if $0 < \theta < \frac{\pi}{4}$. The graph is sketched in Fig. VIII.9.3.

(*)9.4. Find the points on the curve $r = a(1 + \cos\theta)$, where the tangent line is

(a) parallel to the x-axis;
(b) parallel to the y-axis.

$$\frac{dy}{dx} = \text{slope of tangent line} = \frac{\frac{dy}{d\theta}}{\frac{dx}{d\theta}} = \frac{\frac{d(r\sin\theta)}{d\theta}}{\frac{d(r\cos\theta)}{d\theta}} = \frac{\sin\theta\frac{dr}{d\theta} + r\cos\theta}{\cos\theta\frac{dr}{d\theta} - r\sin\theta}.$$

For the curve $r = f(\theta) = a(1 + \cos\theta)$ (Fig. VIII.9.4), $\frac{dr}{d\theta} = -a\sin\theta$. Hence

$$\begin{aligned}
\frac{dy}{d\theta} = 0 &\leftrightarrow -a\sin^2\theta + a\cos\theta(1 + \cos\theta) = 0 \\
&\leftrightarrow \cos\theta + \cos^2\theta - \sin^2\theta = 0 \\
&\leftrightarrow \cos\theta + \cos^2\theta - (1 - \cos^2\theta) = 0 \\
&\leftrightarrow 2\cos^2\theta + \cos\theta - 1 = 0 \\
&\leftrightarrow (\cos\theta + 1)(2\cos\theta - 1) = 0 \\
&\leftrightarrow \cos\theta = -1, \quad \cos\theta = \tfrac{1}{2} \\
&\leftrightarrow \theta = \pi, \pm\frac{\pi}{3}.
\end{aligned}$$

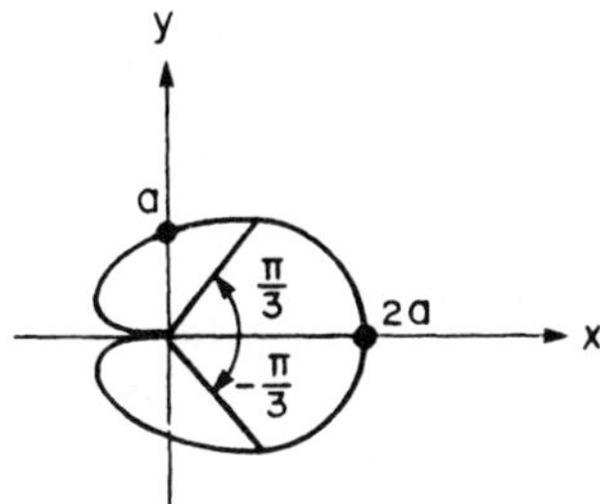

Figure VIII.9.4. Graph of $r = a(1+\cos\theta)$, a cardioid.

$$\begin{aligned}\frac{dx}{d\theta} = 0 &\leftrightarrow -a\cos\theta\sin\theta - a(1+\cos\theta)\sin\theta = 0\\ &\leftrightarrow \sin\theta(2\cos\theta+1) = 0\\ &\leftrightarrow \sin\theta = 0, \quad \cos\theta = -\tfrac{1}{2}\\ &\leftrightarrow \theta = 0, \pi, \pm\frac{2\pi}{3}.\end{aligned}$$

(a) The tangent line is parallel to the x-axis at points where $\frac{dy}{dx} = 0$. At such points it is necessary that $\frac{dy}{d\theta} = 0$. Hence the possible points are $\theta = \pi$, $\pm\frac{\pi}{3}$. Since $\frac{dx}{d\theta} \neq 0$ at $\theta = \pm\frac{\pi}{3}$, $\frac{dy}{dx} = 0$ at $\theta = \pm\frac{\pi}{3}$. Since $\frac{dx}{d\theta} = 0$ at $\theta = \pi$, one must investigate the indeterminate form

$$\begin{aligned}\lim_{\theta\to\pi}\left(\frac{\frac{dy}{d\theta}}{\frac{dx}{d\theta}}\right) &= \lim_{\theta\to\pi}\frac{(2\cos\theta-1)(\cos\theta+1)}{\sin\theta(2\cos\theta+1)}\\ &= \lim_{\theta\to\pi}\frac{2\cos^2\theta+\cos\theta-1}{\sin 2\theta+\sin\theta}\\ &= \lim_{\theta\to\pi} -\frac{\sin\theta(4\cos\theta+1)}{2\cos 2\theta+\cos\theta} = 0, \qquad \text{by l'Hôpital's rule.}\end{aligned}$$

Hence $\frac{dy}{dx} = 0$ at $\boxed{\theta = \pi, \pm\frac{\pi}{3}}$.

(b) The tangent line is parallel to the y-axis at points where $\frac{dy}{dx} = \infty$. At such points $\frac{dx}{d\theta} = 0$. Hence the possible points are $\theta = 0, \pi, \pm\frac{2\pi}{3}$. $\frac{dy}{d\theta} \neq 0$ at $\theta = 0, \pm\frac{2\pi}{3}$; $\frac{dy}{d\theta} = 0$ at $\theta = \pi$, but from part (a), $\frac{dy}{dx} = 0$ at $\theta = \pi$. Hence $\frac{dy}{dx} = \infty$ at $\boxed{\theta = 0, \pm\frac{2\pi}{3}}$.

9.5. Sketch the following polar graph and compute the area A of one leaf:

$$r=1+\sin 2\theta.$$

$r=f(\theta)=1+\sin 2\theta$; $f(\theta+\pi)=f(\theta)$; $\dfrac{dr}{d\theta}=2\cos 2\theta$; $\dfrac{dr}{d\theta}=0$ at $\theta=\pm\dfrac{\pi}{4}, \pm\dfrac{3\pi}{4}$. Then the following table of values is helpful to sketch the graph shown in Figure VIII.9.5.

θ	0	$\frac{\pi}{4}$	$\frac{\pi}{2}$	$\frac{3\pi}{4}$	π
r	1	2	1	0	1

$$A=\frac{1}{2}\int_{-\pi/4}^{3\pi/4} f^2(\theta)\,d\theta=\frac{1}{2}\int_0^{\pi} f^2(\theta)\,d\theta \quad \text{since} \quad f(\theta+\pi)=f(\theta).$$

Hence

$$A=\frac{1}{2}\int_0^{\pi}[1+\sin^2 2\theta+2\sin 2\theta]\,d\theta=\frac{1}{2}\int_0^{\pi}[1+\sin^2 2\theta]\,d\theta$$

$$=\frac{1}{2}\int_0^{\pi}\left[1+\frac{1}{2}\right]d\theta=\boxed{\frac{3\pi}{4}}.$$

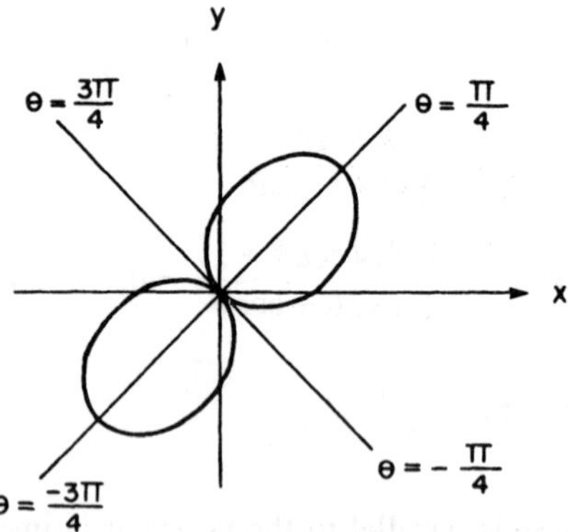

Figure VIII.9.5

9.6. Sketch the region which is inside the curve $r=2+2\cos\theta$ and outside the curve $r=3$. (Here r,θ are polar coordinates.) Find the area A of this region.

The curves intersect when $r = 2+2\cos\theta = 3 \leftrightarrow 2\cos\theta = 1 \leftrightarrow \cos\theta = \frac{1}{2} \leftrightarrow \theta = \pm\pi/3$. Let $f(\theta) = 3$, $g(\theta) = 2+2\cos\theta$ (Fig. VIII.9.6). Then

$$A = \frac{1}{2}\int_{-\pi/3}^{\pi/3} [g^2(\theta) - f^2(\theta)]\, d\theta = \int_0^{\pi/3} [g^2(\theta) - f^2(\theta)]\, d\theta$$

$$= \int_0^{\pi/3} [4 + 8\cos\theta + 4\cos^2\theta - 9]\, d\theta = \int_0^{\pi/3} [8\cos\theta + 4\cos^2\theta - 5]\, d\theta$$

$$= \int_0^{\pi/3} \left[8\cos\theta + 4\left(\frac{1+\cos 2\theta}{2}\right) - 5\right] d\theta = \int_0^{\pi/3} [8\cos\theta + 2\cos 2\theta - 3]\, d\theta$$

$$= [8\sin\theta + \sin 2\theta]\Bigg|_{\theta=0}^{\theta=\pi/3} - \pi = 4\sqrt{3} + \frac{\sqrt{3}}{2} - \pi = \boxed{\frac{9\sqrt{3}}{2} - \pi}.$$

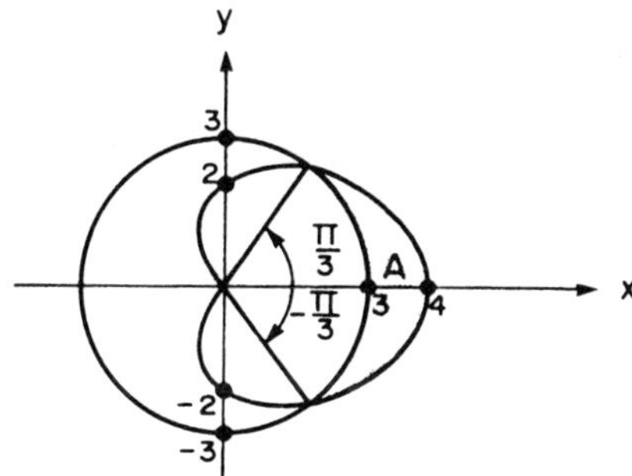

Figure VIII.9.6

9.7. Find the area A of the region interior to the right branch ($x \geq 0$) of the curve described by the equation (see Fig. VIII.9.7) $(x^2 + y^2)^2 = a^2(x^2 - y^2)$, $a > 0$.

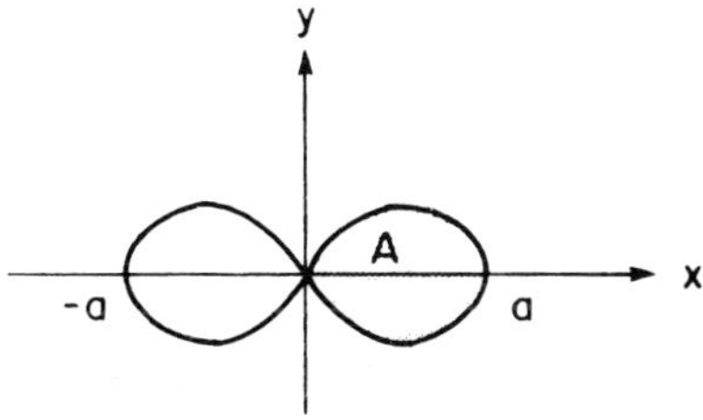

Figure VIII.9.7

The curve is a *lemniscate*. It is convenient to compute A using polar coordinates. Let $x = r\cos\theta$, $y = r\sin\theta$. Then the curve becomes $r^4 = a^2r^2(\cos^2\theta - \sin^2\theta) \leftrightarrow r^2 = a^2\cos 2\theta$. The right branch lies in the region where $\cos 2\theta \geq 0$ on the domain $-\frac{\pi}{2} \leq \theta \leq \frac{\pi}{2}$. Hence $-\frac{\pi}{4} \leq \theta \leq \frac{\pi}{4}$ for the right branch.

$$A = \frac{1}{2}\int_{-\pi/4}^{\pi/4} r^2\, d\theta = a^2 \int_0^{\pi/4} \cos 2\theta\, d\theta = \frac{a^2}{2}\sin 2\theta\bigg|_{\theta=0}^{\theta=\pi/4} = \boxed{\frac{a^2}{2}}.$$

9.8. Consider the curve C described by the polar equation $r = a(1+\cos\theta)$, $0 \leq \theta \leq 2\pi$. Calculate its length l.

$$l = \int_0^{2\pi} \sqrt{r^2 + \left(\frac{dr}{d\theta}\right)^2}\, d\theta \quad \text{where} \quad r = a(1+\cos\theta).$$

$$\frac{dr}{d\theta} = -a\sin\theta. \text{ Thus } r^2 + \left(\frac{dr}{d\theta}\right)^2 = a^2[1 + \cos^2\theta + 2\cos\theta + \sin^2\theta]$$

$$= 2a^2[1+\cos\theta] = 4a^2\cos^2\frac{\theta}{2}.$$

Hence $l = 2a\int_0^{2\pi}\left|\cos\frac{\theta}{2}\right| d\theta = 4a\int_0^{\pi}\cos\frac{\theta}{2}\, d\theta = 8a\sin\frac{\theta}{2}\bigg|_{\theta=0}^{\theta=\pi} = \boxed{8a}$.

10.1. Discuss whether the following series converge or diverge. Give reasons.

(a) $\sum_{n=1}^{\infty} n^{-n}$.

(b) $\sum_{n=1}^{\infty} (-1)^{n+1}\frac{1}{\sqrt{n}}$.

(a) Let $a_n = n^{-n}$. Use the ratio test:

$$\lim_{n\to\infty}\left|\frac{a_{n+1}}{a_n}\right| = \lim_{n\to\infty}\frac{n^n}{(n+1)^{n+1}} = \lim_{n\to\infty}\left(\frac{1}{n+1}\right)\left(\frac{1}{1+\frac{1}{n}}\right)^n$$

$$= \lim_{n\to\infty}\left(\frac{1}{n+1}\right)\lim_{n\to\infty}\left(\frac{1}{1+\frac{1}{n}}\right)^n$$

$$= (0)\left(\frac{1}{e}\right) = 0 < 1.$$

Hence the series $\boxed{\text{converges}}$.

(b) The series is alternating.

$$\lim_{n\to\infty} a_n = 0 \quad \text{where} \quad a_n = \frac{(-1)^{n+1}}{\sqrt{n}} \quad \text{and moreover} \quad |a_{n+1}| < |a_n|,$$

$$n = 1, 2, \ldots.$$

Hence the series $\boxed{\text{converges}}$ by the alternating series test.

10.2. Show whether the series $\sum_{k=1}^{\infty} \frac{1}{k^{1/2}+k^{3/2}}$ converges or diverges.

For $k \geq 1$, $k^{1/2}+k^{3/2} > k^{3/2}$. Hence $0 < \frac{1}{k^{1/2}+k^{3/2}} < \frac{1}{k^{3/2}}$ for $k \geq 1$. Thus $0 < \sum_{k=1}^{\infty} \frac{1}{k^{1/2}+k^{3/2}} < \sum_{k=1}^{\infty} \frac{1}{k^{3/2}}$ which converges. Hence the given series $\boxed{\text{converges}}$ by the comparison test.

10.3. Discuss the series $\sum_{n=1}^{\infty} \frac{\tan^{-1} n}{n^2+1}$.

For each $n \geq 1$, $0 < \frac{\tan^{-1} n}{n^2+1} < \frac{\frac{\pi}{2}}{n^2+1} < \frac{\frac{\pi}{2}}{n^2}$. But $\sum_{n=1}^{\infty} \frac{1}{n^2}$ converges. Hence the given series $\boxed{\text{converges}}$ by the comparison test.

(*)10.4. Show whether each of the following series converges or diverges.

(a) $\sum_{n=1}^{\infty} \frac{\sqrt{n+1}-\sqrt{n}}{\sqrt{n}}$.

(b) $\sum_{n=1}^{\infty} \frac{n^2}{1000n^3+1}$.

(c) $\sum_{n=1}^{\infty} \frac{(-1)^n(n-3)}{(n+1000)^{1.5}}$.

(d) $\sum_{n=1}^{\infty} \left[\frac{(1)(3)(5)\ldots(2n-1)}{(2)(4)(6)\ldots(2n)}\right]$.

(a) $\frac{\sqrt{n+1}-\sqrt{n}}{\sqrt{n}} = \left(\frac{\sqrt{n+1}-\sqrt{n}}{\sqrt{n}}\right)\left(\frac{\sqrt{n+1}+\sqrt{n}}{\sqrt{n+1}+\sqrt{n}}\right) = \frac{1}{\sqrt{n^2+n}+n}$.

If $n \geq 1$, then

$$\sqrt{n^2+n}+n \leq \sqrt{n^2+n^2}+n = (1+\sqrt{2})n \Rightarrow \frac{\sqrt{n+1}-\sqrt{n}}{\sqrt{n}} \geq \frac{1}{(1+\sqrt{2})n}.$$

Hence $\sum_{n=1}^{\infty} \frac{\sqrt{n+1}-\sqrt{n}}{\sqrt{n}} \geq \frac{1}{(1+\sqrt{2})} \sum_{n=1}^{\infty} \frac{1}{n}$. Hence the series $\boxed{\text{diverges}}$ by comparison with the series $\sum_{n=1}^{\infty} \frac{1}{n}$.

(b) $\frac{n^2}{1000n^3+1} > \frac{n^2}{1000n^3+n^3} = \frac{1}{(1001)n}$. Hence the series $\boxed{\text{diverges}}$ by comparison with the series $\sum_{n=1}^{\infty} \frac{1}{n}$.

(c) Let $a_n = (-1)^n \dfrac{(n-3)}{(n+1000)^{1.5}}$. For n large enough, $|a_n|$ $= \dfrac{n-3}{(n+1000)^{1.5}}$ is monotonically decreasing. $\lim_{n\to\infty} a_n = 0$. Hence this alternating series $\boxed{\text{converges}}$ by the alternating series test.

(d) $\dfrac{(1)(3)(5)\ldots(2n-1)}{(2)(4)(6)\ldots(2n)} = (3/2)(5/4)(7/6)\ldots\left(\dfrac{2n-1}{2n-2}\right)\left(\dfrac{1}{2n}\right) > \dfrac{1}{2n}$.

Hence the series $\boxed{\text{diverges}}$ by comparison with the series $\sum_{n=1}^{\infty} \dfrac{1}{n}$.

10.5. Find the sum of the following series:

$$S = \frac{2}{\pi} - \frac{4}{\pi^2} + \frac{8}{\pi^3} - \cdots + (-1)^{n-1}\left(\frac{2}{\pi}\right)^n + \cdots.$$

The given series is the geometric series $-\sum_{n=1}^{\infty} r^n$ where $r = -\dfrac{2}{\pi}$, $|r|<1$. Hence $S = 1 - \sum_{n=0}^{\infty} r^n = 1 - \left(\dfrac{1}{1-r}\right)$. With $r = -\dfrac{2}{\pi}$, $S = 1 - \dfrac{1}{1+\dfrac{2}{\pi}} = \dfrac{2/\pi}{1+\dfrac{2}{\pi}}$

$= \boxed{\dfrac{2}{\pi+2}}$.

(*)10.6. (a) Obtain a closed form expression for the nth partial sum of the series $\sum_{n=1}^{\infty} \dfrac{1}{2^n}\tan\left(\dfrac{x}{2^n}\right)$. (Hint: Note that $\tan\theta = \cot\theta - 2\cot 2\theta$.)

(b) Hence show that

$$\sum_{n=1}^{\infty} \frac{1}{2^n}\tan\left(\frac{x}{2^n}\right) = \frac{1}{x} - \cot x \quad \text{for} \quad x \neq 0.$$

(a) $\dfrac{1}{2^k}\tan\left(\dfrac{x}{2^k}\right) = \dfrac{1}{2^k}\left(\cot\left(\dfrac{x}{2^k}\right) - 2\cot\left(\dfrac{x}{2^{k-1}}\right)\right)$

$= \dfrac{1}{2^k}\cot\left(\dfrac{x}{2^k}\right) - \dfrac{1}{2^{k-1}}\cot\left(\dfrac{x}{2^{k-1}}\right)$, leading to a telescoping property. Hence the nth partial sum of the series is

$$S_n = \sum_{k=1}^{n} \frac{1}{2^k}\tan\left(\frac{x}{2^k}\right)$$

$$= -\left\{\left(\cot x - \frac{1}{2}\cot\frac{x}{2}\right) + \left(\frac{1}{2}\cot\frac{x}{2} - \frac{1}{4}\cot\frac{x}{4}\right) + \left(\frac{1}{4}\cot\frac{x}{4} - \frac{1}{8}\cot\frac{x}{8}\right)\right.$$

$$\left. + \cdots + \left(\frac{1}{2^{n-1}}\cot\left(\frac{x}{2^{n-1}}\right) - \frac{1}{2^n}\cot\left(\frac{x}{2^n}\right)\right)\right\}$$

$= \boxed{\dfrac{1}{2^n}\cot\left(\dfrac{x}{2^n}\right) - \cot x}$, since all terms cancel except for the end terms.

(b) $\lim_{n\to\infty} \frac{1}{2^n}\cot\left(\frac{x}{2^n}\right) = \lim_{u\to 0} \frac{u}{\tan ux}$ where $u = \frac{1}{2^n}$. But $\lim_{u\to 0} \frac{\tan ux}{u}$ $= \left[\frac{d}{du}(\tan ux)\right]\Big|_{u=0} = [x\sec^2 ux]|_{u=0} = x$. Hence for $x \neq 0$, $\lim_{n\to\infty} \frac{1}{2^n}\cot\left(\frac{x}{2^n}\right) = \frac{1}{x}$. Thus $\lim_{n\to\infty} S_n = \boxed{\frac{1}{x} - \cot x}$.

10.7. Prove that $S = \sum_{n=1}^{\infty} \frac{1}{n^2}$ converges and that $1.5 < \sum_{n=1}^{\infty} \frac{1}{n^2} < 1.8$.

The function $f(x) = \frac{1}{x^2}$ is decreasing for $x > 1$. Moreover $\int_1^{\infty} \frac{1}{x^2}dx = \lim_{b\to+\infty} \frac{-1}{x}\Big|_{x=1}^{x=b} = 1$. Hence the series $\boxed{\text{converges}}$ by the integral test. From Fig. VIII.10.7,

$$\int_2^{\infty} \frac{1}{x^2}dx + 1 < S < 1 + \frac{1}{2^2} + \int_2^{\infty} \frac{1}{x^2}dx \leftrightarrow 1.5 < S < 1.75.$$

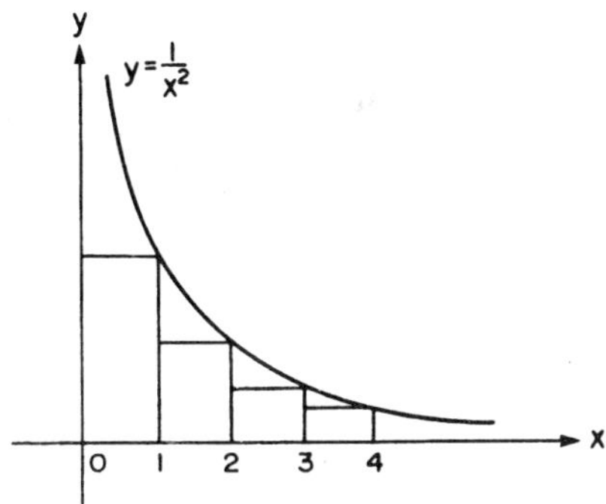

Figure VIII.10.7

(*)**10.8.** Evaluate $l = \lim_{n\to\infty} \frac{1 + \sqrt[n]{e} + \sqrt[n]{e^2} + \cdots + \sqrt[n]{e^{n-1}}}{n}$.

$$l = \lim_{n\to\infty} \frac{1}{n}\left(1 + e^{1/n} + e^{2/n} + \cdots + e^{(n-1)/n}\right)$$

$$= \lim_{n\to\infty} \sum_{k=0}^{n-1} \frac{1}{n}e^{k/n}.$$

Now consider the function $f(x) = e^x$ over the interval $[0,1]$ (Fig. VIII.10.8). If one subdivides $[0,1]$ into n equal subintervals, then $f\left(\frac{k-1}{n}\right) = e^{(k-1)/n}$ is the value of $f(x)$ at the initial point of the kth subinterval. Let $\Delta x = \frac{1}{n}$.

Then

$$l = \lim_{\substack{n\to\infty \\ (\Delta x\to 0)}} \sum_{k=1}^{n} f((k-1)\,\Delta x)\,\Delta x$$

$$= \int_0^1 f(x)\,dx = \int_0^1 e^x\,dx = \boxed{e-1}.$$

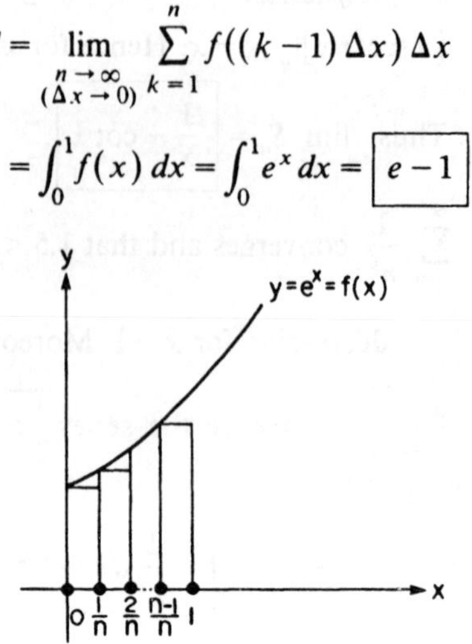

Figure VIII.10.8

(*)10.9. An ant crawls at the rate of 10 centimetres per minute along a rubber band which can be stretched uniformly. Suppose the rubber band is initially 1 metre long and it is stretched an additional metre at the end of each minute.

If the ant begins at one end of the band, does it ever reach the other end? If so, how long does it take the ant?

At the end of N minutes, the ant has travelled $l(N)$ centimetres and the band is $(100)(N+1)$ centimetres long. At the end of $(N+1)$ minutes $l(N+1)=[l(N)+10]\left(\dfrac{N+2}{N+1}\right)$ where $\left(\dfrac{N+2}{N+1}\right)$ is the stretch factor for the lengthening of the band in the $(N+1)$th minute.

$l(0)=0$; $\quad l(1)=10(2)=20$; $\quad l(2)=[10(2)+10]\left(\tfrac{3}{2}\right)=(3)(10)\left(1+\tfrac{1}{2}\right)$;
$l(3)=\left[(3)(10)\left(1+\tfrac{1}{2}\right)+10\right]\left(\tfrac{4}{3}\right)=(4)(10)\left(1+\tfrac{1}{2}+\tfrac{1}{3}\right)$.

Claim. $l(N)=(N+1)(10)\left(1+\dfrac{1}{2}+\cdots+\dfrac{1}{N}\right)$, $N\geq 1$.

Proof of claim by induction. One need only check for $N+1$;

$$l(N+1)=[l(N)+10]\left(\frac{N+2}{N+1}\right)$$

$$=(N+2)(10)\left(1+\frac{1}{2}+\cdots+\frac{1}{N}\right)+(N+2)(10)\left(\frac{1}{N+1}\right)$$

$$=(N+2)(10)\left(1+\frac{1}{2}+\cdots+\frac{1}{N}+\frac{1}{N+1}\right) \qquad \text{(end of proof)}.$$

At the end of N minutes the band is $(100)(N+1)$ centimetres in length. The ant reaches the other end in n minutes for the first integer n such that $(n+1)\left(1+\frac{1}{2}+\cdots+\frac{1}{n}\right)>10(n+1)\leftrightarrow\left(1+\frac{1}{2}+\cdots+\frac{1}{n}\right)>10$. n is a finite number since $\sum_{k=1}^{\infty}\frac{1}{k}=+\infty$. Now estimate n. $\log(n+1)=\int_1^{n+1}\frac{dx}{x}<1+\frac{1}{2}+\cdots+\frac{1}{n}$. Clearly the ant has already reached the other end of the band at a time when $\log(n+1)>10\leftrightarrow n>22{,}026\leftrightarrow$ $\boxed{\text{15 days}}$. (Do this problem for a continuously stretching band. Have the band stretch α metres/minute and the ant crawl β metres/minute.)

(*)10.10. The sequence $\{a_n\}$ is defined by

$$a_n=\frac{1}{n+1}+\frac{1}{n+2}+\frac{1}{n+3}+\cdots+\frac{1}{n+n}.$$

(a) Prove that $\lim_{n\to\infty} a_n$ exists.

(b) If you have not already done so in (a), show that $\lim_{n\to\infty} a_n>\frac{1}{2}$.

(a)

$$a_{n+1}=\frac{1}{n+2}+\frac{1}{n+3}+\cdots+\frac{1}{n+n}+\frac{1}{n+n+1}+\frac{1}{(n+1)+(n+1)}.$$

$$a_{n+1}-a_n=\frac{1}{2n+1}+\frac{1}{2n+2}-\frac{1}{n+1}=\frac{1}{2n+1}-\frac{1}{2n+2}>0.$$

$0<a_n\le\frac{n}{n+1}<1$ since $\frac{1}{n+k}\le\frac{1}{n+1}$ if $k\ge1$. Hence $\{a_n\}$ is bounded and monotonically increasing. Thus $\lim_{n\to\infty} a_n$ exists.

(b) For each $n\ge2$, $a_n\ge a_2$. Hence $\lim_{n\to\infty} a_n\ge a_2$. But $a_2>a_1=\frac{1}{2}$. Hence $\lim_{n\to\infty} a_n>\frac{1}{2}$.

10.11. Determine the interval of convergence for the series

$$S(x)=\sum_{k=1}^{\infty}\frac{(x-4)^k}{\sqrt{3^k}}.$$

Let $a_k(x)=\frac{(x-4)^k}{\sqrt{3^k}}$. Then $\left|\frac{a_{k+1}(x)}{a_k(x)}\right|=\frac{|x-4|}{\sqrt{3}}$; $\lim_{k\to\infty}\left|\frac{a_{k+1}(x)}{a_k(x)}\right|=\frac{|x-4|}{\sqrt{3}}$. Hence by the ratio test $S(x)$ converges for $|x-4|<\sqrt{3}$ and diverges for $|x-4|>\sqrt{3}$. If $|x-4|=\sqrt{3}$, then $\lim_{k\to\infty}|a_k(x)|=1\ne0$. Hence the series $S(x)$ diverges when $x=4\pm\sqrt{3}$. The interval of convergence is $\boxed{4-\sqrt{3}<x<4+\sqrt{3}}$. (Alternatively, note that $S(x)$ is a geometric series.)

(*)10.12. Find the Taylor polynomial $p(x)$ of degree 6 in x which approximates $\sqrt{1+x}$ on the interval $[0, \frac{1}{2}]$. Estimate $|p(x)-\sqrt{1+x}|$ using the standard error formula.

Let $f(x)=\sqrt{1+x}$; $p(x)=\sum_{n=0}^{6}\frac{f^{(n)}(0)}{n!}x^n$.

$$f'(x)=\frac{1}{2}(1+x)^{-1/2},\quad f''(x)=-\frac{1}{2^2}(1+x)^{-3/2},$$

$$f'''(x)=\frac{3}{2^3}(1+x)^{-5/2},\quad f^{(4)}(x)=\frac{-15}{2^4}(1+x)^{-7/2},$$

$$f^{(5)}(x)=\frac{(3)(5)(7)}{2^5}(1+x)^{-9/2},\quad f^{(6)}(x)=\frac{-(3)(5)(7)(9)}{2^6}(1+x)^{-11/2}.$$

Hence $\boxed{p(x)=1+\frac{x}{2}-\frac{x^2}{8}+\frac{x^3}{16}-\frac{5}{128}x^4+\frac{7}{256}x^5-\frac{21}{1024}x^6}$.

$|p(x)-\sqrt{1+x}|=\left|\frac{f^{(7)}(c)x^7}{7!}\right|$ for some c, $0\le c\le x\le\frac{1}{2}$. $\left|\frac{f^{(7)}(c)}{7!}\right|$ $=\left(\frac{21}{1024}\right)\left(\frac{11}{2}\right)\left(\frac{1}{7}\right)(1+c)^{-13/2}<\left(\frac{21}{1024}\right)\left(\frac{11}{2}\right)\left(\frac{1}{7}\right)=\frac{33}{2^{11}}$. Hence $|p(x)-\sqrt{1+x}|\le\boxed{\frac{33}{2^{11}}x^7}$ for $0\le x\le\frac{1}{2}$.

10.13. Let $P(x)=x^4-4x^3+5x$. Express $P(x)$ in the form

$$P(x)=a_0+a_1(x-1)+a_2(x-1)^2+a_3(x-1)^3+a_4(x-1)^4.$$

(Evaluate a_0, a_1, a_2, a_3, a_4.)

Since $P(x)$ is a polynomial of degree 4, $P(x)=\sum_{n=0}^{4}\frac{P^{(n)}(1)}{n!}(x-1)^n$; $P'(x)=4x^3-12x^2+5$, $P''(x)=12x^2-24x$, $P'''(x)=24x-24$, $P^{(4)}(x)=24$; $P(1)=2$, $P'(1)=-3$, $P''(1)=-12$, $P'''(1)=0$, $P^{(4)}(1)=24$. $P(x)=a_0+a_1(x-1)+a_2(x-1)^2+a_3(x-1)^3+a_4(x-1)^4$; $a_0=P(1)=2$, $a_1=P'(1)=-3$, $a_2=\frac{P''(1)}{2!}=-6$, $a_3=\frac{P'''(1)}{3!}=0$, $a_4=\frac{P^{(4)}(1)}{4!}=1$. Hence

$$\boxed{P(x)=2-3(x-1)-6(x-1)^2+(x-1)^4}.$$

10.14. A function $f(x)$ satisfies the two conditions

$$f'(x)=1+\{f(x)\}^{10}\quad\text{and}\quad f(0)=1.$$

Calculate the first 4 coefficients of its Taylor series expansion about $x=0$.

The 3rd degree Taylor polynomial for $f(x)$ about $x = 0$ is

$P(x) = a_0 + a_1 x + a_2 x^2 + a_3 x^3$, where $a_0 = f(0) = \boxed{1}$; $a_1 = f'(0) = 1 + 1 = \boxed{2}$; $f''(x) = 10f'(x)[f(x)]^9$, $a_2 = \dfrac{f''(0)}{2!} = \dfrac{20}{2!} = \boxed{10}$; $f'''(x) = 10f''(x)[f(x)]^9 + 90[f'(x)]^2[f(x)]^8$, $a_3 = \dfrac{f'''(0)}{3!} = \dfrac{(10)(20)+(90)(4)}{3!} = \boxed{\dfrac{280}{3}}$. Hence

$$\boxed{P(x) = 1 + 2x + 10x^2 + \frac{280}{3}x^3}.$$

(*)10.15. Sum the series $S = \dfrac{1}{2} + \dfrac{2}{4} + \dfrac{3}{8} + \dfrac{4}{16} + \dfrac{5}{32} + \dfrac{6}{64} + \cdots$.

A possible representation for this series is

$$S = \sum_{n=1}^{\infty} \frac{n}{2^n}.$$

Within its interval of convergence a power series can be differentiated term-by-term, i.e.,

if $f(x) = \sum_{n=0}^{\infty} a_n x^n$, then $f'(x) = \sum_{n=1}^{\infty} n a_n x^{n-1}$, with the same radius of convergence as $f(x)$. Thus $\sum_{n=0}^{\infty} x^n = \dfrac{1}{1-x}$ if $|x| < 1 \Rightarrow \sum_{n=1}^{\infty} n x^{n-1} = \dfrac{1}{(1-x)^2}$ if $|x| < 1$. Now let $x = \dfrac{1}{2}$. Then $\sum_{n=1}^{\infty} \dfrac{n}{2^{n-1}} = \dfrac{1}{\left(1 - \frac{1}{2}\right)^2} = 4 \Rightarrow$

$$S = \frac{1}{2} \sum_{n=1}^{\infty} \frac{n}{2^{n-1}} = \boxed{2}.$$

10.16. Find the first five terms of the Maclaurin series for $f(x) = \dfrac{\sin x}{x}$.

The Maclaurin series expansion of a function is unique. Since the Maclaurin series for $\sin x = x - \dfrac{x^3}{3!} + \dfrac{x^5}{5!} - \dfrac{x^7}{7!} + \dfrac{x^9}{9!} + \cdots$, the Maclaurin series for

$$\frac{\sin x}{x} = \boxed{1 - \frac{x^2}{3!} + \frac{x^4}{5!} - \frac{x^6}{7!} + \frac{x^8}{9!} + \cdots}.$$

(*)10.17. The Taylor series for $\dfrac{1}{1-x}$ about $x = 0$ is

$$\frac{1}{1-x} = 1 + x + x^2 + x^3 + \cdots.$$

(a) Determine the series for $\frac{1}{1+x}$.

(b) Determine the series for $\frac{1}{1+x^2}$.

(c) Determine the series for arctan x.

(d) For each of the power series in this question, state the radius of convergence.

The uniqueness of the Taylor series expansion of a function about a point is exploited in solving this problem.

(a) Substitute $-x$ for x in the series for $\frac{1}{1-x}$:

$$\frac{1}{1+x}=\frac{1}{1-(-x)}=1-x+x^2-x^3+\cdots.$$

(b) Substitute x^2 for x in the series for $\frac{1}{1+x}$:

$$\frac{1}{1+x^2}=1-x^2+x^4-x^6+\cdots.$$

(c) $\arctan x=\int_0^x \frac{1}{1+t^2}dt$. Since a power series can be integrated term-by-term,

$$\arctan x=\int_0^x\left[1-t^2+t^4-t^6+\cdots\right]dt$$

$$=x-\frac{x^3}{3}+\frac{x^5}{5}-\frac{x^7}{7}+\cdots.$$

(d) All series have radius of convergence 1, i.e., converge absolutely for $|x|<1$.

(*)10.18. Find the complete Taylor series expansions and corresponding regions of convergence for the following functions about the indicated points.

(a) x^2+5x+2, about $x=1$.

(b) $\frac{1}{x-5}$, about $x=3$.

(c) $\int_0^x \frac{dt}{\sqrt{1+t^2}}$, about $x=0$.

(a) $x^2+5x+2=[(x-1)+1]^2+5[(x-1)+1]+2=$

$$\boxed{8+7(x-1)+(x-1)^2\text{, converging for all } x}.$$

(b) $\frac{1}{x-5} = \frac{1}{x-3-2} = \frac{-1}{2-(x-3)} = \frac{-1/2}{1-\left(\frac{x-3}{2}\right)} =$

$-\frac{1}{2}\sum_{n=0}^{\infty}\left(\frac{x-3}{2}\right)^n = \boxed{-\sum_{n=0}^{\infty}\frac{(x-3)^n}{2^{n+1}}}$, converging for $\left|\frac{x-3}{2}\right| < 1$

$\leftrightarrow \boxed{|x-3| < 2}$.

(c) $(1+t^2)^{-1/2} = 1 + (-1/2)t^2 + \frac{(-1/2)(-3/2)t^4}{2!} + \cdots +$
$\frac{(-1)^n(1)(3)(5)\cdots(2n-1)}{2^n n!}t^{2n} + \cdots = \sum_{n=0}^{\infty}\frac{(-1)^n(2n)!}{(2^n n!)^2}t^{2n}$. Hence

$$\int_0^x \frac{dt}{\sqrt{1+t^2}} = \sum_{n=0}^{\infty}\frac{(-1)^n(2n)!}{(2^n n!)^2}\int_0^x t^{2n}\,dt$$

$$= \boxed{\sum_{n=0}^{\infty}\frac{(-1)^n(2n)!x^{2n+1}}{(2^n n!)^2(2n+1)}}, \qquad \text{converging for } \boxed{|x| \le 1}.$$

(*)10.19. Let $f(x) = (x-1)^7 e^x$. Find $f^{(20}(1)$.

$f(x) = \sum_{n=0}^{\infty}\frac{f^{(n)}(1)}{n!}(x-1)^n$. $e^x = ee^{(x-1)} = e\sum_{k=0}^{\infty}\frac{(x-1)^k}{k!}$. Then $f(x)$ $= \sum_{k=0}^{\infty}\frac{e(x-1)^{k+7}}{k!} = \sum_{n=7}^{\infty}\frac{e(x-1)^n}{(n-7)!}$. Since the power series expansion of $f(x)$ about $x=1$ is unique, $\frac{f^{(20)}(1)}{20!} = \frac{e}{13!} \Rightarrow f^{(20)}(1) = \boxed{\frac{e(20!)}{13!}}$.

11.1. Find $L = \lim_{x\to\infty}\frac{4x+\sqrt{1+4x^2}}{x}$.

If $x > 0$, then

$\frac{4x+\sqrt{1+4x^2}}{x} = \frac{x\left[4+\sqrt{\frac{1}{x^2}+4}\right]}{x} = 4 + \sqrt{4+\frac{1}{x^2}}$. Hence $L = 4$ $+ \lim_{x\to\infty}\sqrt{4+\frac{1}{x^2}} = 4+2 = \boxed{6}$.

11.2. Find $L = \lim_{x \to \infty} (\sqrt{x^2+5x} - x)$.

If $x > 0$, then

$$\sqrt{x^2+5x} - x = \frac{(\sqrt{x^2+5x} - x)(\sqrt{x^2+5x} + x)}{\sqrt{x^2+5x} + x} = \frac{x^2+5x-x^2}{\sqrt{x^2+5x}+x}$$

$$= \frac{5x}{\sqrt{x^2+5x}+x} = \frac{5x}{x\left(\sqrt{1+\frac{5}{x}}+1\right)} = \frac{5}{1+\sqrt{1+\frac{5}{x}}}.$$

Hence $L = \dfrac{5}{1 + \lim_{x \to \infty} \sqrt{1+\frac{5}{x}}} = \boxed{\dfrac{5}{2}}$.

11.3. Compute the following limit: $L = \lim_{x \to -\infty} \left(\dfrac{5^x + 3^x}{4^x}\right)$.

$L = \lim_{x \to -\infty} \left[\left(\frac{5}{4}\right)^x + \left(\frac{3}{4}\right)^x\right]$. Let $u = -x$. Then

$L = \lim_{u \to +\infty} \left[\left(\frac{5}{4}\right)^{-u} + \left(\frac{3}{4}\right)^{-u}\right] = \lim_{u \to +\infty} \left[\left(\frac{4}{5}\right)^u + \left(\frac{4}{3}\right)^u\right] = \boxed{+\infty}$

since $\lim_{u \to +\infty} \left(\frac{4}{3}\right)^u = +\infty$, $\lim_{u \to +\infty} \left(\frac{4}{5}\right)^u = 0$.

(*)11.4. Evaluate $L = \lim_{x \to +\infty} e^{x^2} \int_0^x e^{-t^2}\, dt$.

If $x > 1$, $f(x) = \int_0^x e^{-t^2}\, dt > \int_0^1 e^{-t^2}\, dt > e^{-1} \int_0^1 1\, dt = e^{-1}$. $\lim_{x \to +\infty} e^{x^2} = +\infty$.

Thus $L = \boxed{+\infty}$.

11.5. Evaluate $L = \lim_{h \to 0} \dfrac{\sin 3h \sin 7h}{\sin 5h \sin 9h}$.

$$\frac{\sin 3h \sin 7h}{\sin 5h \sin 9h} = \frac{21}{45}\left[\frac{\left(\frac{\sin 3h}{3h}\right)\left(\frac{\sin 7h}{7h}\right)}{\left(\frac{\sin 5h}{5h}\right)\left(\frac{\sin 9h}{9h}\right)}\right].$$

$\lim_{x \to 0} \dfrac{\sin x}{x} = \lim_{x \to 0} \dfrac{\sin x - \sin 0}{x} = \left[\dfrac{d}{dx} \sin x\right]\Big|_{x=0} = \cos 0 = 1$. Thus for any constant k, $\lim_{h \to 0} \dfrac{\sin kh}{kh} = \lim_{x \to 0} \dfrac{\sin x}{x} = 1$ where $x = kh$. Hence

$$L = \frac{7}{15}\left[\frac{\lim_{h \to 0}\left(\frac{\sin 3h}{3h}\right) \lim_{h \to 0}\left(\frac{\sin 7h}{7h}\right)}{\lim_{h \to 0}\left(\frac{\sin 5h}{5h}\right) \lim_{h \to 0}\left(\frac{\sin 9h}{9h}\right)}\right] = \boxed{\frac{7}{15}}.$$

11.6. Find $L = \lim_{t\to 0} \dfrac{\sin t \cos t}{t - t^2}$.

$2\sin t\cos t = \sin 2t$. Hence $L = \dfrac{1}{2}\lim_{t\to 0}\dfrac{\sin 2t}{t - t^2} = \lim_{t\to 0}\left[\dfrac{\dfrac{\sin 2t}{2t}}{1-t}\right] =$

$$\frac{\lim_{t\to 0}\left(\dfrac{\sin 2t}{2t}\right)}{\lim_{t\to 0}(1-t)} = \frac{1}{1} = \boxed{1}.$$

(*)11.7. Find $L = \lim_{x\to\infty} x\left(\dfrac{\pi}{2} - \tan^{-1}x\right)$.

One must distinguish between the cases $x \to -\infty$, $x \to +\infty$. $\lim_{x\to-\infty}\tan^{-1}x = -\dfrac{\pi}{2} \Rightarrow \lim_{x\to-\infty} x\left(\dfrac{\pi}{2} - \tan^{-1}x\right) = \boxed{-\infty}$. But $\lim_{x\to+\infty}\tan^{-1}x = \dfrac{\pi}{2}$. In this case $\lim_{x\to+\infty} x\left(\dfrac{\pi}{2} - \tan^{-1}x\right)$ is an indeterminate form $\left((\infty)(0) = \dfrac{(0)}{(1/\infty)} = \dfrac{(0)}{(0)}\right)$. By l'Hôpital's rule,

$$L = \lim_{x\to+\infty}\left(\frac{\pi/2 - \tan^{-1}x}{1/x}\right) = \lim_{x\to+\infty}\left[\frac{\dfrac{d}{dx}\left(\dfrac{\pi}{2} - \tan^{-1}x\right)}{\dfrac{d}{dx}\left(\dfrac{1}{x}\right)}\right]$$

$$= \lim_{x\to+\infty}\left(\frac{-\dfrac{1}{1+x^2}}{-\dfrac{1}{x^2}}\right) = \lim_{x\to+\infty}\frac{x^2}{1+x^2}$$

$$= \lim_{x\to+\infty}\frac{x^2}{x^2\left(1+\dfrac{1}{x^2}\right)} = \frac{1}{\lim_{x\to+\infty}\left(1+\dfrac{1}{x^2}\right)} = \boxed{1}.$$

11.8. If $f(1) = 1$ and $f'(1) = 2$, compute $\lim_{x\to 1}\dfrac{[f(x)]^2 - 1}{x^2 - 1}$.

Method 1.

$$\frac{[f(x)]^2-1}{x^2-1} = \left(\frac{f(x)-1}{x-1}\right)\left(\frac{f(x)+1}{x+1}\right) = \left(\frac{f(x)-f(1)}{x-1}\right)\left(\frac{f(x)+1}{x+1}\right).$$

Hence

$$\lim_{x\to 1}\frac{[f(x)]^2-1}{x^2-1} = \lim_{x\to 1}\left(\frac{f(x)-f(1)}{x-1}\right)\lim_{x\to 1}\left(\frac{f(x)+1}{x+1}\right)$$

$$= \left[\frac{f'(1)}{2}\right]\left[1 + \lim_{x\to 1} f(x)\right] = f'(1) = \boxed{2}.$$

($f(1) = \lim_{x\to 1} f(x)$ since $f'(1)$ exists.)

Method 2.

By l'Hôpital's rule,

$$\lim_{x\to 1}\frac{[f(x)]^2-1}{x^2-1}=\lim_{x\to 1}\frac{2f(x)f'(x)}{2x}=f(1)f'(1)=\boxed{2}.$$

11.9. Evaluate $L=\lim_{x\to 0^+} x^{2x}$ if the limit exists.

Let $y=x^{2x}$. Then $\log y=2x\log x$.

$$\lim_{x\to 0^+}\log y=2\lim_{x\to 0^+}x\log x=2\lim_{x\to 0^+}\frac{\log x}{1/x}=2\lim_{x\to 0^+}\frac{1/x}{-1/x^2},$$

by l'Hôpital's rule. Hence $\lim_{x\to 0^+}\log y=-2\lim_{x\to 0^+}x=0$. Thus $\lim_{x\to 0^+}x^{2x}$ $=\lim_{x\to 0^+}y=\exp\left[\lim_{x\to 0^+}\log y\right]=e^0=\boxed{1}$.

11.10. Evaluate $L=\lim_{n\to\infty}\left(1-\frac{2}{3n}\right)^{n+1}$.

Let $y=\left(1-\frac{2}{3x}\right)^{1+x}$. Then $\log y=(1+x)\log\left(1-\frac{2}{3x}\right)$;

$$\lim_{x\to+\infty}\log y=\lim_{x\to+\infty}\left[\frac{\log\left(1-\frac{2}{3x}\right)}{\frac{1}{1+x}}\right]=\lim_{x\to+\infty}\left[\frac{\left(\frac{1}{1-\frac{2}{3x}}\right)\left(\frac{2}{3x^2}\right)}{-\left(\frac{1}{1+x}\right)^2}\right]$$

$$=-\frac{2}{3}\,\frac{\lim_{x\to+\infty}\left(\frac{1}{1-\frac{2}{3x}}\right)}{\lim_{x\to+\infty}\left(\frac{x}{1+x}\right)^2}=-\frac{2}{3}.$$

Hence $L=\lim_{n\to\infty}\left(1-\frac{2}{3}\right)^{n+1}=\boxed{e^{-2/3}}$.

In general one can show that for constant a, $\boxed{\lim_{x\to\infty}\left(1+\frac{a}{x}\right)^x=e^a.\ (*)}$

11.11. Evaluate $L=\lim_{x\to 0}(1+3x)^{1/(2x)}$.

By an appropriate substitution this limit reduces to equation (*) of Solved Problem 11.10. Let $y=\frac{1}{2x}$. Then $L=\lim_{y\to\infty}\left(1+\frac{3}{2y}\right)^y=\boxed{e^{3/2}}$.

(*)11.12. Evaluate $L = \lim_{x \to 0} (1+3x)^{(1+5x)/x}$.

$(1+3x)^{(1+5x)/x} = (1+3x)^{1/x}(1+3x)^5$. Hence

$$L = \lim_{x \to 0} (1+3x)^{1/x} \lim_{x \to 0} (1+3x)^5 = \lim_{x \to 0} (1+3x)^{1/x} = \lim_{y \to \infty} \left(1+\frac{3}{y}\right)^y = \boxed{e^3}.$$

(*)11.13. Determine the limit $L = \lim_{x \to +\infty} (e^x + e^{-x})^{2/x}$.

$(e^x + e^{-x})^{2/x} = e^2(1+e^{-2x})^{2/x}$; $\lim_{x \to +\infty} (1+e^{-2x})^{2/x} = 1^0 = 1$. Hence $L = \boxed{e^2}$.

(*)11.14. Evaluate

$$L = \lim_{t \to 1^+} [1-(t-1)x]^{2/(t^2-1)},$$

x fixed.

Write $[1-(t-1)x]^{2/(t^2-1)} = [1-(t-1)x]^{2/(t-1)(t+1)}$. Let $u = t-1$. Then $L = \lim_{u \to 0^+} [1-xu]^{2/u(u+2)} = \lim_{u \to 0^+} [(1-xu)^{2/u}]^{1/(u+2)}$. But $\lim_{u \to 0^+} (1-xu)^{2/u} = \lim_{y \to +\infty} \left(1-\frac{2x}{y}\right)^y$ where $y = \frac{2}{u}$. Thus equation (*) of Solved Problem 11.10 $\Rightarrow \lim_{u \to 0^+} (1-xu)^{2/u} = e^{-2x}$. Hence $L = [e^{-2x}]^{\left(\lim_{u \to 0^+} \frac{1}{u+2}\right)} = (e^{-2x})^{(1/2)} = \boxed{e^{-x}}$.

(*)11.15. Evaluate $L = \lim_{x \to 0^+} \frac{x - \sin x}{(x \sin x)^{3/2}}$.

$$L = \lim_{x \to 0^+} \frac{x - \sin x}{(x \sin x)^{3/2}} = \lim_{x \to 0^+} \left[\frac{\frac{x^3}{3!} - \frac{x^5}{5!} + \cdots}{\left(x^2 - \frac{x^4}{3!} + \cdots\right)^{3/2}} \right]$$

$$= \lim_{x \to 0^+} \left[\frac{x^3\left(\frac{1}{3!} - \frac{x^2}{5!} + \cdots\right)}{x^3\left(1 - \frac{x^2}{3!} + \cdots\right)^{3/2}} \right] = \frac{1}{3!} = \boxed{\frac{1}{6}}.$$

(*)11.16. Evaluate $\lim_{x\to 0^+}[x^{x^x}-1]$.

As stated this question is ambiguous. x^{x^x} could mean $x^{(x^x)}$ or $(x^x)^x = x^{(x^2)}$.

Case I. $L_1 = \lim_{x\to 0^+} x^{(x^x)}$. Let $y = x^{(x^x)}$. Then $\log y = x^x \log x$. By the method of Solved Problem 11.10, it is easy to see that $\lim_{x\to 0^+} x^x = 1$. Thus $\lim_{x\to 0^+}\log y = \left(\lim_{x\to 0^+} x^x\right)\left(\lim_{x\to 0^+}\log x\right) = (1)(-\infty) = -\infty$. Hence $L_1 = e^{-\infty} = 0$. Thus $\lim_{x\to 0^+}[x^{(x^x)}-1] = \boxed{-1}$.

Case II. $L_2 = \lim_{x\to 0^+}(x^x)^x = \lim_{x\to 0^+} x^{x^2}$. Since $\lim_{x\to 0^+} x^x = 1$, $L_2 = 1^0 = 1$. Hence $\lim_{x\to 0^+}[(x^x)^x - 1] = \boxed{0}$.

12.1. Solve the differential equation $\dfrac{dy}{dx} = \dfrac{x}{\sqrt{1-x^4}}$, given that $y = \dfrac{\pi}{3}$ when $x = \dfrac{1}{\sqrt{2}}$.

First one finds the general solution of $\dfrac{dy}{dx} = g(x)$ where $g(x) = \dfrac{x}{\sqrt{1-x^4}}$, i.e., one determines the most general function $y = f(x)$ whose derivative is $g(x)$. Hence $y = f(x) = \int g(x)\,dx = \int \dfrac{x}{\sqrt{1-x^4}}dx$. In this integrand, let $u = x^2$; $x\,dx = \dfrac{du}{2}$. Thus $y = \dfrac{1}{2}\int \dfrac{du}{\sqrt{1-u^2}} = \dfrac{\sin^{-1}u}{2} + C = \dfrac{\sin^{-1}(x^2)}{2} + C$, where C is an arbitrary constant. Now one determines C from the given data: $y\left(\dfrac{1}{\sqrt{2}}\right) = \dfrac{\pi}{3} \leftrightarrow \dfrac{\sin^{-1}(1/2)}{2} + C = \dfrac{\pi}{3} \leftrightarrow C = \dfrac{\pi}{3} - \dfrac{\pi}{12} = \dfrac{\pi}{4}$. Hence the solution is $y = \boxed{\dfrac{1}{2}\sin^{-1}(x^2) + \dfrac{\pi}{4}}$.

(*)12.2. A mothball exposed to the air evaporates in such a way that the amount of substance evaporating per unit time is proportional to the surface area. If the mothball remains spherical at all times and its radius after one year is only one-half its original radius, when will it completely disappear?

Let $r(t)$ be the radius of the mothball at time t (in years); $V(t) = \dfrac{4}{3}\pi r^3$ and $S(t) = 4\pi r^2$ are its volume and surface area at time t. $\dfrac{dV}{dt} = kS$ for some constant k; $r(1) = \dfrac{1}{2}r(0)$.

$$\frac{dV}{dt} = \frac{4\pi}{3}\frac{dr^3}{dt} = \frac{4\pi}{3}\frac{dr^3}{dr}\frac{dr}{dt} = 4\pi r^2\frac{dr}{dt} = kS = k(4\pi r^2).\ \text{Thus}\ \frac{dr}{dt} = k.$$

Hence $r(t)=\int k\,dt = kt + C$ for some constant C. $r(1)=\frac{1}{2}r(0) \leftrightarrow k + C = \frac{C}{2} \Rightarrow C = -2k$. Thus $r(t)=k(t-2)$. The mothball completely disappears at the time T when $r(T)=0=k(T-2) \Rightarrow \boxed{T = 2 \text{ years}}$.

12.3. Find the equation of the curve which contains the points $(0,1)$ and $(2, e^4)$, and whose slope at (x, y) is proportional to the product xy.

Let $y=f(x)$ be the equation of the curve tc be found; $\frac{dy}{dx}=kxy$ for some constant k. Separating variables for this first order differential equation, $\frac{dy}{y}=kx\,dx \leftrightarrow \int \frac{dy}{y} = k\int x\,dx \leftrightarrow \log y = \frac{kx^2}{2}+C$ for some constant C. Hence $y = Ae^{kx^2/2}$ for some constants A, k. One is given the data $y(0)=1$, $y(2)=e^4$ to determine A and k. $y(0)=1 \leftrightarrow A=1$; $y(2)=e^4 \leftrightarrow e^4 = e^{2k} \leftrightarrow k = 2$. Hence the curve is $\boxed{y = f(x) = e^{x^2}}$.

(*)12.4. The portion of a tangent line l to a curve included between the x-axis and the point of tangency is bisected by the y-axis. If the curve passes through $(1,2)$, find its equation.

Let $y=f(x)$ be the equation of the curve to be found. Let l be the tangent line passing through $(x_0, y_0)=(x_0, f(x_0))$. Then l is given by the equation $y-f(x_0)=f'(x_0)(x-x_0)$. One is given that l passes through $\left(0, \frac{y_0}{2}\right)$ (Fig. VIII.12.4). Hence for any x_0 on the curve, $f(x_0)$ satisfies the relation

$$\frac{f(x_0)}{2} - f(x_0) = -x_0 f'(x_0)$$

$$\leftrightarrow x_0 f'(x_0) = \frac{f(x_0)}{2},$$

i.e. the curve $y=f(x)$ satisfies the differential equation $\frac{dy}{dx}=\frac{y}{2x}$. Separating variables, $\frac{dy}{y}=\frac{dx}{2x} \leftrightarrow \int \frac{dy}{y} = \frac{1}{2}\int \frac{dx}{x} \leftrightarrow \log y = \frac{1}{2}\log x + C$ for some

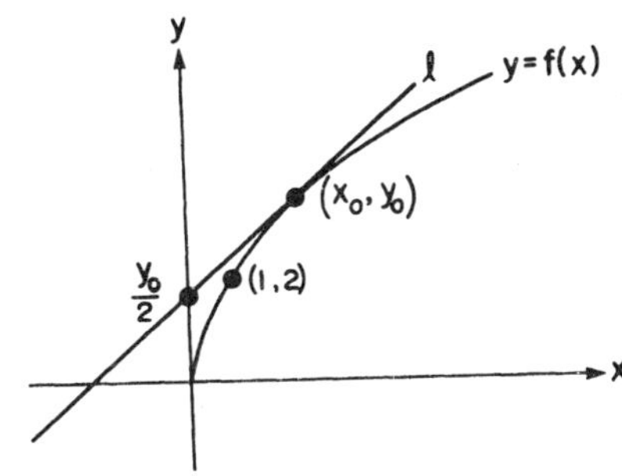

Figure VIII.12.4

constant $C \leftrightarrow \log y = \log x^{1/2} + C \leftrightarrow y = A\sqrt{x}$ for some constant A. Since $y(1) = 2$, $A = 2$. Hence the curve is $\boxed{y = 2\sqrt{x}}$.

12.5. A function $y = f(x)$ has a second derivative of $f''(x) = 6(x-1)$. Find the function, if its graph passes through the point $(2,1)$ and at that point is tangent to the line $y = 3x - 5$.

The aim is to solve the second order differential equation $\dfrac{d^2y}{dx^2} = 6(x-1)$ where $y(2) = 1$, $y'(2) = 3$. $\dfrac{d^2y}{dx^2} = \dfrac{d}{dx}\left(\dfrac{dy}{dx}\right) = 6(x-1)$. Hence $\dfrac{dy}{dx} = \int 6(x-1)\,dx = 3x^2 - 6x + C_1$ for some constant C_1.
$y = \int (3x^2 - 6x + C_1)\,dx = x^3 - 3x^2 + C_1x + C_2$ for some constants C_1, C_2. Now one determines C_1 and C_2 from the data specified at $x = 2$: $y'(2) = 3 = 12 - 12 + C_1 \leftrightarrow C_1 = 3$; $y(2) = 1 \leftrightarrow 8 - 12 + 6 + C_2 = 1 \leftrightarrow C_2 = -1$. Hence $f(x)$

$= \boxed{x^3 - 3x^2 + 3x - 1 = (x-1)^3}$.

12.6. Show that the second order equation $\dfrac{d^2y}{dx^2} + y\dfrac{dy}{dx} = 0$ can be transformed into the first order equation $\dfrac{dv}{dy} = -y$, where $v = \dfrac{dy}{dx} \neq 0$.

$$\frac{d^2y}{dx^2} = \frac{d}{dx}\left(\frac{dy}{dx}\right) = \frac{dv}{dx} = \frac{dv}{dy}\frac{dy}{dx} = v\frac{dv}{dy}.$$

Hence $\dfrac{d^2y}{dx^2} + y\dfrac{dy}{dx} = 0 \leftrightarrow v\dfrac{dv}{dy} + yv = 0 \leftrightarrow \dfrac{dv}{dy} = -y$.

(*)12.7. Consider $\alpha\dfrac{dx}{dt} + x = F(t)$, $t > 0$, where the input $F(t)$ is a periodic step function with period 2τ:

$$F(t) = \begin{cases} \beta & \text{if } \; 2n\tau \le t < (2n+1)\tau \\ -\beta & \text{if } \; (2n+1)\tau \le t < (2n+2)\tau, \; \beta = \text{constant}, \; n = 0,1,2,\ldots. \end{cases}$$

Question. Say $x(0) = x_0$. What must the value of x_0 be in order that the output $x(t)$ is a periodic function with period 2τ, i.e., $x(t+2\tau) = x(t)$, for any value of t? Sketch the periodic solution(s).

The problem reduces to determining x_0 so that $x(2\tau) = x(0) = x_0$. (To see this let $\tilde{t} = t - 2\tau$. Then $F(t) = F(\tilde{t})$ and $\dfrac{dx}{dt} = \dfrac{dx}{d\tilde{t}}\dfrac{d\tilde{t}}{dt} = \dfrac{dx}{d\tilde{t}}$, so $x(\tilde{t})$ solves the same differential equation as $x(t)$. To ensure that these functions are identical, one need only require that their initial conditions be identical, i.e., $x_0 = x(\tilde{t} = 0) = x(2\tau)$.) Therefore one considers

$$\alpha\frac{dx}{dt} + x = \left\{\begin{array}{ll} \beta & \text{if } \; 0 \le t < \tau \\ -\beta & \text{if } \; \tau \le t < 2\tau \end{array}\right\}. \tag{1}$$

Hence $\frac{dx}{dt}$ has a finite jump of $-\frac{2\beta}{\alpha}$ at $t=\tau$ provided $x(t)$ is continuous at $t=\tau$, i.e., $\lim_{t\to\tau^-} x(t) = \lim_{t\to\tau^+} x(t)$. $\lim_{t\to\tau^+} x'(t) - \lim_{t\to\tau^-} x'(t) = -\frac{2\beta}{\alpha}$.

Case I: $\boxed{0<t<\tau}$. In this case the differential equation is $\alpha\frac{dx}{dt}+x=\beta$. $\alpha\frac{dx}{dt}=\beta-x \leftrightarrow \frac{dx}{x-\beta}=-\frac{dt}{\alpha}$. Then $\log(x-\beta)=-\frac{t}{\alpha}+C$ for some constant C. Thus the general solution is $x-\beta=Ae^{-t/\alpha}$ for some constant A to be determined.

Case II: $\boxed{\tau<t<2\tau}$. Here the differential equation is $\alpha\frac{dx}{dt}+x=-\beta$. To obtain the general solution of this differential equation one simply replaces β by $-\beta$ in the general solution for *Case I*. Hence $x+\beta=Be^{-t/\alpha}$ for some constant B to be determined.

Now A and B are determined so that $x(0)=x_0$ and $\lim_{t\to\tau^-} x(t) = \lim_{t\to\tau^+} x(t)$. $x(0)=x_0 \leftrightarrow x_0-\beta=A$. Thus $x(t)=\beta+(x_0-\beta)e^{-t/\alpha}$ if $0<t<\tau$, and $\lim_{t\to\tau^-} x(t)=\beta+(x_0-\beta)e^{-\tau/\alpha}$. If $\tau<t<2\tau$, $x(t)=-\beta+Be^{-t/\alpha}$; $\lim_{t\to\tau^+} x(t)=-\beta+Be^{-\tau/\alpha}$. Hence $Be^{-\tau/\alpha}=2\beta+(x_0-\beta)e^{-\tau/\alpha} \leftrightarrow B=2\beta e^{\tau/\alpha}+(x_0-\beta)$. Thus

$$x(t)=\begin{cases} \beta+(x_0-\beta)e^{-t/\alpha}, & 0<t<\tau, \\ -\beta+\left[2\beta e^{\tau/\alpha}+(x_0-\beta)\right]e^{-t/\alpha}, & \tau\le t<2\tau. \end{cases}$$

Now x_0 is determined so that $x(2\tau)=x_0$ where $x(2\tau)=\lim_{t\to 2\tau} x(t) = -\beta + [2\beta e^{\tau/\alpha}+(x_0-\beta)]e^{-2\tau/\alpha}$. Thus $x_0[1-e^{-2\tau/\alpha}]=\beta[2e^{-\tau/\alpha}-e^{-2\tau/\alpha}-1] \leftrightarrow$

$$x_0=\frac{-\beta(1-e^{-\tau/\alpha})^2}{(1-e^{-\tau/\alpha})(1+e^{-\tau/\alpha})}=\frac{\beta(e^{-\tau/\alpha}-1)}{(e^{-\tau/\alpha}+1)}=\boxed{-\beta\tanh\frac{\tau}{2\alpha}}.$$

(Note that another way to compute x_0 is to see that from symmetry, $x_0+x(\tau)=0$.)

Let $x_s(t)$ be the unique periodic solution. For any other initial condition where $x(0)=x_0\neq-\beta\tanh\frac{\tau}{2\alpha}$ with corresponding solution $x(t)$, one can

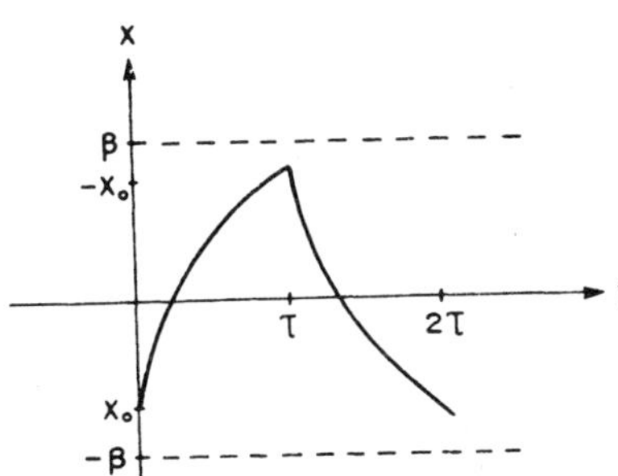

Figure VIII.12.7. Sketch of the periodic solution for one period.

show that $\lim_{t \to +\infty} [x(t) - x_s(t)] = 0$, i.e., *all* solutions tend to the periodic solution eventually. In this sense one sees that in general the output $x(t)$ *eventually* "forgets" about its given initial value x_0 and behaves as if it started with the initial value $-\beta \tanh \dfrac{T}{2\alpha}$.

Supplementary Problems

1.1. Solve each of the following for x:

(a) $|x-3|+|x-7|=2$.
(b) $|x-3|+|x-7|=4$.
(c) $|x-3|+|x-7|=6$.

1.2. Solve for x: $x^2 \le 5x-6$.

1.3. Solve for x: $|3x-4| \le 1$.

1.4. Solve for x: $|x-5| \ge |x+2|$.

1.5. Solve the following inequality: $2-x \ge |3x+1|$.

1.6. Solve the following inequality: $\dfrac{x^4-x^2}{x^2+1} < 0$.

1.7. Solve for x: $\dfrac{1-2x}{x+2} \le 3$.

1.8. Solve the following inequality: $\dfrac{-2x+2}{x+2} < 5-x$.

(*)1.9. For which real numbers x is the following inequality valid?

$$\frac{x}{x+1} \ge \frac{2}{x+2}.$$

1.10. A line whose inclination is 120° passes through the intersection of the lines $2x-3y=4$ and $x+2y+5=0$. Find its equation.

1.11. Find k if the line through the points $(1,4)$ and $(6,k)$ is parallel to the line $5x-y=3$.

1.12. Find the coordinates of the centre and the length of the radius of the circle whose equation is $x^2+y^2+6x-8y+21=0$.

1.13. Find an equation of the circle having $(5,-6)$ and $(-1,4)$ as ends of a diameter.

1.14. Graph the curve given by

$$4x^2-8x+y^2+4y-8=0$$

and locate the focal points.

1.15. Find the length of the common chord of the parabolas $y=x^2$ and $y^2=8x$. Illustrate graphically.

1.16. Consider the equation $(1+k)x^2+(2-k)y^2+x+2y=3$ (k constant). For what values of k is the graph of this equation in the xy-plane a parabola? an ellipse? a hyperbola?

1.17. Write the equation

$$y=2\sqrt{3}\cos x+2\sin x$$

in the form $y=A\sin(x+\theta)$, and hence find the period, amplitude, and phase displacement of $y=2\sqrt{3}\cos x+2\sin x$.

1.18. Express $\sin\left(2\tan^{-1}\dfrac{3}{4}\right)$ as a rational number $\dfrac{a}{b}$.

1.19. Determine the values of x which satisfy the equation $\log 2x(3\log x-2)=0$.

1.20. If $f(x)=2x^3-1$ with domain of $f=(-\infty,\infty)$, and $g(x)=3x+1$ with domain of $g=(-\infty,\infty)$, determine

(a) $\dfrac{f(1)}{g(1)}$;

(b) $f[g(1)]$;

(c) the inverse function of $f(x)$.

1.21. Find the inverse of $f(x)$ for the following:

(a) $f(x)=x^2+1,\ -1\le x<0$;

(b) $f(x)=e^{2x}$.

1.22. Consider the function $f(x)=e^{x^3-8}$. What are the domain and range of f? Find the inverse of f, and specify the domain and range of this inverse function.

(*)1.23. Show that the function $f(x)=\dfrac{x-2}{x+k}$ has an inverse for all constants $k\ne-2$. Find the value of k so that $f(x)\equiv f^{-1}(x)$.

2.1. Find y' where $y=\dfrac{u^2-1}{u^2+1}$ and $u=\sqrt[3]{x^2+2}$.

2.2. Find $\dfrac{dy}{dx}$ if $y=\log(x+\sqrt{x^2-1})$.

2.3. Find y' where $y=\log|3+5x-x^2|^{3/2}$.

2.4. Find $f'(x)$ where $f(x)=x|x|$.

2.5. Find the second derivative of $y=\dfrac{x-1}{x+1}$.

2.6. Find y' if $y=\dfrac{2x^2-x}{(1-x)^3}$.

(*)2.7. Find $\dfrac{dy}{dx}$ for the following cases:

(a) $y=\log_{10}(1+x\sqrt{x})$, (b) $y=x^x$.

2.8. Find $\frac{d}{dv}\left[\sin^{-1}\left(\frac{3v-1}{2}\right)\right]$ when $v = 0$.

2.9. Let $f(g(x)) = x$. Suppose $g(0) = 1$ and $g'(0) = 2$. Find $f'(1)$.

2.10. Use implicit differentiation to find $\frac{dy}{dx}$ at the point $(2,2)$ for the equation $\frac{1}{x} + \frac{1}{y} = 1$.

2.11. Find y' where $3^{xy} = \cot x + y$.

2.12. Find $\frac{dy}{dx}$ where $x \cosh y + \sqrt{xy} = 7$.

2.13. Find $\frac{dy}{dx}$ and $\frac{d^2y}{dx^2}$, given that $a^y + b^x = a^2b^2$ (a and b are constants).

2.14. (a) For $x \in [0, \infty)$, let $f(x) = e^{x^2}$. Show that $f(x)$ is increasing on its domain.

(b) Explain why $f(x)$ has an inverse function $f^{-1}(x)$.
(c) What is the domain of the inverse?
(d) What is $f^{-1}(e)$?
(e) Sketch the graph of $f^{-1}(x)$.

2.15. Let $f(x) = 2x + \log x$ for $x > 0$.
(a) Explain why $f(x)$ has an inverse function $g(x)$.
(b) For which values of x is $g(x)$ defined?
(c) Find $g'(2)$.

2.16. (a) One of the following two functions has an inverse for $x \geq 0$. Decide which one has an inverse and find the inverse function:

(i) $f(x) = e^{\sqrt{x+1}}$.
(ii) $g(x) = \tan^{-1}((x-1)^2)$.

(b) Both functions have inverses for $x \geq \frac{3}{2}$. Find the derivative of the inverse function of g.

2.17. For (a) $f(x) = \sqrt{x}$, $x > 0$, and (b) $f(x) = \cot x$, $0 < x < \pi$:
(i) Sketch the graph of $y = f(x)$.
(ii) Sketch the graph of the inverse function $y = f^{-1}(x)$.
(iii) Find the domain of $f^{-1}(x)$.
(iv) Find the range of $f^{-1}(x)$.
(v) Compute $\frac{d}{dx}f^{-1}(x)$ at $x = 1$.

2.18. (a) Sketch $y = \tanh x$.
(b) $x = \tanh y$ defines $y = \tanh^{-1}x$. Sketch $y = \tanh^{-1}x$.
(c) Find $\frac{d}{dx}\tanh^{-1}x$ as a function of x.

2.19. Find $\frac{dy}{dx}$ where $y = \int_0^x (2^t + t^2 - 3)\,dt$.

2.20. Let $f(x)=\int_0^{\sin x} g(t)\,dt$. Find $f'(x)$.

2.21. Find $f'(x)$ in the case $f(x)=\int_3^{e^{x^2}}(t^2+1)\,dt$.

2.22. Find $\dfrac{dy}{dx}$ where $y=\int_2^{x^2}\sin(t^2)\,dt$.

2.23. Let $f(x)=\int_{x^3}^{x^2}\dfrac{1}{(1+t^2)^3}\,dt$. Find $f'(x)$.

2.24. Evaluate $\dfrac{d}{dx}\int_x^{e^x}\sqrt{1+t^2}\,dt$.

2.25. Find a function f such that $\int_0^t f(x)\,dx=\dfrac{\sin t}{1+t^2}$ for all t.

(*)**2.26.** Prove that the function f, where $f(x)=\left(1+\dfrac{1}{x}\right)^x$ for $x>0$, is always increasing.

(*)**3.1.** Find the angle between the curves $y=\cos x$ and $y=\sin x$ at one of their points of intersection.

3.2. Find the equation of the tangent line to the curve $y=\log x$ at $x=1$.

3.3. Find a, b and c so that the parabola $y=ax^2+bx+c$ passes through the points $(3,0)$ and $(-5,0)$, and so that its tangent at $(3,0)$ has slope -3.

3.4. Determine the constants a, b, c and d such that the cubic polynomial $y=ax^3+bx^2+cx+d$ passes through the points $(0,0)$, $(1,1)$ and has zero slope at these points.

(*)**3.5.** Of all the straight lines through the point $(0,-2)$, which ones intersect the cubic curve $y=x^3-x$ thrice, twice, once, or not at all? You should use a picture to help you to think, but your answer must be algebraic —in terms of the equation of the line.

3.6. If $f(x)=\tan^{-1}[A(x-B)^2]$ for all x, determine the constants A and B so that the following conditions are both satisfied:

(a) $f(2)=\dfrac{\pi}{4}$;

(b) the graph of the function f has a horizontal tangent line at the point where $x=1$.

3.7. Determine the points on the curve $x^2+xy+y^2=12$ at which the slope of the tangent is 1. Also, write the equations of the tangent lines at these points.

(*)**3.8.** Find all points on the graph of $f(x)=x^3-3x^2+3x-1$ at which the tangent lines to the graph of $f(x)$ pass through the origin.

(*)**3.9.** A tangent line to the curve $x^{2/3}+y^{2/3}=a^{2/3}(a>0\text{ const.})$ cuts the coordinate axes at the points P and Q, respectively. Prove that the segment PQ has constant length a, independent of the point of tangency. Sketch.

3.10. Prove that the tangent line to the curve $y = x + 2x^2 - x^4$ at the point $(-1,0)$ is also tangent to the curve at the point $(1,2)$.

3.11. Find the value of x in the interval $0 < x < 4$ at which the tangents to the curves $y = x^3 + 6x^2 + x + 1$ and $y = 3x^2 + 25x + 10$ are parallel.

(*)3.12. Prove that the line tangent to the curve $y = -x^4 + 2x^2 + x$ at the point $(1,2)$ is also tangent to the curve at another point. Find this point.

(*)3.13. Find the equation of the line which is tangent to both the curves $y = x^2$ and $y = x^2 - 2x + 3$. Sketch.

3.14. Find the equation of the line normal to the curve $y = x^3 + 2x - 5$ at the point $(1, -2)$.

3.15. Show that the line normal to the curve $y = x^3 - 3x^2 + 2x + 1$, at the point where $x = 1$, passes through the origin.

3.16. If $(x + y)^3 - 5x + y = 1$ defines y as a differentiable function of x, find y' in terms of x and y. Find the equations of the tangent and normal lines to the curve at the point where it meets the line $x + y = 1$.

3.17. (a) By taking natural logarithms, find the slope of the tangent to the graph of $y = \sqrt[3]{2x+1}\,(4x-1)^2$ at the point $(0,1)$.

(b) Write down, in the form $y = mx + c$, the equation of the normal line to this graph at the given point.

(*)3.18. A smooth curve $y = f(x)$ is given which does not pass through the origin O. If Q is the point on the curve which is nearest to O, prove that the segment OQ is orthogonal to the curve at Q.

(*)3.19. Find the equation of the line in the xy-plane which makes an angle of $\frac{\pi}{3}$ with the tangent to the curve $x^3 + y^3 - 2xy = 0$ at the point $(1,1)$. Sketch.

(*)3.20. A circle is centred at $(3, \frac{9}{2})$ and just touches the curve $xy = 1$. What is the radius of the circle? Sketch the situation.

3.21. Determine the intervals of increase and of decrease of the function $f(x) = \dfrac{x+1}{(x-4)^2 x^{1/3}}$. Find the local extrema of $f(x)$.

3.22. Determine the value of the constant A for which the function $f(x) = x^2 + \dfrac{A}{x}$ has a relative minimum at $x = 2$.

3.23. Find the absolute maximum of $f(x) = 2x^3 - 3x^2 - 12x + 1$ on the interval $[0,4]$.

3.24. Determine the minimum value of $y = 2x^3 - 3x^2$ on $[-1,1]$.

3.25. Find the absolute minimum value of the function $f(x) = 2x - x^2$ on the interval $[2,3]$. Give reasons.

3.26. Find the maximum value of $f(p)=p^3+(1-p)^3$, $\frac{1}{4}\le p\le\frac{3}{4}$.

3.27. Find the absolute maximum and absolute minimum of $f(x)=x^3-\dfrac{3x^2}{2}-6x+2$ on the interval $[-2,4]$.

3.28. Determine the maximum and minimum values of the function $f(t)=t^3-3t$ on the interval $[0,2]$.

3.29. A student has determined that his mark in calculus will be

$$M(t)=\frac{t^3}{5}-\frac{21t^2}{5}+27t+25$$

where t is the number of hours per week that he spends studying calculus. How many hours per week should he spend studying calculus, if he cannot devote more than a total of 10 hours a week to it?

(*)3.30. Find the maximum and minimum values of the function $f(x)=\sqrt{|2x-3|}$ on $[0,2]$.

(*)3.31. For each constant $a>0$, a function f_a is defined for all $x>0$ by

$$f_a(x)=\begin{cases}\dfrac{1}{x}+4x+3 & \text{if } 0<x<a,\\[2ex] \dfrac{1}{x}+4x+2 & \text{if } x\ge a.\end{cases}$$

Find the value of $x>0$ which minimizes $f_a(x)$, and find the corresponding minimum value of $f_a(x)$, in the cases (a) $a=\frac{1}{4}$, (b) $a=\frac{3}{4}$, (c) $a=\frac{5}{4}$.

(*)3.32. Find the maximum and minimum values of $y=2\sin x+\cos 2x$.

(*)3.33. Find the maximum possible slope for a tangent line to the curve

$$y=\frac{8}{1+3e^{-x}}.$$

3.34. What is the largest possible value of $x^m(1-x)^n$ as x goes from 0 to 1? (m,n are positive constants).

(*)3.35. Let $f(x)=\dfrac{\log(x)}{1+[\log(x)]^2}$ for $x>0$.

(a) Consider $F(u)=\int_0^{u^2} f(x)\,dx$, where $f(x)$ is as above. Evaluate $F'(u)$ and show that $F'(e)=\dfrac{4}{5}e$.

(b) Express $f(e)$ and $f\left(\dfrac{1}{e}\right)$ in the form $\dfrac{p}{q}$ where p and q are integers.

(c) Show that f has a relative maximum when $x=e$.

(d) Show that $|f(x)|\le\frac{1}{2}$ for $x>0$.

4.1. Evaluate the following integral: $\displaystyle\int\frac{1}{x+\sqrt{x}}dx.$

4.2. Find all anti-derivatives of $\dfrac{x}{\sqrt{4-x^2}}$.

4.3. Evaluate

(a) $\int \sin^3\theta \cos^4\theta \, d\theta$.

(b) $\int \sin^2\theta \cos^2\theta \, d\theta$.

4.4. Evaluate $\int \dfrac{3x^2+2x}{(x^3+x^2+7)^{2/3}}dx$.

4.5. Evaluate $\int \dfrac{x}{\sqrt{1-4x^4}}dx$.

4.6. Evaluate $\int \cos(\log x)\, dx$.

4.7. Evaluate $\int \dfrac{dx}{x(x+1)}$.

(*)4.8. Determine:

(a) $\int (4-x^2)^{1/2}\, dx$;

(b) $\int \dfrac{x+1}{x^4+2x^2}dx$.

4.9. Find the indefinite integrals:

(a) $\int \dfrac{3x+1}{x^2+2x-3}dx$.

(b) $\int \dfrac{x-2}{x^2+4x+8}dx$.

4.10. (a) Resolve the function $\dfrac{9x^2+2x-3}{x^4-1}$ into a sum of partial fractions.

(b) Hence evaluate $\int \dfrac{9x^2+2x-3}{x^4-1}dx$.

4.11. Evaluate $\int \dfrac{x+4}{(x+2)(x^2+4x+5)}dx$.

4.12. Calculate $\int \dfrac{3x^3+x^2+x-1}{x^4-1}dx$.

4.13. Evaluate $\int \dfrac{1}{(x^2+1)(x-1)^2}dx$.

4.14. Find $\int \dfrac{5x^3-5x^2+8x-13}{(x^2+4)(x-1)^2}dx$.

4.15. Evaluate $\int \dfrac{x^2-3x+2}{x^3+6x^2+5x-12}dx$.

(*)4.16. Find $\int \frac{dx}{2+\tan x}$.

(*)4.17. Evaluate $\int \frac{dx}{1+\cos x+\sin x}$.

(*)4.18. Evaluate $\int \frac{2\,dx}{\sin x+\tan x}$.

5.1. Compute $\int_0^{\pi/4} \sin 2x\,dx$.

5.2. Evaluate the integral $\frac{1}{2}\int_0^{\pi/2} 36\sin^2\theta\,d\theta$ by interpreting it geometrically.

5.3. Evaluate the integral $\int_0^{\pi/4} \sec^4\theta\tan^2\theta\,d\theta$.

5.4. Compute $\int_0^{\pi/2} \sin^5 x\,dx$.

5.5. Determine $\int_0^3 |x^2-4|\,dx$.

5.6. Evaluate $\int_{-1}^1 \sqrt{|x|+x}\,dx$.

5.7. Evaluate $\int_{-2}^2 x|x-1|dx$.

5.8. Integrate:

(a) $\int_0^{\pi/4} (\sin x)\log(\cos x)\,dx$;

(b) $\int_1^2 x\log\sqrt{x}\,dx$ (use integration by parts);

(c) $\int_0^1 \frac{1}{(4-x^2)^{3/2}}dx$.

5.9. Evaluate the following definite integral:

$$\int_0^1 \frac{y^2\,dy}{y^2+2y+5}.$$

5.10. By taking $\log 1.5 \simeq 0.41$, find an approximate value for the integral

$$\int_1^2 \left(\frac{3x^4-15x^3+2x-5}{x^2-5x}\right)dx.$$

(*)5.11. Compute $\int_{-\pi/3}^{\pi/3} \frac{dx}{1-\sin x}$.

5.12. Evaluate $\int_0^1 \frac{dx}{\sqrt{(1+x^2)^3}}$.

5.13. If $a>1$ and $b>1$ show that

$$\int_1^a \frac{dx}{x}+\int_1^b \frac{dx}{x}=\int_1^{ab} \frac{dx}{x}.$$

5.14. (a) Show that the integral $I_n = \int_0^1 (1-x^2)^n \, dx$ satisfies the reduction formula $I_n = \frac{2n}{2n+1} I_{n-1}$ for any positive integer n.

(Hint: Write $(1-x^2)^n = (1-x^2)^{n-1} - x^2(1-x^2)^{n-1}$, thus expressing I_n as the difference of two integrals, and evaluate the second integral by parts.)

(b) Using part (a), evaluate $\int_0^1 (1-x^2)^4 \, dx$.

5.15. Show that $\int_0^1 x^n(1-x)^m \, dx = \int_0^1 x^m(1-x)^n \, dx$ where n and m are positive integers.

5.16. Show, by integration by parts, the equality

$$\int_0^1 x^2(1-x^2)^{1/2} \, dx = \frac{1}{3} \int_0^1 (1-x^2)^{3/2} \, dx.$$

6.1. Test for convergence, giving reasons: $I = \int_0^\infty xe^{-x} \, dx$.

6.2. Evaluate the following improper integral: $\int_0^\infty t^2 e^{-2t} \, dt$.

6.3. In each case evaluate the expression if it exists and if it does not exist explain why.

(a) $\int_1^2 x^4 \log x \, dx$;

(b) $\int_0^3 (x-1)^{-3} \, dx$.

6.4. Determine if the following improper integral converges:

$$I = \int_2^\infty \frac{x^3}{\sqrt{x^7+1}} \, dx.$$

6.5. Determine whether the following improper integrals converge or diverge:

(a) $\int_1^\infty \frac{x}{x^2+4} \, dx$;

(b) $\int_0^2 \frac{1}{(x-1)^{2/3}} \, dx$.

(*)6.6. Criticize the computation

$$I = \int_{-1}^2 \frac{dx}{x^2} = -x^{-1}\Big|_{-1}^{2} = -3/2$$

and show the correct solution.

6.7. Evaluate $\int_0^\infty \frac{dx}{x^2+2x+2}$.

(*)6.8. Evaluate $\int_1^\infty \frac{dx}{(1+x^2)^2}$.

6.9. Without attempting to evaluate, state whether the following integrals converge or diverge and give your reasoning.

(a) $\int_0^1 \frac{e^{-2x}}{x^{1/3}} dx$;

(b) $\int_2^{\infty} \frac{\log x}{\sqrt{x}} dx$.

(*)6.10. Let $F(x) = \int_0^{\infty} t^{x-1} e^{-t} dt$.

(a) Prove that $F(x)$ converges if and only if $x > 0$.

(b) Show that $F(x+1) = xF(x)$. Hence show that $F(n) = (n-1)!$ for any integer $n \geq 1$.

7.1. Sketch the region bounded by the graphs of the two functions $y = x^3 - 3x^2 + 3x - 1$, $y = 4x - 4$, and find the area A of this region.

7.2. Find the area bounded by the curves $y = x(x-2)^2$ and $y = x$.

7.3. Determine the value of $a > 0$ so that the area bounded by the two parabolas $y = ax^2 - 1$ and $y = 1 - ax^2$ is $\sqrt{8}$.

7.4. Sketch the region bounded by the graphs of $y = \frac{3}{x^2}$ and $y = 4 - x^2$, for $x > 0$, then find the area A of this region.

7.5. Find the value of k satisfying $0 < k < 1$ such that the area under $y = x^2$ from 0 to k is equal to the area under the same curve from k to 1.

7.6. Sketch the region lying in the half-plane $x > 0$ and bounded by the curves $y = x + \frac{1}{x}$ and $y = 3 - \frac{1}{x}$. Find the area A of this region.

7.7. Find the area of the region enclosed by the curves whose equations are $y^2 = x$ and $y = x - 2$.

7.8. Find the area of the region bounded by the graphs of the equations $x = 2y - y^2$ and $y = 2 + x$ (Fig. S.7.8).

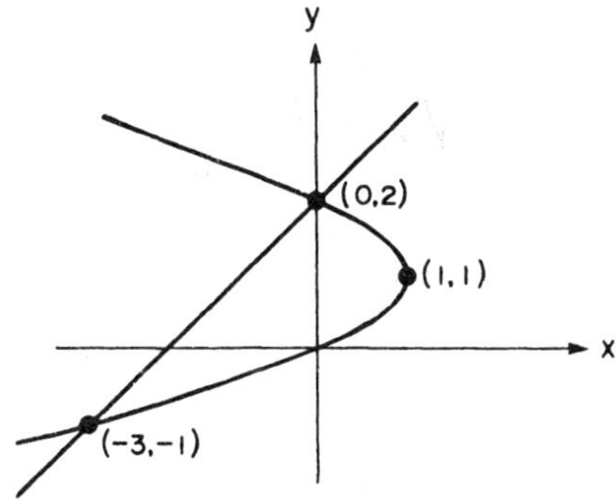

Figure S.7.8

7.9. Sketch the plane region consisting of all points whose coordinates (x, y) satisfy both

$$0 \le x \le \pi \quad \text{and} \quad 1 - \sin x \le y \le \sin x;$$

and find the area A of the region.

7.10. Find the area of the region bounded by the curve $x^{1/2} + y^{1/2} = a^{1/2}$, the x-axis, and the lines $x = a$, $x = b$ where $0 < b < a$.

7.11. Find the area in the first quadrant of the region bounded by the x-axis and the curves $x^2 + y^2 = 10$ and $y^2 = 9x$.

7.12. Find the area A bounded by the curve $x = 1 - \dfrac{1}{y^2}$ and lines $x = 1$, $y = 1$, and $y = 4$. Draw a rough sketch of the area you are computing.

7.13. Find thc area of the region in the first quadrant bounded by the y-axis and the curves $y = \cos x$, $y = \sin x$.

7.14. Over a certain portion of the interval $0 \le x \le 2\pi$, the graph of the function $y = \sin x$ lies above the graph of the function $y = \cos x$. Find the area between these two curves on that portion of the interval.

7.15. Find the area of the triangle bounded by the lines $y = 4x$, $y = -2x$ and $y = 8 - 4x$.

7.16. Find the area of the region bounded by $y = (\log x)^2$, $x = e$ and the x-axis.

(*)7.17. A manufacturer is designing square floor tiles of unit length on a side with two lines separating two colours as shown in Figure S.7.17. What should the value of the constant n be in order that equal amounts of each colour are used in every tile?

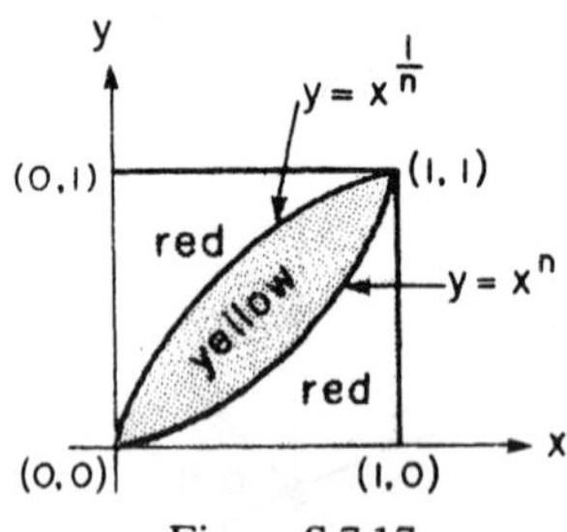

Figure S.7.17

(*)7.18. Make a sketch of the region R in the first quadrant bounded by the curves $y = \dfrac{4x^3}{\sqrt{x^4 - x^2 + 1}}$ and $y = \dfrac{2x}{\sqrt{x^4 - x^2 + 1}}$ between $x = 0$ and $x = 1$. Show that the area of R is $4 - 2\sqrt{3}$.

(*)7.19. A rectangle is constructed with its base on the x-axis and the two upper vertices on the curve $y = e^{-x^2}$.

(a) Find the area of the largest rectangle that can be so constructed.

(b) Show that the upper vertices of this rectangle of maximum area are at the points of inflection of the graph of $y = e^{-x^2}$.

7.20. Find the volume of the ellipsoid obtained by rotating the region bounded by the curve $y = \frac{1}{2}\sqrt{4 - x^2}$ and the x-axis, about the x-axis.

7.21. A football has a shape approximately the same as that generated by rotating the upper half of the area inside the ellipse

$$\frac{x^2}{9} + \frac{y^2}{4} = 1,$$

about the x-axis. Find the volume so generated.

7.22. Sketch the finite region bounded by $y = \sqrt{x}$, $y = 2$ and the y-axis. Find the volume V of the solid generated by rotating this region about the x-axis.

7.23. The curve $y = e^x$, $1 \leq x \leq 2$, is rotated about the x-axis. Make a sketch showing the resulting solid, and find the volume V of the solid.

7.24. Find the volume of the solid generated by revolving the area in the first quadrant bounded by the curve $y = x^2 + 1$, the y-axis and the line $y = 5$ about the x-axis.

7.25. Let R be the region bounded by the axes, $x = 1$ and $y = (4 - x^2)^{1/4}$. Find the volume of the solid obtained by rotating R about

(a) the x-axis;

(b) the y-axis.

7.26. Find the volume V of the solid generated when the region R bounded by the graphs $y = x^2 + 1$, $x = 0$, $x = 2$ and $y = 0$ is revolved about the y-axis. Make a clear sketch of R.

7.27. Consider the ellipse $\dfrac{x^2}{a^2} + \dfrac{y^2}{b^2} = 1$. Find the volume of the solid obtained by revolving this ellipse

(a) about the y-axis;

(b) about the x-axis.

7.28. Find the volume of the solid generated by revolving about the x-axis the area of the triangle having vertices $(0,0)$, $(0,2)$ and $(4,2)$.

7.29. Suppose that the region bounded below by the curve $y = x^2$, above by $y = 2x$ and on the left by $x = 1$ is revolved about the line $x = \frac{1}{2}$. Find the volume of the resulting solid.

7.30. The curve $y = e^x$, $1 \leq x \leq 2$, is rotated about the line $y = 1$. Sketch the resulting solid, and find its volume V.

7.31. Use the cylindrical shell method to show that the volume of the solid generated when the region

$$R = \{(x, y) | \ 0 \le x \le y^{3/2}, 0 \le y \le 3\}$$

is revolved through 2π radians about the line $y = -1$ is

$$\frac{792\sqrt{3}\,\pi}{35}.$$

7.32. Using the techniques of integral calculus, derive a formula for the volume of a right circular cone of radius r and height h.

7.33. (a) Sketch the region bounded by the graphs of

$$f(x) = |x| \quad \text{and} \quad g(x) = 2 - x^2.$$

(b) Find the area of this region.

(c) Find the volume of the solid which is generated by revolving the region about the x-axis.

7.34. The area bounded by the curve

$$y = \sqrt{x}\, e^{-x/2} (x \ge 0), \qquad x = M (M > 0), \quad \text{and}$$

$y = 0$ is rotated about the x-axis. Set up the integral that represents the volume of the solid which is generated, including an appropriate rough sketch. Find the volume V. This volume depends on M; what is its limiting value as $M \to +\infty$?

(*)7.35. Consider the region between the curve $y = e^{-x}$ and the x-axis, as x runs from zero to infinity.

(a) Find the area A of the region, if it is finite.

(b) If the region is revolved about the line $y = -1$, will the volume V of the solid swept out by the region be finite? Explain your answer.

(*)7.36. (a) A solid of revolution is formed by revolving the area bounded by the parabola $y = 16 - x^2$ and the x-axis about the y-axis. Find the volume of this solid.

(b) Find the radius of the right circular cylinder of maximum volume which can be inscribed in the solid body in (a) (Fig. S.7.36).

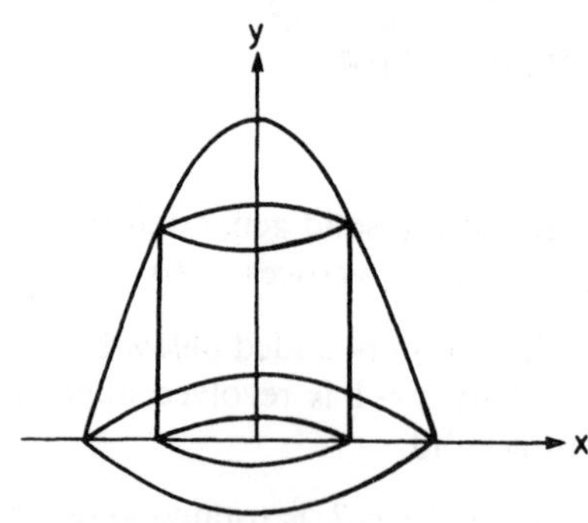

Figure S.7.36

7.37. Find the length of the curve $9x^2 = 4y^2$ from $(0,0)$ to $(2\sqrt{3}, 3)$.

(Hint: The problem is most easily solved when y is used as the variable of integration.)

(*)7.38. Find the length of the loop of the curve $9y^2 = x(x-3)^2$.

(*)7.39. An electric wire connecting a telephone pole to a house (Fig. S.7.39) hangs in the shape of a catenary $y = c\cosh\frac{x}{c}$, where the units are metres. Find the length of the wire.

(Hint: First find the value of c.)

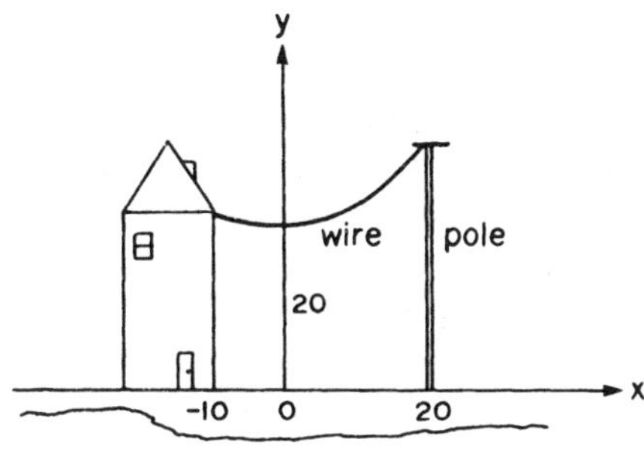

Figure S.7.39

(*)7.40. Find the area of the surface of revolution generated by revolving about the x-axis the arc of $4x = 2\log|y| - y^2$ from $y=1$ to $y=3$.

7.41. Sketch the arc of the parabola $y^2 = 4 - x$ that lies in the first quadrant.

(a) Write an integral that describes the volume of the figure obtained by rotating about the y-axis the area bounded by the above arc, the x-axis, and the y-axis.

(b) Write an integral that describes the volume of the figure obtained by rotating about the y-axis the area bounded by the above arc, the x-axis, and the line $x=1$.

(c) Write an integral that represents the length of the arc between the point $(0,2)$ and the point $(4,0)$.

(*)7.42. Let $f, g\colon [1,4] \to \mathbb{R}$ be differentiable functions satisfying $f(3) = g(3)$. Suppose the regions J, H, K are bounded by the curves $y = f(x)$, $y = g(x)$, $x = 1$, $x = 4$, and the x-axis, as shown in Figure S.7.42. Can you express the following quantities by means of definite integrals?

(a) The area of J.

(b) The total length of the boundary of H.

(c) The volume of the solid generated by the revolution of K about the x-axis.

(d) The total surface area of the solid generated by the revolution of H about the line $y = -3$.

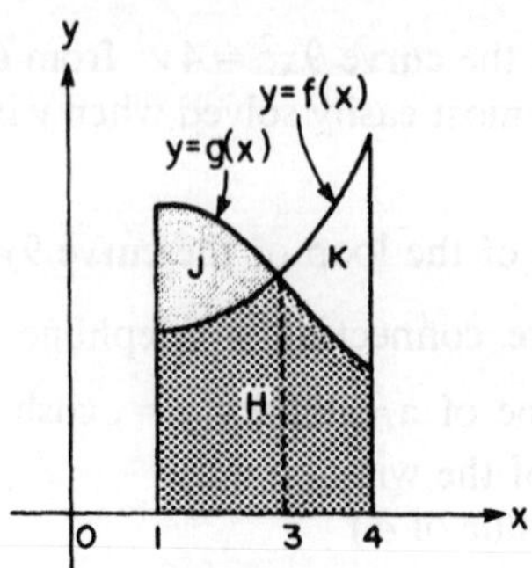

Figure S.7.42

(*)**7.43.** A sphere of radius 2 and a circular cylinder of radius 1 with axis of symmetry along the y-axis are both centred at the origin and meet in a circle. Consider the ice cream cone so formed that its cone has vertex at the origin and base on this circle and the scoop of ice cream is bounded by the portion of the hemisphere cut off by the cylinder. Calculate

(a) the volume of the entire ice cream cone (cone plus ice cream top);

(b) the surface area of the entire ice cream cone (cone plus ice cream).

8.1. Parametrize the ellipse E given by the equation

$$\frac{x^2}{a^2}+\frac{y^2}{b^2}=1.$$

8.2. Parametrize the ellipse E given by the equation

$$\frac{x^2}{a^2}+\frac{y^2}{b^2}-\frac{2x}{a}-\frac{2y}{b}=0.$$

8.3. Graph the curve given parametrically by

$$\left\{\begin{array}{l} x=\dfrac{1}{\cos t} \\ y=\cos t \end{array}\right\}, \qquad 0\le t<\frac{\pi}{2}.$$

8.4. Consider the curve given by $x=\cos t$, $y=\sin t$, $0\le t\le 2\pi$. Find the slope of this curve at $t=\dfrac{\pi}{6}$.

8.5. Find $\dfrac{dy}{dx}$ for $\left\{\begin{array}{l} x=\sin t+t \\ y=e^t+t^2+2 \end{array}\right\}$.

8.6. A curve C has parametric representation $x=6t+1$, $y=t^3-2t$.

(a) Find points on C at which the tangent line is perpendicular to the line $3x+5y-8=0$.

(b) Does the curve have a vertical tangent at any point?

(c) Find all the relative maximum values of y.

8.7. Find $\dfrac{d^2y}{dx^2}$ for the following parametric equations: $x = t^3 - 3t$, $y = 2 - t - t^2$.

8.8. Find $\dfrac{d^2y}{dx^2}$ *in terms of* t, if $\left\{\begin{matrix} x = \log t \\ y = e^t \end{matrix}\right\}$.

8.9. Find $\dfrac{dy}{dx}$ and $\dfrac{d^2y}{dx^2}$ if $y = \sin t$, $x = t^2 - \cos t$.

(*)**8.10.** In the xy-plane consider a circle of radius R centred at the origin O and consider a point $A = (a, 0)$. At any point T on the circle construct the tangent line l to the circle. Let $M = (x, y)$ be the point on l where the line AM is perpendicular to l. Let t be the angle which OT makes with the x-axis. Find the coordinates of M in terms of t.

(*)**8.11.** Calculate the area A of the region R formed by the curve C parametrized by the equations $x = 4a\cos t$, $y = a\sqrt{2}\sin t\cos t$, $-\dfrac{\pi}{2} \le t \le \dfrac{\pi}{2}$.

(*)**8.12.** Since $\cosh^2 t - \sinh^2 t = 1$, the curve described parametrically by $\left\{\begin{matrix} x = \cosh t \\ y = \sinh t \end{matrix}\right\}$ is the hyperbola $x^2 - y^2 = 1$.

(a) Show that the area of the shaded region (Fig. S.8.12) bounded by the line joining the origin to $P(\cosh T, \sinh T)$, the hyperbola $x^2 - y^2 = 1$ and the positive x-axis is $\dfrac{T}{2}$.

(b) What is the (well known) analogous result for $\cos t$, $\sin t$?

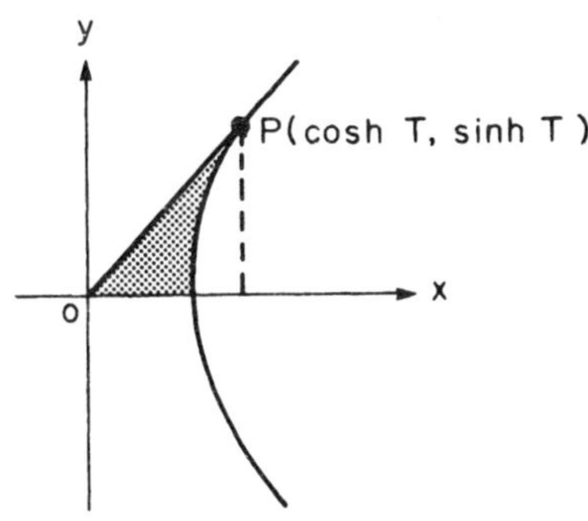

Figure S.8.12

8.13. Find the length of the arc described parametrically by $x = 8t^3$, $y = 4t^2$, from $t = 0$ to $t = 1$.

8.14. (a) Give the formula for finding the length of a curve $x = x(t)$, $y = y(t)$ for $t_0 \le t \le t_1$.

(b) Find the length of the curve $\left\{\begin{matrix} x(t) = 2(4t + 5)^{3/2} \\ y(t) = 3(2t + 2)^2 \end{matrix}\right\}$ for $-1 \le t \le 1$.

8.15. Find the length of the arc of the curve

$$\begin{Bmatrix} x = e^t \sin t \\ y = e^t \cos t \end{Bmatrix}$$

from $t = 0$ to $t = \pi$.

8.16. Consider the parametric curve

$$\begin{cases} x = t^2 + 2, \\ y = t^3, \end{cases} \qquad -\infty < t < \infty.$$

(a) Find $\dfrac{dy}{dx}$, $\dfrac{d^2y}{dx^2}$ and the equation of the tangent line at the point (3, 1).
(b) Find the length of the curve from the point (2,0) to (3,1).

(*)**8.17.** The equations $x = t - \sin t$, $y = 1 - \cos t$ $(0 \le t \le 2\pi)$ are the parametric equations of an arc of a cycloid (Fig. S.8.17).
(a) Find the length of the arc.
(b) Find the area of the region between the arc and the x-axis.

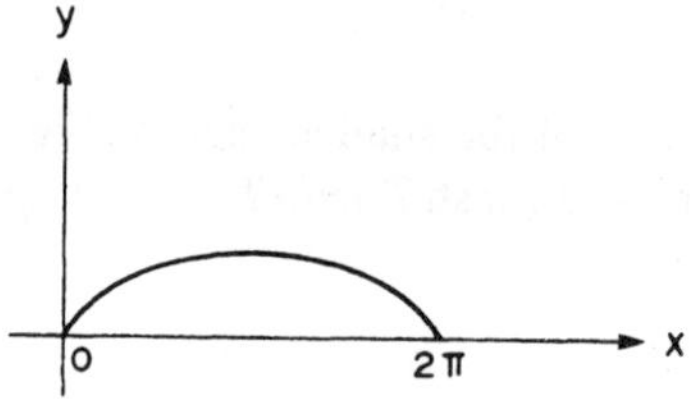

Figure S.8.17

9.1. (a) Express the equation $r = 3 + 2\sin\theta$ in rectangular coordinates.
(b) Is (0,0) a point on this curve? Explain.

9.2. Graph $r = \dfrac{1}{\sqrt{2}}(\sin\theta + \cos\theta)$.

9.3. Sketch $r = \dfrac{1}{2 - \cos\theta}$.

(*)**9.4.** Compute $\dfrac{dy}{dx}$ given $r = \dfrac{2}{1 - \sin\theta}$, $0 \le \theta < 2\pi$. At what points (if any) on the graph of this function does the curve have horizontal tangents? vertical tangents?

9.5. Compute the area of the region inside $r = \sqrt{\sin\theta}$.

9.6. Find the area enclosed by the cardioid $r = 2 - 2\sin\theta$.

9.7. Find the area bounded by the curve which, in polar coordinates, is given by the equation $r = 3 + 2\cos\theta$.

9.8. Set up the integral (do not evaluate it) which would give the area outside the cardioid $r = 1 + \sin\theta$ but inside the circle $r = 1$.

9.9. Calculate the area of the crescent-shaped region inside the circle $r = 3\cos\theta$, and outside the cardioid $r = 1 + \cos\theta$. Sketch the curves.

(*)9.10. (a) Express in cartesian coordinates the curves $r = 2\sin\theta$ and $r = 2\cos\theta$.

(b) Find the area included by these curves.

(*)9.11. Find the area of the region that is outside $r = \cos 3\theta$ and inside $r = 2\cos\theta$.

9.12. Sketch the graph of the curve $r = \sin^2\theta$ and evaluate the area A interior to the curve.

9.13. Consider the two curves of the respective polar equations $r^2 = 2a^2\cos 2\theta$ and $r = a$. Calculate the area of the region R between the two curves as indicated in Figure S.9.13.

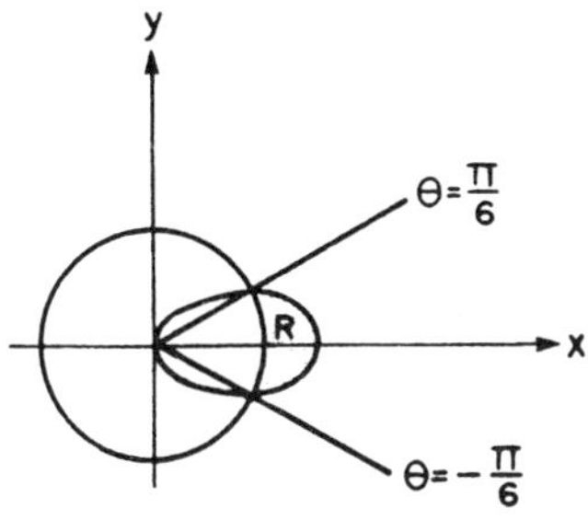

Figure S.9.13

9.14. Calculate the area of the region T bounded by the circle $x^2 + y^2 = 1$, $x \geq 0$ and the line $x + y = 1$, $y \geq 0$.

9.15. Find the area of the region interior to the cardioid $x^2 + y^2 = 2[x + \sqrt{x^2 + y^2}]$ and exterior to the circle $x^2 + y^2 = 4$. Draw a sketch.

(*)9.16. (a) What is the polar equation $r = f(\theta)$ for the ellipse whose cartesian form is $\dfrac{x^2}{a^2} + \dfrac{y^2}{b^2} = 1$?

(b) Find the points of intersection for the two ellipses $\dfrac{x^2}{a^2} + \dfrac{y^2}{b^2} = 1$ and $\dfrac{x^2}{b^2} + \dfrac{y^2}{a^2} = 1$.

(c) Find the area common to these two curves. (Draw a sketch.)

(*)9.17. Find the area of the surface generated by revolving the lemniscate $r^2 = 2a^2\cos 2\theta$ about the y-axis.

10.1. Show whether $\sum\limits_{n=1}^{\infty} ne^{-n^2}$ converges absolutely, conditionally, or is divergent.

10.2. Determine whether the given series converges or diverges. Explain your answers.

(a) $\sum_{k=1}^{\infty} (-1)^k \frac{k^2}{k^4+2k^2+1}$.

(b) $\sum_{k=1}^{\infty} (-1)^k \frac{\log k}{k}$.

10.3. Determine the convergence or divergence of the following series. Give reasons.

(a) $\sum_{n=2}^{\infty} \frac{1}{n(\log n)^2}$;

(b) $\sum_{n=1}^{\infty} \frac{1}{n^2-8}$;

(c) $\sum_{n=1}^{\infty} n \sin\left(\frac{1}{n}\right)$;

(d) $\sum_{n=1}^{\infty} \frac{(-1)^n}{\sqrt[6]{n}}$;

(e) $\sum_{n=1}^{\infty} \frac{5^n}{(2n+1)!}$;

(f) $\sum_{n=1}^{\infty} \log\left(1+\frac{1}{n}\right)$.

(*)10.4. Discuss the convergence of $\sum_{n=0}^{\infty} \sin^n k$ for various values of the constant k.

(*)10.5. Show that $\sum_{n=1}^{\infty} \left\{\frac{4}{2^n} - \frac{2}{n(n+1)}\right\}$ converges and find its sum S.

10.6. (a) Use a partial sum and the integral test to find bounds for $S = \sum_{k=1}^{\infty} \frac{1}{k^6}$.

(b) Find a value for S which has an error of no more than 0.001.

(*)10.7. Show that $\lim_{n\to\infty} \sum_{i=0}^{n} \frac{1}{n+i} = \log 2$ by writing the sum under the limit as an nth upper sum of an appropriate function.

(*)10.8. Suppose $\lim_{n\to\infty} a_n = A$.

(a) Find a formula for the nth partial sum of the series $\sum_{k=1}^{\infty} (a_k - a_{k+1})$.

(b) Find $\sum_{k=1}^{\infty} (a_k - a_{k+1})$.

(c) Find $\sum_{k=1}^{\infty} (a_k - a_{k+2})$.

10.9. Give an example of each of the following:

(a) A sequence which is unbounded.

(b) A sequence which is bounded, but does not converge.

(c) A sequence which converges, but for which the terms are those of a divergent series.

(d) A sequence whose terms are terms of a convergent series.

(*)10.10. (a) Find a divergent sequence $\{a_n\}_{n=1}^{\infty}$ such that $-1<a_n<1$ for all n.

(b) Find two divergent series $\sum_{n=1}^{\infty} a_n$ and $\sum_{n=1}^{\infty} b_n$ such that $\sum_{n=1}^{\infty} (a_n+2b_n)$ is convergent.

(c) Find a series $\sum_{n=1}^{\infty} a_n$ such that $\sum_{n=1}^{\infty} a_n=2$ and $\sum_{n=1}^{\infty} a_n^2=10$.

10.11. Determine the interval of convergence for the series $\sum_{k=1}^{\infty} \frac{(x+3)x^k}{k!}$.

10.12. What is the Taylor polynomial of degree 3 of $x+\cos x$ about $x=0$?

10.13. Let $f(x)=x^5$. Find the Taylor polynomial $p_5(x)$ of degree 5 for f around $x=1$. Show for $x=-1,0,1$, that $p_5(x)=f(x)$. (In fact, this is true for all x.)

(*)10.14. A certain function f has the property that $f''(x)=\frac{1}{1+x^3}$ whenever $x>-1$. It is known also that $f(1)=1$ and $f'(1)=2$.

(a) Write down an appropriate Taylor's formula and

(b) use it to find a second degree polynomial that is an approximation to $f(x)$ when x is close to 1. Hence

(c) obtain an approximate value for $f(\frac{3}{2})$ and

(d) (by using Taylor's theorem) estimate the error in your result. (Do not attempt to determine $f(x)$ by integration.)

10.15. Calculate the quadratic polynomial approximation (first three terms of Taylor's formula) for the function $f(x)=x^2\log x$ about the point $x=e$.

(*)10.16. Use the Taylor series for e^x about $x=0$ to write down the first few terms of the Taylor series for $f(x)=xe^{-x^3}$ about $x=0$. Use this Taylor series for $f(x)$ to determine $f^{(4)}(0)$ and $f^{(5)}(0)$.

10.17. (a) Calculate the first two nonvanishing terms of the Taylor series for $\sin x$ about $x=0$.

(b) Use the results of part (a) to obtain the first two nonvanishing terms of the Taylor series for $\sin(x^2)$ about $x=0$.

(c) Use your result to obtain an approximation to $F(x)=\int_0^x \sin(t^2)\,dt$ that is valid for small x.

10.18. Exhibit the first six terms of the Maclaurin series for $f(x)=\sin(x)+\cos(2x)$.

(*)10.19. Use the Maclaurin series of $\dfrac{1}{1-x}$ to find the Maclaurin series of $\dfrac{1}{(1-x)^2}$.

(*)10.20. Find the Maclaurin series of $f(x) = \dfrac{x^{20}}{(1-x)^4}$.

(*)10.21. For each of the following functions find the full Taylor series expansion and region of convergence about the indicated point:

(a) $\dfrac{x}{1+x^2}$, about $x = 0$.

(b) $\dfrac{1}{x}\operatorname{erf}(x)$ where $\operatorname{erf}(x) = \dfrac{2}{\sqrt{\pi}}\int_0^x e^{-t^2}\,dt$, about $x = 0$.

(c) $\sinh x$, about $x = 2$.

(*)10.22. Find $\lim_{n\to\infty} (a + 2ar + 3ar^2 + \cdots + nar^{n-1})$ where $|r| < 1$.

(*)10.23. If $f(x) = \dfrac{x^5}{1+x^2}$, find $f^{(31)}(0)$.

(*)10.24. Find the power series representation about $x = 0$ for

(a) $\dfrac{1}{2-x}$;

(b) $\dfrac{x}{1+x-2x^2}$.

11.1. Evaluate $\lim_{x\to\infty} \dfrac{(2x-1)(x^2+1)^3(x^2-4)}{x^4(x+1)^2(x^3+9)}$.

11.2. Evaluate the following limits:

(a) $\lim_{x\to 2} \dfrac{3x^2 - x - 10}{x^2 + 5x - 14}$;

(b) $\lim_{x\to+\infty} \dfrac{x}{2x - \dfrac{7}{x}}$.

11.3. Compute $\lim_{x\to\infty} \dfrac{3x^2 + 6\sqrt{x}}{x^2 + 8\sqrt{x}}$.

11.4. Evaluate $\lim_{x\to-\infty} (\sqrt{2-x} - \sqrt{1-x})$.

11.5. Find $\lim_{x\to+\infty} \left[\sqrt{9x^2+4x} - 3x\right]$.

(*)11.6. Let $f(k) = \lim_{x\to 2} \dfrac{x^2 - 5x + 3k}{x^2 - 3x + k}$. Determine $f(k)$ for all k.

(*)11.7. Evaluate $\lim_{x\to 0^-} \dfrac{x}{x^2 + |x|}$.

(*)11.8. Determine whether the following limits exist and evaluate when possible:

(a) $\lim_{x \to -3} \frac{|x+3|}{x+3}$;

(b) $\lim_{x \to \infty} \frac{(x^3+2x)\sin(1/x)}{2x^2-1}$.

11.9. Evaluate $\lim_{x \to 0} \left(\frac{1-\cos x}{\sin x}\right)$.

11.10. Evaluate $\lim_{x \to +\infty} \frac{\log x}{x^{1/2}}$.

11.11. Find $\lim_{x \to 0} \left(\frac{e^x - 1}{xe^x}\right)$.

11.12. If $f(1) = 1$ and $f'(1) = 2$, compute $\lim_{x \to 1} \left(\frac{\sqrt{f(x)} - 1}{\sqrt{x} - 1}\right)$.

11.13. Find $\lim_{x \to \infty} e^{-x}\sqrt{x}$.

(*)11.14. (a) Prove that $\lim_{x \to +\infty} \frac{\log x}{x} = 0$.

(b) From part (a) deduce that $\lim_{x \to 0^+} x \log x = 0$.

(*)11.15. Evaluate $\lim_{x \to 0} \left[\frac{\int_{-x}^{x} f(t)\,dt}{\int_0^{2x} f(t+4)\,dt}\right]$ where f is continuous.

11.16. Evaluate $\lim_{x \to 1} \frac{1}{\log x} \int_1^x e^{t^2}\,dt$.

11.17. Find $\lim_{x \to 1} \left[\frac{\int_a^x \cos(t^2)\,dt}{\sin \pi x}\right]$.

11.18. Evaluate $\lim_{x \to +\infty} e^{-x^2} \int_0^x e^{t^2}\,dt$.

11.19. Find $\lim_{x \to 0} \frac{\int_0^x \sin(t^2)\,dt}{x^3}$.

(*)11.20. Find $\lim_{x \to +\infty} \left(\frac{1}{x} \int_0^x e^{t^2 - x^2}(t^2+1)\,dt\right)$.

(*)11.21. Show that, if f is continuous on $[0, 1]$, then

$$\lim_{x \to 0^+} \left[x \int_x^1 \frac{f(u)}{u^2}\,du\right] = f(0).$$

11.22. Find $\lim_{x \to 0} \left(\frac{1}{\sin x} - \frac{1}{x}\right)$.

11.23. Evaluate $\lim_{n\to\infty}\left(1-\frac{1}{2n}\right)^n$.

(*)11.24. Evaluate the following limit: $\lim_{x\to\infty}\left(1+\frac{1}{2x}\right)^{x^2}$.

(*)11.25. Evaluate the limit: $\lim_{x\to 0^+}(e^x+2x)^{1/x}$ if it exists. If it fails to exist, specify whether the answer $+\infty$ or $-\infty$ is suitable, giving a reason.

(*)11.26. Evaluate the following:

(a) $\lim_{x\to\infty}\left(1+\frac{5}{x}\right)^{3x}$;

(b) $\lim_{x\to 0^+}(\tan x)^{\sin x}$;

(c) $\lim_{x\to 1^+}\left(\frac{\log[\sin(x-1)]}{2x^2}\right)$;

(d) $\lim_{x\to 5}\left[\frac{(x-5)^2 e^{x^2}\cos(2\pi x)\log x}{\sin^2(x-5)}\right]$.

(*)11.27. Compute:

(a) $\lim_{n\to\infty}\left[\frac{2n^2-3\sqrt{n}+5}{(3n+1)^2+\sqrt[3]{n}}\right]$;

(b) $\lim_{n\to\infty}\sqrt[n]{n^2+n}$;

(c) $\lim_{x\to 0}\left[\frac{\log(1+x)-x+\frac{x^2}{2}-\frac{x^3}{3}}{x^3}\right]$.

(*)11.28. Compute the following limits:

(a) $\lim_{x\to 0}\left(\frac{2-\sin x}{x+1}\right)$;

(b) $\lim_{x\to 0}\left[\frac{e^x-1-x-\sin x^2}{(\sin x)^3}\right]$;

(c) $\lim_{x\to 1^+}\left[x^{1/(1-x^2)}\right]$.

(*)11.29. (a) Give the first three non-zero terms of the Maclaurin series for each of the functions $\sin(2x)$, $\cos(3x)$ and e^{-x^2}.

(b) Using your results from (a), evaluate:

(i) $\int_0^{0.1} e^{-x^2}\,dx$ correct to four decimal places.

(ii) $\lim_{x\to 0}\left[\frac{\cos(3x)-e^{-x^2}}{x\sin(2x)}\right]$.

(*)11.30. The height above ground of an object that is thrown up with an initial velocity of 1000 feet per second, in a medium whose resistance is proportional to the velocity, is

$$y(t)=\frac{-32t}{k}+\frac{32+1000k}{k^2}(1-e^{-kt}).$$

Show that as $k\to 0$, this formula approaches that for zero resistance ($k=0$ case), namely, $-16t^2+1000t$.

12.1. If $\dfrac{dy}{dt}=\dfrac{9}{9+t^2}$ and $y(0)=\pi/2$, find $y(\sqrt{3})$.

12.2. If $\dfrac{dy}{dt}=\dfrac{t}{\sqrt{t^2+9}}$ and $y(0)=1$, find $y(4)$.

12.3. If $\dfrac{dy}{dt}=\dfrac{1}{\sqrt{4-t^2}}$ and $y(1)=0$, find $y(2)$.

12.4. A psychological model of stimulus and response (the Weber Fechner Law) states that the rate of change of response, R, with respect to stimulus, S, is inversely proportional to the stimulus.

(a) State this law as a differential equation.

(b) Solve the differential equation with the initial condition that $R=0$ for $S=S_0$.

12.5. At each point (x, y) of a certain curve the slope is $(2\log x)/y$. If the graph contains the point $(1,2)$ find the equation of the curve.

12.6. Find the equation of the curve passing through the point $(3,2)$ and having the slope $m=\dfrac{x^2}{y^3}$ at any point (x, y).

12.7. Find the general solution of the differential equation $\dfrac{d^2y}{dx^2}=4x+3$.

12.8. At any point (x, y) on a curve, $\dfrac{d^2y}{dx^2}=1-x^2$. At the point $(1,1)$ on this curve the equation of the tangent line to the curve is $y=2-x$. Find the equation of the curve.

12.9. Verify that $y=x\sin x$ is a solution of the differential equation $x^2y''-2xy'+(x^2+2)y=0$.

(*)12.10. If $x=e^t$, show that the differential equation $x^2\dfrac{d^2y}{dx^2}+x\dfrac{dy}{dx}+y=0$ may be written as $\dfrac{d^2y}{dt^2}+y=0$.

13.1. Represent $y=2x/(1-x^2)$ as a sum of two simpler functions and find the nth derivative $y^{(n)}(x)$.

(*)13.2. Consider the function $f(x)=\dfrac{1-\sqrt{1+x}}{1+\sqrt{1+x}}$.

(a) Calculate the derivative $Df(x)$ and $\lim_{x\to\infty} f(x)$;
(b) give the domain of f and the domain of its inverse f^{-1};
(c) calculate the derivative $Df^{-1}(0)$;
(d) sketch the graph of f and that of f^{-1}.

(*)13.3. The equation $y = k_1 x(x - k_2)$ determines a family of parabolas. (Thus, a particular member of the family is obtained by assigning values to the constants k_1 and k_2.) Find the member of this family (i.e., determine the appropriate k_1 and k_2) which passes through the point $(4,1)$ and which bounds the least area with the x-axis. Assume $k_1 < 0$.

(*)13.4. (a) Sketch that portion of the graph of $x^k + y^k = 1$ which lies in the first quadrant, where $k > 0$. Distinguish three cases: $0 < k < 1$, $k = 1$, $k > 1$.

(b) Which points on the curve sketched in (a) lie
(i) closest to the origin?
(ii) furthest from the origin?

(c) Find the arc length of the curve in (a) for the particular cases when $k = 2/3$, $k = 1$, $k = 2$. Find a good approximation to the arc length when k is very large.

(d) Let (u, v) be a point on the curve in (a) (so that $u \geq 0$, $v \geq 0$). Find the equation of the tangent to the curve through (u, v). What is the length of the segment of this tangent which lies in the first quadrant? Give two values of k for which the length of this segment is independent of the point (u, v) chosen.

13.5. (a) Find all anti-derivatives of

(i) $2x\tan^{-1}x$,
(ii) $\dfrac{\sec^2 x}{1+\tan x}$.

(b) Hence solve the differential equation $\dfrac{dy}{dx} = 2x\tan^{-1}x + \dfrac{\sec^2 x}{1+\tan x}$, if $y = 2$ when $x = 0$.

(*)13.6. Let f be the function defined by the following:

$$f(t) = \begin{cases} (1+t)^{1/t} & \text{for } -1 < t < 0, \text{ and } t > 0 \\ e & \text{for } t = 0. \end{cases}$$

(a) Prove that f is continuous for $t > -1$.
(b) Show that $f'(t) = f(t)t^{-2}[t(1+t)^{-1} - \log(1+t)]$ for $t \neq 0$, and hence that $f'(t) < 0$ for $t > -1$, except at $t = 0$.
(Hint: Show that the quantity in square brackets has a maximum value, 0, for $t = 0$.)
(c) Conclude from (a) and (b) that for $0 < t < 1$,

$$(1+t)^{1/t} < e < (1-t)^{-1/t}.$$

13.7. Choose a non-mathematician:

(a) John von Neumann.
(b) Mick Jagger.
(c) Georg Cantor.
(d) Pablo Casals.
(e) Stanley I. Grossman.
(f) Rene Descartes.
(g) Guy Lafleur.

Chapter IX

Answers to Supplementary Problems

Chapter I

1. Equation: $\dfrac{(x+3)^2}{4}+\dfrac{(y-1)^2}{16}=1$; ellipse, centre $(-3, 1)$; horizontal semi-axis length $=2$; vertical semi-axis length $=4$; if $y=0$, then $x=-3 \pm \frac{1}{2}\sqrt{15}$; if $x=0$, y does not exist. See Figure SI.1.

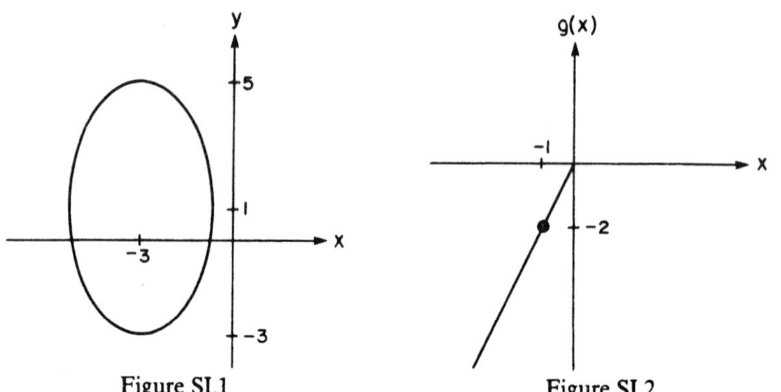

Figure SI.1

Figure SI.2

2. See Figure SI.2.

3. For $x \leq \frac{1}{2}$, $x=-\frac{1}{2}y^2-\frac{1}{2}$, $y \geq 0$; for $x \geq \frac{1}{2}$, $x=+\frac{1}{2}y^2+\frac{3}{2}$, $y \geq 0$. Note symmetry with respect to the line $x=\frac{1}{2}$ (Fig. SI.3).

4. Relative maximum at $(-1, 13)$; relative minimum at $(3, -19)$; increasing for $-\infty < x < -1$, $3 < x < \infty$; inflection point at $(1, -3)$; concave up for $1 < x < \infty$. (Use synthetic division to evaluate polynomials.) See Figure SI.4.

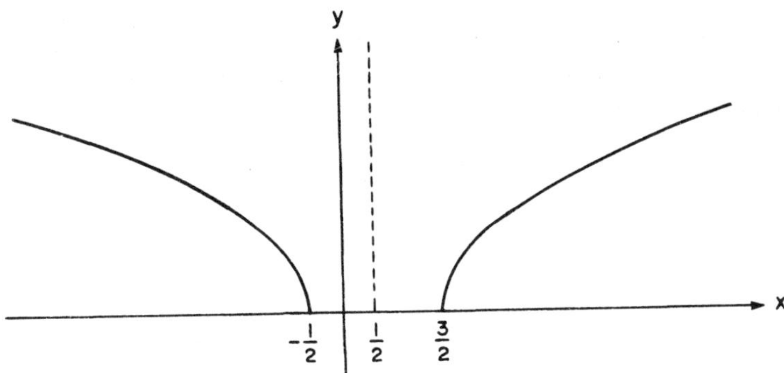

Figure SI.3

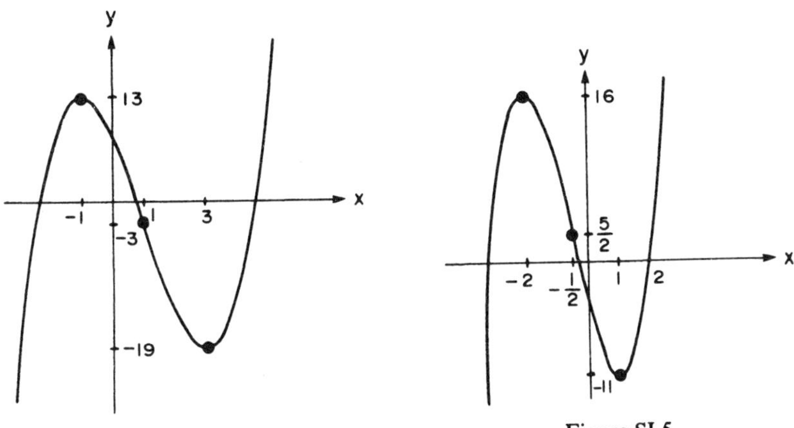

Figure SI.4

Figure SI.5

5. Relative maximum at $(-2, 16)$; relative minimum at $(1, -11)$; increasing for $-\infty < x < -2$, $1 < x < \infty$; inflection point at $x = -\frac{1}{2}$; concave up for $-\frac{1}{2} < x < \infty$. See Figure SI.5.

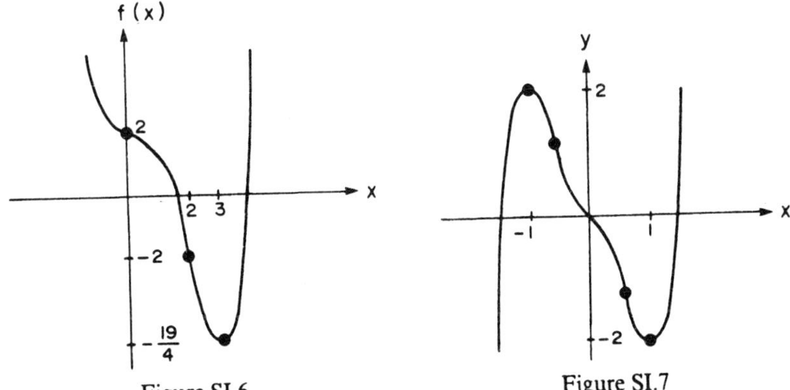

Figure SI.6

Figure SI.7

6. Minimum at $\left(3, -\frac{19}{4}\right)$; increasing for $3 < x < \infty$; inflection points at $x = 0, 2$; concave up for $-\infty < x < 0$, $2 < x < \infty$. (Observe that no relative maxima exist.) See Figure SI.6.

7. (Note that y has odd symmetry.) Relative minimum at $(1, -2)$; relative maximum at $(-1, 2)$; increasing for $-\infty < x < -1$, $1 < x < \infty$; inflection points at $\left(\mp\sqrt{\frac{3}{10}}, \pm\frac{221}{100}\sqrt{\frac{3}{10}}\right)$, $(0,0)$; concave up for $-\sqrt{\frac{3}{10}} < x < 0$, $\sqrt{\frac{3}{10}} < x < \infty$. See Figure SI.7.

8. (a) Local maximum at $\left(-\frac{1}{3}, \frac{32}{27}\right)$; local minimum at $(1,0)$;

(b) inflection point at $\left(\frac{1}{3}, \frac{16}{27}\right)$;

(c) (i) increasing for $-\infty < x < -\frac{1}{3}$, $1 < x < \infty$;
(ii) decreasing for $-\frac{1}{3} < x < 1$;
(iii) concave up for $\frac{1}{3} < x < \infty$;
(iv) concave down for $-\infty < x < \frac{1}{3}$.

See Figure SI.8.

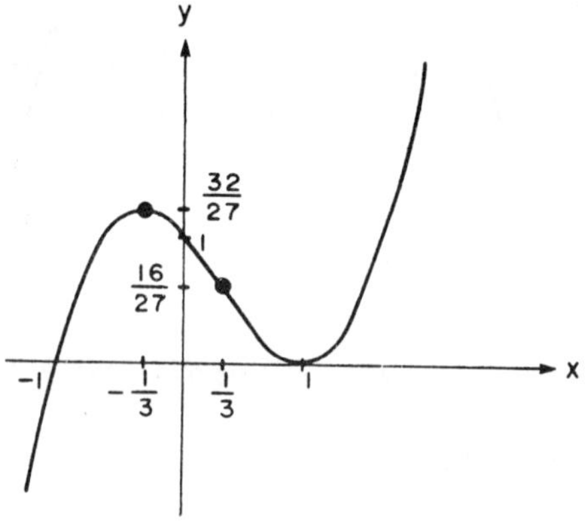

Figure SI.8

9. Relative maximum at $(1,0)$; relative minimum at $(5, -256)$; increasing for $-\infty < x < 1$, $5 < x < \infty$; inflection point at $(4, -162)$; concave up for $4 < x < \infty$; roots at $x = 1, 6$. See Figure SI.9.

10. Relative maximum at $\left(\frac{3}{2}, \frac{9}{16}\right)$; minima at $\left(\frac{3}{2} \pm \frac{\sqrt{5}}{2}, -1\right)$; increasing for $\frac{3}{2} - \frac{\sqrt{5}}{2} < x < \frac{3}{2}$, $\frac{3}{2} + \frac{\sqrt{5}}{2} < x < \infty$; inflection points at $x = \frac{3}{2} \pm \frac{\sqrt{15}}{6}$; concave up for $-\infty < x < \frac{3}{2} - \frac{\sqrt{15}}{6}$, $\frac{3}{2} + \frac{\sqrt{15}}{6} < x < \infty$; roots at $x = 0, 1, 2, 3$.

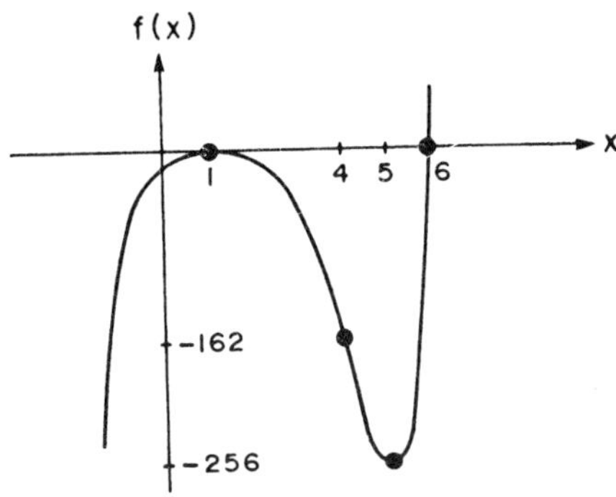

Figure SI.9

(Note that if $f(x) = x(x-1)(x-2)(x-3)$, then

$$f\left(\tfrac{3}{2}-x\right) = \left(\tfrac{3}{2}-x\right)\left(\tfrac{1}{2}-x\right)\left(-\tfrac{1}{2}-x\right)\left(-\tfrac{3}{2}-x\right)$$
$$= \left[x^2-\left(\tfrac{3}{2}\right)^2\right]\left[x^2-\left(\tfrac{1}{2}\right)^2\right]$$

and

$$f\left(\tfrac{3}{2}+x\right) = \left(\tfrac{3}{2}+x\right)\left(\tfrac{1}{2}+x\right)\left(-\tfrac{1}{2}+x\right)\left(-\tfrac{3}{2}+x\right)$$
$$= \left[x^2-\left(\tfrac{3}{2}\right)^2\right]\left[x^2-\left(\tfrac{1}{2}\right)^2\right],$$

i.e., $f(x)$ has even symmetry about $x = \frac{3}{2}$.) See Figure SI.10.

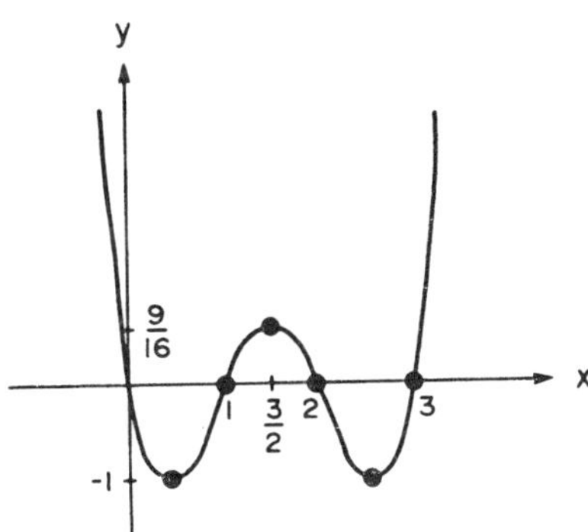

Figure SI.10

11. Maximum at $\left(1, \dfrac{1}{4}\right)$; inflection point at $\left(2, \dfrac{2}{9}\right)$. As $x \to -1$, $f(x) \to \dfrac{-1}{(x+1)^2}$. As $|x| \to \infty$, $f(x) \to \dfrac{1}{x}$. $f(x)$ concave down for $-\infty < x < -1$, $-1 < x < 2$, $f(x)$ concave up for $2 < x < \infty$. Intercept at $(0,0)$.

(Note that this graph, Figure SI.11, is the same as that for Solved Problem 6 (cf. Fig. I.6) after an appropriate translation and reflection.)

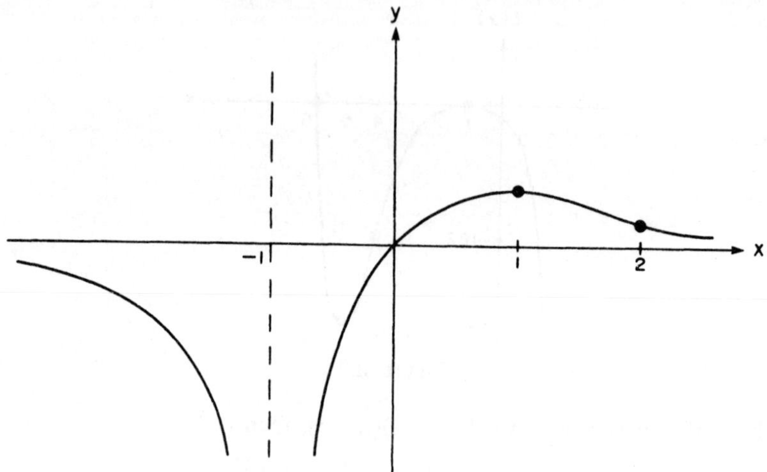

Figure SI.11

12. It is helpful to rewrite $y=\frac{x-1}{x+1}=\frac{(x+1)-2}{x+1}=1-\frac{2}{x+1}$. No critical points or inflection points; singular point at $x=-1$. See Figure SI.12.

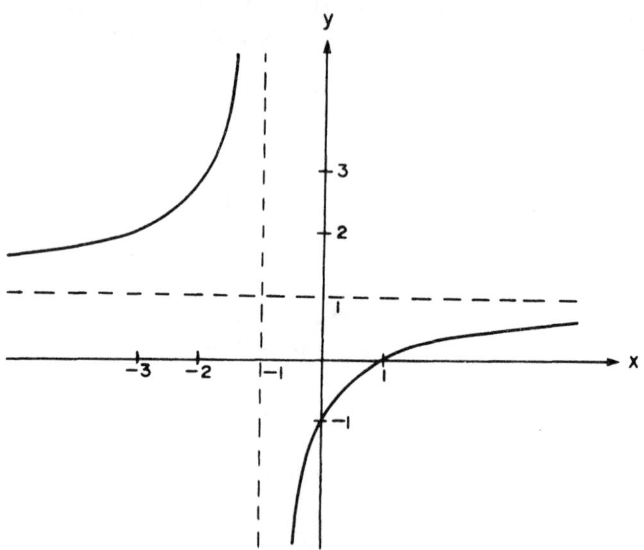

Figure SI.12

13. Singular point: $x=0$. Zeros: $x=\pm 1$. As $x\to 0$, $f(x)=\frac{x^2-1}{x(x^2+1)}\to -\frac{1}{x}$. As $|x|\to\infty$, $f(x)\to\frac{1}{x}$.

(Note that $f(x)$ is an odd function, i.e., $f(-x)=-f(x)$.) See Figure SI.13.

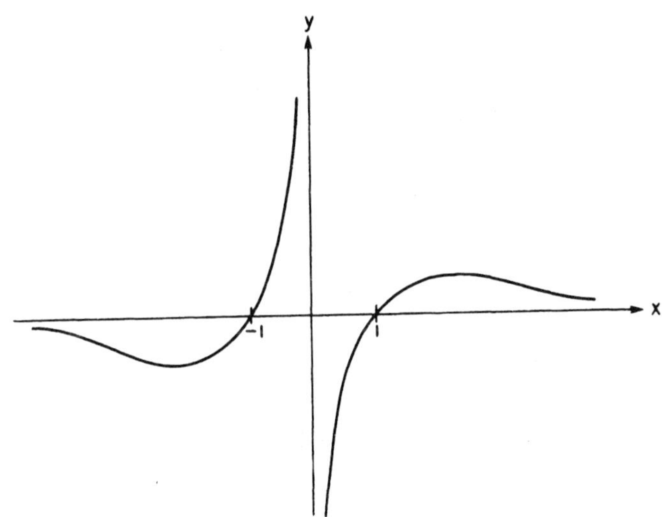

Figure SI.13

14. $y=\dfrac{x^2+x-5}{x-2}=x+3+\dfrac{1}{x-2}$.

(a) As $x \to 2$, $y \to \dfrac{1}{x-2}$. As $|x| \to \infty$, $y \to x+3$.

(b) Local maximum at $x=1$, local minimum at $x=3$.

(c) Intercepts at $\left(0, \dfrac{5}{2}\right)$, $\left(\dfrac{-1 \pm \sqrt{21}}{2}, 0\right)$.

(d) See Figure SI.14.

15. (a) $\lim\limits_{x \to -\infty} xe^x = 0$.

(b) $f(x)=xe^x$: increasing for $x>-1$, decreasing for $x<-1$, concave up for $x>-2$, concave down for $x<-2$. Minimum at $(-1, -e^{-1})$, inflection point at $(-2, -2e^{-2})$. $\lim\limits_{x \to +\infty} xe^x = +\infty$. See Figure SI.15.

16. (a) (Note that $f(x)$ has odd symmetry.) $f(x)$: increasing for $-\dfrac{1}{\sqrt{2}} < x < \dfrac{1}{\sqrt{2}}$; decreasing for $-\infty < x < -\dfrac{1}{\sqrt{2}}$, $\dfrac{1}{\sqrt{2}} < x < \infty$; absolute minimum at $\left(-\dfrac{1}{\sqrt{2}}, -\dfrac{1}{\sqrt{2e}}\right)$; absolute maximum at $\left(\dfrac{1}{\sqrt{2}}, \dfrac{1}{\sqrt{2e}}\right)$.

(b) $f(x)$: concave up for $-\sqrt{\dfrac{3}{2}} < x < 0$, $\sqrt{\dfrac{3}{2}} < x < \infty$; concave down for $-\infty < x < -\sqrt{\dfrac{3}{2}}$, $0 < x < \sqrt{\dfrac{3}{2}}$; inflection points at $\pm\sqrt{\dfrac{3}{2}}$.

(c) See Figure SI.16.

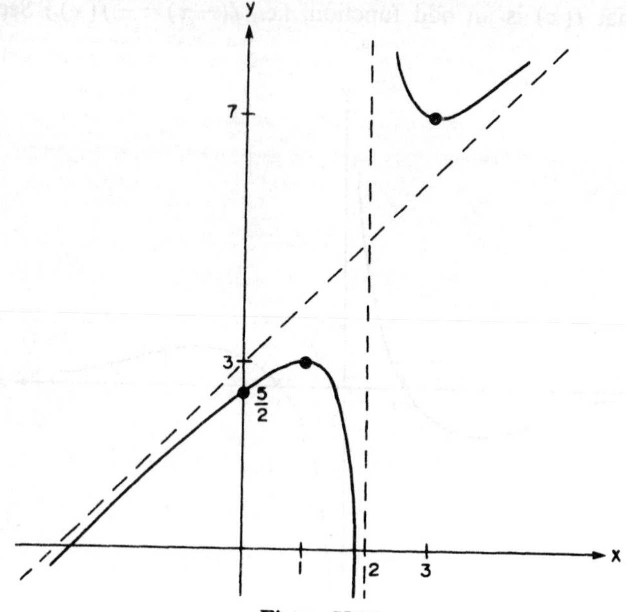

Figure SI.14

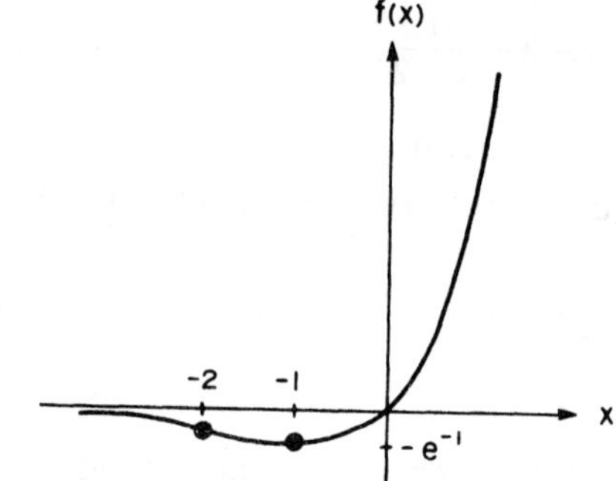

Figure SI.15

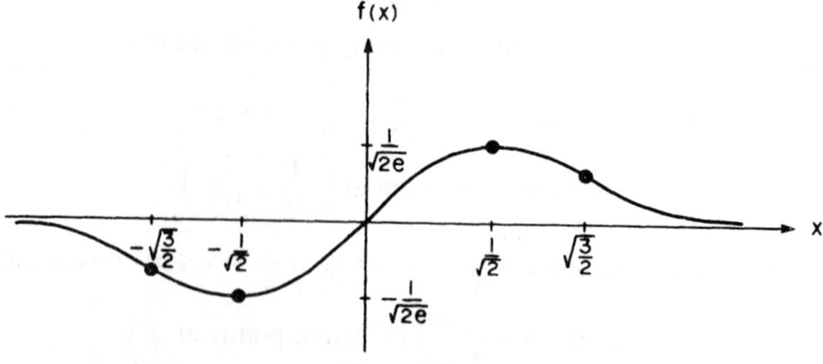

Figure SI.16

17. $\cosh x = \dfrac{e^x + e^{-x}}{2}$, $\sinh x = \dfrac{e^x - e^{-x}}{2}$; $f(x) = 10\cosh x + 6\sinh x = 8e^x + 2e^{-x}$. Minimum value of $f(x)$ is 8. See Figure SI.17.

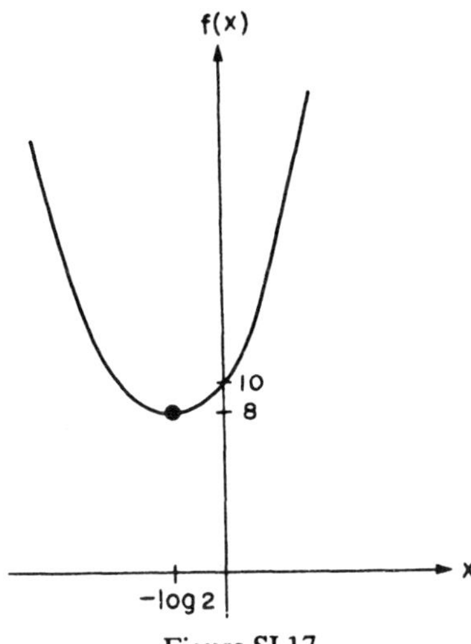

Figure SI.17

18. (a) Horizontal asymptote: $y = 1$. Vertical asymptote: $x = 6$.

(b) Relative minimum at (0,0). Inflection point at $\left(1, \dfrac{1}{\sqrt[3]{25}}\right) \simeq (1, 0.3)$.

See Figure SI.18.

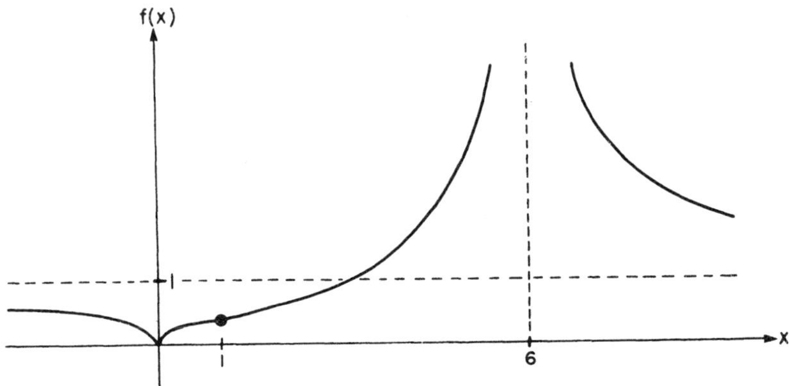

Figure SI.18

19. (a) 0.

(b) 0; $+\infty$.

(c) Relative maximum at (0,0). Relative minima at $(\pm e^2, -e^4)$.

(d) $(\pm e, -3e^2)$.

(e) Concave up for $-\infty < x < -e$, $e < x < +\infty$. Concave down for $-e < x < e$. (Note that $g(x)$ is even.) See Figure SI.19.

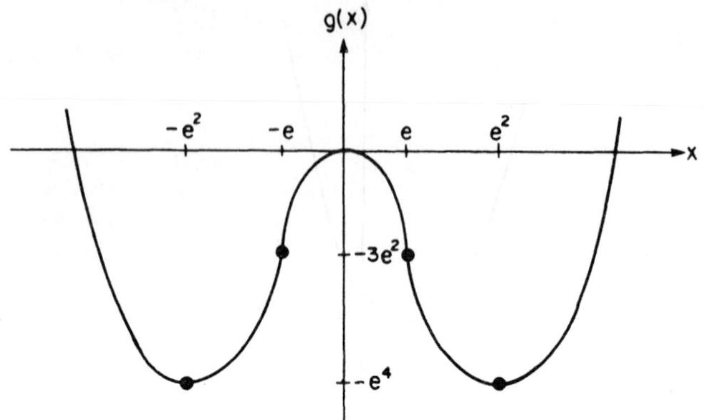

Figure SI.19

20. Vertical asymptote: $x = -1$. Local maximum at $\left(-\frac{2}{3}, \frac{2}{3} - \log 3\right) \approx \left(-\frac{2}{3}, -0.432\right)$. Local minimum at $\left(-\frac{1}{2}, \frac{1}{4} - \log 2\right) \approx \left(-\frac{1}{2}, -0.443\right)$. Inflection point at $\approx \left(-1 + \frac{1}{\sqrt{6}}, -0.437\right)$. Root at $x = 0$ only. See Figure SI.20.

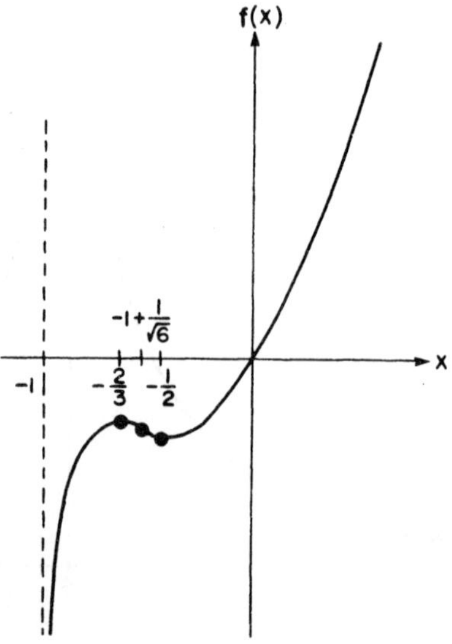

Figure SI.20

22. Note symmetry: if (x, y) lies on graph then so do $(x+2\pi, y-2\pi), (\pi - x, -\pi - y)$. Since $\sin x \geq 0$, one need only consider domain $0 \leq x \leq \frac{\pi}{2}$ to obtain whole graph. See Figure SI.22.

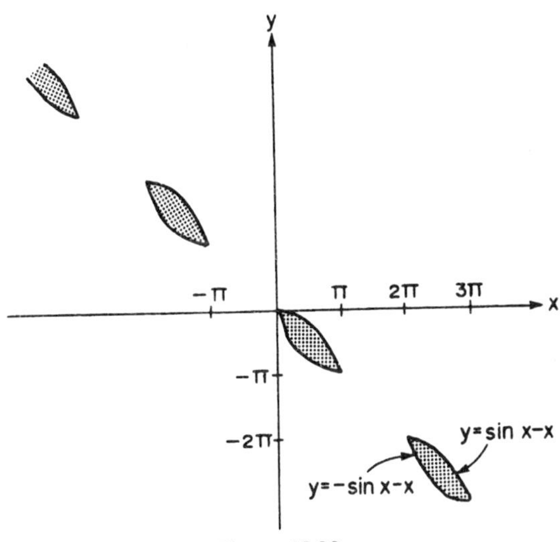

Figure SI.22

23. Hints

(a) Let $h(x) = f(x) - g(x)$. Justify the following:

(i) $h(0) = 0$.

(ii) For $x < -1$, $h(x) \simeq f(x) -$ constant.

(iii) For large $x > 0$, $h(x) \simeq f(x)$.

(b) Say $h(x) = f(x+1)$. Then in h, x takes the value that $(x+1)$ had in f. Plot points. ($h(x)$ is a horizontal *translation of* $f(x)$.)

(c) Say $h(x) = f(x)+1$. Then $h(x)$ is a vertical translation of $f(x)$.

(d) Say $h(x) = f(-x)$. Then $h(x)$ is a reflection of $f(x)$ about the y-axis. Check: Which points remain the same?

(e) Say $h(x) = \int_0^x f(t)\,dt$. $h(x)$ should increase where $f(x) > 0$. What is $h(0)$?

Solution

I $\leftrightarrow f(x)+1$;
II $\leftrightarrow f(-x)$;
III $\leftrightarrow \int_0^x f(t)\,dt$;
IV $\leftrightarrow f(x) - g(x)$;
V $\leftrightarrow f(x+1)$;
VI $\leftrightarrow f'(x)$.

24. (a) The concavity of $f(x)$ is left up to you. You can include as many inflection points as you wish on the intervals $0 < x < \frac{1}{2}$ and $\frac{1}{2} < x < 1$, so long as $f(x)$ is increasing. Obviously there must be at least one inflection point on the interval $0 < x < \frac{1}{2}$. But there must also be at least one inflection point on the interval $\frac{1}{2} < x < 1$. Why must this be true?

(b) When $x = \dfrac{1}{4}$, $f(x)$ is somewhere between 0 and $\dfrac{1}{2}$ and is increasing. When x increases from 0 to $\dfrac{1}{2}$, $g(x)$ decreases. Hence $g(f(x))$ is *decreasing* when $x = \dfrac{1}{4}$. Similarly, $g(f(x))$ is *increasing* when $x = \dfrac{3}{4}$.

(c) It is given that $D_x[g(x)] = 0$ on $[0,1]$ only when $x = \dfrac{1}{2}$. Then $D_f[g(f(x))] = 0$ only when $f(x) = \dfrac{1}{2}$, i.e., at $x = \dfrac{1}{2}$. Therefore $D_x[g(f(x))] \equiv D_f[g(f(x))] \cdot D_x f(x)$ is zero when $\dfrac{dg}{df} = 0$ or when $\dfrac{df}{dx} = 0$, i.e. at $x = 0$ and $x = \dfrac{1}{2}$. See Figure SI.24.

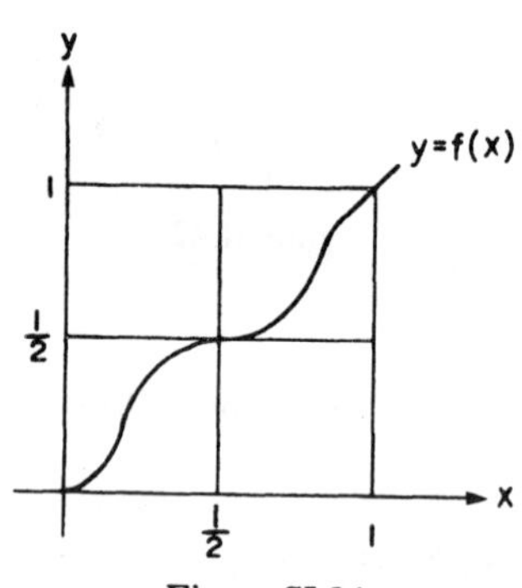

Figure SI.24

25. Hints

(a) You should have a "preliminary" sketch as shown in Figure SI.25*a*.

(b) It may be helpful to rewrite each piece of concavity information in words, as we have done in Figure SI.25*b*.

(c) Remember $f(x)$ is continuous—i.e. unbroken. See Figure SI.25*c*.

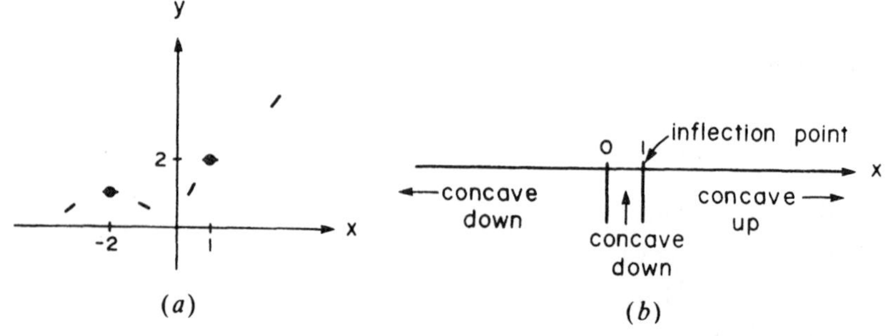

Figures SI.25*a* and *b*.

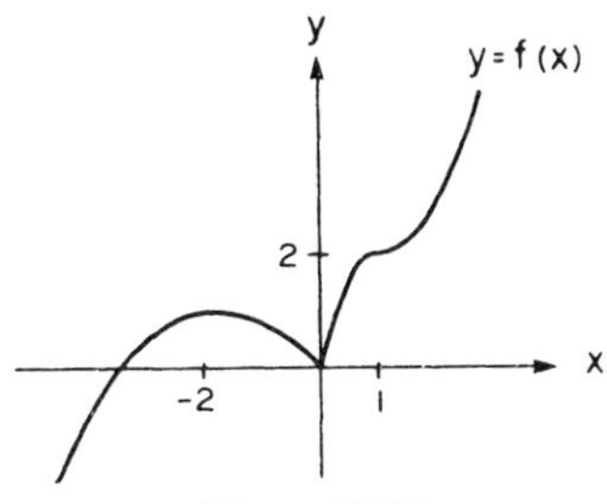

Figure SI.25c

26. (a) A function which is not differentiable at all points; for example, one with a "cusp", as shown in Figure SI.26a.

(b) A function having a minimum whose derivative is either infinite or different on either side of the minimum point (Fig. SI.26b).

(c) Conditions for continuity of f at $x = a$ are:

(i) $f(a)$ exists;

(ii) $\lim_{x \to a} f(x)$ exists;

(iii) $\lim_{x \to a} f(x) = f(a)$.

One is given (ii); therefore one must leave either (i) or (iii) not satisfied. Fig. SI.26c $\leftrightarrow$ (i) not satisfied; Fig. SI.26d $\leftrightarrow$ (iii) not satisfied.

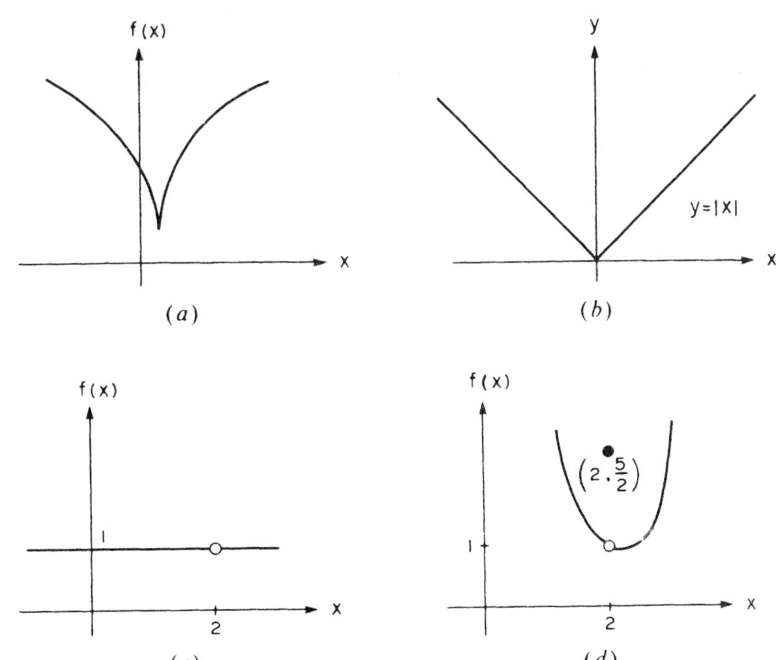

Figures SI.26$a-d$.

Chapter II

1. 20 inches × 25 inches.

2. 640 cm^2 (dimensions are 16 cm × 40 cm).

3. 6 cm × 6 cm × 3 cm.

4. $x = 5$ cm.

5. (b) $\frac{5}{3}$ cm × $\frac{5}{3}$ cm.

6. 12 ft^2.

7. radius $= r^* = \sqrt[3]{\frac{V}{2\pi}}$, height $= 2r^*$.

8. height $= \frac{40}{\pi}$ cm, radius = 5 cm.

9. height of rectangle = radius of semi-circle $= \frac{12}{\pi + 4}$ ft.

10. $\frac{250{,}000}{\pi}$ ft^2.

11. 100.

12. 1.

13. length of each side of equilateral triangle $= \left(\frac{6+\sqrt{3}}{33}\right)L$ feet; width of rectangle $= \left(\frac{5-\sqrt{3}}{22}\right)L$ feet.

14. $10(1+\sqrt{2})$ units.

15. 5 inches × 10 inches.

16. $\frac{4\sqrt{5}\,\pi}{375}a^3$.

17. $4\sqrt{2} \times 3\sqrt{2}$.

18. $\frac{2\sqrt{3}\,\pi}{27}R^3$.

19. $y = \frac{L}{\sqrt{3}}$, $x = \sqrt{2}\,y$.

20. $2[\sqrt{5} - 1]$.

21. (a) $S = \frac{5}{3}\pi r^2 + \frac{2V}{r}$, (b) $r = \sqrt[3]{\frac{3V}{5\pi}}$.

22. width $= \frac{1}{2}$ m, depth $= \dfrac{\sqrt{3}}{2}$ m.

23. 7500 ft^2.

24. $x = 10\sqrt{2}$ feet, $y = 15\sqrt{2}$ feet.

25. (a) See Figure SII.25.
(b) 15,000 m^2.

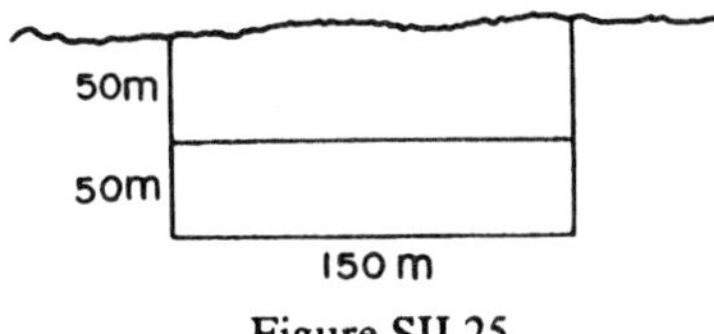

Figure SII.25

26. 3 m.

28. 8 inches.

29. $x = 12$ inches, $y = 24$ inches.

30. width of window $= \dfrac{48}{8-\pi}$ feet.

31. height $= \dfrac{3}{\sqrt{2}}$ feet, length (lying on diameter of semi-circle) $= 3\sqrt{2}$ feet.

32. 4 miles.

33. a point $\dfrac{5}{\sqrt{39}}$ miles from P.

34. $\dfrac{25}{12}$ miles from A.

35. a distance $\dfrac{SD}{\sqrt{v^2 - S^2}}$ from the point on the track closest to the maiden.

36. $x = 1$.

37. 80 metres from the line joining the posts to the road.

38. 12 feet.

39. 125 feet.

40. $x = 2$ inches.

41. $P = \left(-\dfrac{3}{2}, \dfrac{1}{2}\right)$.

42. $4\sqrt{3}$.

43. $\sqrt{3}$.

45. $\sqrt{\dfrac{2}{5\pi}}$ cm/s.

46. increasing at a rate of 5 cm/s.

47. $2\pi r$, $\dfrac{\pi d}{2}$, $\dfrac{c}{2\pi}$. (r = radius, d = diameter, c = circumference)

48. decreasing at a rate of $\dfrac{3}{25}$ radians/min.

49. $\dfrac{16}{5}$ m/s.

50. 20 ft/sec.

51. $\frac{1}{2}$ m/s.

52. $3\sqrt{5}$ mph.

53. increasing at rate of $\dfrac{\sqrt{39}}{104}$ radians/sec.

54. $11 + \dfrac{32}{\sqrt{13}}$ knots.

55. decreasing at rate of $\dfrac{34}{5}$ mph.

56. $128\sqrt{\dfrac{5}{13}}$ miles/hour.

57. 40 ft/sec.

58. 68 ft/sec.

59. (a) decreasing at a rate of $\dfrac{\sqrt{1066}}{41}$ miles per minute;

(b) 6 minutes later when distance apart is $18\sqrt{13}$ miles.

60. increasing at a rate of $\dfrac{60}{\sqrt{13}}$ ft/sec.

61. increasing at a rate of 3 cm/s.

62. increasing at a rate of $\dfrac{19}{12}\sqrt{10}$ units/sec.

63. (a) 30π ft^2/sec.
(b) $18\pi t$ ft^2/sec.

64. 7960π cm^2/s.

65. $\dfrac{5}{18\pi}$ ft/min.

66. $\dfrac{1}{2(\sqrt[3]{9\pi})}$ in/sec.

67. increasing at a rate of $\dfrac{10}{3}$ cm^2/s.

68. $\dfrac{1}{48\pi}$ in/min.

69. $\dfrac{1}{16\pi}$ in/hr.

70. $\dfrac{1}{2\pi}$ ft/min.

71. $\dfrac{25}{L\sqrt{20h - h^2}}$ ft/min; L = length of tank in feet, h = depth of water in feet.

72. increasing at a rate of 6 cm/s.

73. $\dfrac{9}{4\pi}$ ft/min.

74. (a) $\dfrac{1}{12}$ ft/min.
(b) $\dfrac{1}{20}$ ft/min.

75. $\dfrac{1}{\pi}$ ft/min.

76. (a) $\dfrac{1}{\pi}$ m/min.
(b) 6 m^2/min.

77. rising at a rate of $\dfrac{5}{8}$ ft/sec.

78. (a) $\dfrac{1600}{3}$ ft/min.
(b) $\dfrac{1600}{3}$ ft/min.

79. increasing at a rate of $\dfrac{3}{5}$ metres per second.

80. at a rate of $\dfrac{5}{9}$ metres/second.

81. $\dfrac{15}{4}$ ft/sec.

82. decreasing at a rate of $\dfrac{2\sqrt{2}}{5}$ radians/sec.

83. $-\dfrac{3}{4}$ metres per second. (top end is falling)

84. at a rate of $\dfrac{1}{10}$ radians/sec.

85. increasing at a rate of $\frac{1}{50}$ radians/sec.

86. 5π miles/minute.

87. 100π feet/second.

88. $\frac{16\pi}{15}$.

89. π.

90. $\frac{64}{5}$.

91. $\frac{1}{3}a^2h$.

92. $\frac{1}{3}abh$.

93. $\frac{3}{10}$.

94. $\frac{1023}{25}$.

95. $\frac{16}{3}$.

96. $\frac{3\pi}{2}$.

97. $\frac{\pi R^3}{6}$.

98. $\frac{11}{15}$.

99. $\frac{4}{3}$.

100. $\frac{4}{3}(8-3\sqrt{3})\pi a^3$.

Chapter III

1. After 1 hour the hare has travelled 20 metres; the tortoise 10 metres. After 4 hours the hare has travelled 160 metres; the tortoise 160 metres. After 9 hours the hare has travelled 540 metres; the tortoise 810 metres.

2. 15 sec.

3. (a) deceleration is 20 m/s^2; (b) 100 km.

4. (a) velocity = 496 units/sec, position = 1472 units;
(b) particle moves to the left if $0 \le t < 2$ and moves to the right if $t > 2$.

5. (a) after 2 seconds; (b) 64 feet; (c) 4 seconds; (d) 64 feet/second.

6. (a) position $= \frac{1}{2}(F-g)T^2$ metres, velocity $= (F-g)T$ metres/second; (b) $\frac{T}{g}[F+\sqrt{F(F-g)}\,]$ seconds.

7. $V = 400[\log 100 - \log(100-10t)] - 32t$; $ub - Mg \geq 0$ is necessary for positive acceleration when $t > 0$.

8. The shell rises to 360,000 ft. The gun crew has 5 minutes to make a getaway.

10. At time t seconds the height (in feet) of the rock above the surface of the moon is $x(t) = -\frac{8}{3}t^2 + 80t + 576$. The rock's maximum height is 1176 feet. The rock hits the surface of the moon after 36 seconds.

11. distance rolled is $5\left[1-\left(\frac{3}{5}\right)^t\right]$ feet where t is measured in seconds.

12. vertex angle $\theta = \frac{\pi}{2}$; to show one has a maximum consider $\frac{d^2A}{d\theta^2} = -\frac{225}{2}\sin\theta < 0$ for $0 < \theta < \pi$. ($A(\theta)$ is the cross-sectional area of the trough.)

13. $\sqrt{\frac{3}{2}}$.

14. (a) 90 metres; (b) 50 metres.

15. $\frac{24}{5}$ units from A.

17. $\frac{5x}{\sqrt{x^2-4}}$ km/s.

18. 60 kilopascals/second.

19. 4 miles.

20. (a) 100 feet after $4\sqrt{3}$ seconds; (b) $\frac{1}{16}$ radians/sec.

21. (a) $\frac{dx}{dt} = \frac{4\pi x \sin\theta}{l\cos\theta - x}$ metres/second where x is measured in metres; (b) when B lies on the x-axis.

22. $\frac{160n\pi}{4\sqrt{2}-1}$ cm/min.

23. $\log 2$ radians/minute.

24. $\bar{x} = \frac{3}{4}, \bar{y} = \frac{8}{5}$.

25. $\bar{x} = -1, \bar{z} = 0, \bar{y} = \frac{2}{5}$.

26. $\bar{x} = \bar{y} = \dfrac{4a}{3\pi}$.

27. $\bar{x} = \dfrac{14}{15\pi}, \bar{y} = \dfrac{9}{10}$.

28. $\bar{x} = \dfrac{\int_{-1/2}^{\sqrt{3}/2} x\cos^{-1}x\,dx}{\int_{-1/2}^{\sqrt{3}/2}\cos^{-1}x\,dx}, \bar{y} = \dfrac{\frac{1}{2}\int_{-1/2}^{\sqrt{3}/2}(\cos^{-1}x)^2\,dx}{\int_{-1/2}^{\sqrt{3}/2}\cos^{-1}x\,dx}$.

29. $\bar{x} = 0, \bar{y} = \dfrac{-20a}{3(\pi+8)}$.

31. $\rho(4-2\sqrt{2})$.

32. 5000 lb.

33. 37,687.5 lb, assuming a density of 62.5 lb/ft^3 for water.

34. $\dfrac{640}{3}$ lb.

35. 400,000 foot-pounds.

36. 6.6×10^7 Joules.

37. 5625π Joules.

38. 7.595×10^6 J, assuming a density of 9800 N/m^3 for water.

39. 5.61×10^7 Joules.

40. 1.11×10^9 Joules.

41. 450 foot-pounds.

42. $\frac{1}{2}(1-e^{-100})$ ft-lb.

43. $\dfrac{8704}{5}\pi w$ foot-pounds.

44. 81 ft-lb.

45. 2 Joules.

46. (a) $\dfrac{5}{3}$ lb/in; (b) $\dfrac{80}{3}$ in-lb.

47. 80 in-lb.

48. $\dfrac{1}{\log\frac{\pi}{2}}\left[\left(\dfrac{2}{\pi}\right)^{1939}-\left(\dfrac{2}{\pi}\right)^{913}\right]$ Einstein-years.

50. ≈ 14 minutes.

51. ≈ 10.0 hours.

52. $\simeq$ 9:00 p.m., assuming body temp. of 98.6°F;
$\simeq$ 9:10 p.m., assuming body temp. of 98.4°F;
$\simeq$ 9:20 p.m., assuming body temp. of 98.2°F.

53. 50 rpm; 8 minutes.

54. $\frac{8}{3}$ hours.

55. 12 units.

57. $\simeq$ 3.1 minutes.

58. length of curve is $10(e - e^{-1}) = 5\ \sinh(1)$ metres.

59. (a) $\dfrac{3600v}{18 + v + \dfrac{v^2}{32}}$ cars; (b) 24 ft/sec.

Chapter IV

1. 3 cm × 3 cm × 6 cm.

2. length of glass wall is 100 metres, length of adjacent wall is 144 metres.

3. (a) The cost function is $C(x) = 6x^2 + 60x + \frac{432}{x} + \frac{2592}{x^2}$;

(c) $x = 4$ metres, $y = 5.4$ metres.

4. The plot should be 30 metres × 40 metres where the 40 metre side is along the neighbour's property.

5. along a straight line from the pumping station to the shore at a distance $\left(2 - \frac{\sqrt{3}}{3}\right)$ km from the refinery, and then along the shore to the refinery.

6. There are an infinite number of equivalent solutions. One solution involves laying a straight-line underwater cable from P to a point on the other bank $2\alpha - \frac{\alpha}{\sqrt{3}}$ metres from the factory and then laying a land cable direct to the factory.

7. If $\frac{a^2}{b^2} \leq 26$, the pipeline should be laid directly from $(1, -5)$ to $(2,0)$; if $\frac{a^2}{b^2} > 26$, the pipeline should follow the polygonal path consisting of a straight line path from $(1, -5)$ to $\left(1 + \frac{5b}{\sqrt{a^2 - b^2}}, 0\right)$ and then a straight line path from $\left(1 + \frac{5b}{\sqrt{a^2 - b^2}}, 0\right)$ to $(2,0)$.

8. $x = 500$.

9. \$70.

10. 2 widgets.

11. 90¢.

12. a reduction of \$2/month.

13. 250 passengers give the company the maximum proceeds of \$3125.

14. 10¢.

15. 20 km/hr.

16. 50 blocks.

17. 10.25%.

18. Here 8% corresponds to the effective annual interest rate for some unknown interest rate compounding continuously. The current value of the bonds is \$51,629.34.

19. 6.9% compounded continuously.

20. in 2.63 years.

21. (a) $I_n = 100\left[\left(1+\frac{r}{100n}\right)^n - 1\right]\%$;
(b) $I_\infty = 100[e^{r/100} - 1]\%$;
(d) for $r = 2$, $I_\infty = 2.02\%$; for $r = 10$, $I_\infty = 10.52\%$; for $r = 20$, $I_\infty = 22.14\%$.

22. (a) $\$(1.04)^{2N}$; (b) $\$(1.02)^{2N}$; (c) $\$[(1.02)^{2N} - 1]$.

23. (a) $\int_1^2 10^5(t+1)^{1/2}\,dt = \$157{,}848$;
(b) \$157,848/year;
(c) $\left\{(e^r - 1)P + \alpha\left[\int_0^1 \sqrt{\tau+2}\,(e^{r(1-\tau)} - 1)\,d\tau\right]\right\}$ dollars where P is the profit in dollars during the first year, $r = 0.08$, $\alpha = 10^5$, $P = \$121{,}895$.

24. (a) $\bar{p} = 3$, $\bar{x} = 4\sqrt{3}$; (b) $\frac{28}{5}\sqrt{3}$.

25. (a) \$184; (b) \$134.67.

26. 60 items should be produced; maximum profit is \$49,000.

27. The profit is positive for $10 < x < 30$; maximum profit for $x = 20$.

28. (a) Maximum at $(0, 15)$; inflection points at $(\pm 3, 15e^{-1/2})$; increasing for $x < 0$; decreasing for $x > 0$; concave up for $x < -3$ and for $x > 3$; concave down for $-3 < x < 3$; $f(-x) = f(x)$. See Figure SIV.28.
(b) $x = 3$.

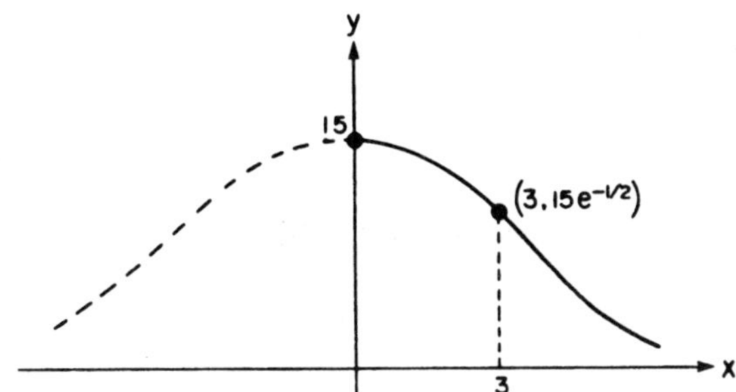

Figure SIV.28. Graph of $y = 15e^{-x^2/18}$.

29. Production should be fixed at $x = 10$; maximum profit is 2600; maximum profit is attained when price/unit is 280.

30. (a) 2842; (b) $x \simeq 25 - \frac{25}{28}, y \simeq 25 + \frac{25}{28}$.

31. (a) $x_0 = \frac{a-d-t}{2b}, p_0 = \$\left(\frac{a+d+t}{2}\right)$;
(b) $\$\left[\frac{(a-d-t)^2}{8b}\right]$.

32. average daily inventory = 200 cases; total daily holding cost = \$60.

33. (a) average cost/week = $\$\left(50T + \frac{800}{T}\right)$; (b) every 4 weeks; (c) \$400.

34. $c(x) = \log\left[\frac{x^2}{2} + x + 1\right] + 50$.

35. (a) $x = 2500$, price = \$75, total profit = \$52,500;
(b) $x = 2000$, price = \$80, total profit = \$30,000.

36. (b) $t = 2$.

37. (a) producer's surplus = 1000, consumer's surplus = $\frac{2000}{3}$;
(b) $\eta = \frac{200}{x^2} - \frac{1}{2}$; (c) $0 \le x < \frac{20}{\sqrt{3}}$.

38. When $p = 10$, the elasticity of demand is 1; change in demand is approximately $-\frac{1}{2}\%$.

39. $x > 2$.

40. $\frac{dc}{dx} = 34 - \frac{5}{3(5x+4)}$.

41. (a) always decreasing for $x > 0$.

42. (a) $p = e^{(3.3 - q/2)}$;
(b) $4.6 - 2\log p$;

(c) $\left(1-\frac{q}{2}\right)e^{(3.3-q/2)}$;

(d) $p=e^{2.3}$, $q=2$.

43. \$2037.

44. (a) \$2015; (b) \$2.62½ per unit; (c) \$2017.63.

45. The demand equation is $p=D(x)=\frac{4}{x}\left[\sqrt{x+1}\left(\log\sqrt{x+1}-1\right)+1\right]$; $D(15)\simeq 0.68$.

46. $R=3$ miles; the result is independent of the population density.

Chapter V

1. 5040 bacteria.

2. 57.6 mg.

3. (a) $20\frac{\log 5}{\log 3}$; (b) $6000\frac{\log 3}{\log 5}$ people/year.

4. 1888. (This is the first year when the estimated population is less than one.)

5. $1975+\left(\frac{\log 1.5}{\log 1.05}\right)T$ where T is the number of years since August 1975 when a student does this problem. If $T=3$ years (August 1978) then the answer is the year 2000.

6. in the year 2108.

7. in the year 1975.

8. (a) 34.7 years; (b) 6%; (c) 8.

9. 3.2×10^7.

10. After 16 weeks the colony of mould will have 16 times its original weight. As $t\to\infty$ the colony will tend to have 32 times its original weight.

11. $10\frac{\log 5}{\log 2}\simeq 23.2$.

12. $\frac{dS}{dt}=-\frac{S}{50}$, $S(0)=50$.

13. $\simeq 273$ lb.

14. (a) $S(t)=\frac{100}{\sqrt{100+t}}$ pounds where t is time in minutes. (b) after 5 hours.

15. 5.

16. (a) 1905; (b) 195.

17. (a) in 8 weeks; (b) in 6 weeks.

18. (a) 1.55 days; (b) 3.11 days.

19. in three hours.

20. $x = a + \dfrac{b-a}{1+e^{(b-a)t}}$; $\lim_{t \to +\infty} x(t) = b$.

21. (a) $c_1 = \dfrac{\alpha}{\beta}$, $c_2 = 2c_1$, $c_3 = 2\alpha\beta$; (b) equilibrium is reached since $\lim_{t \to \infty} p(t) = c_1$; the only inflection point occurs at $t = 0$. There are no extrema. See Figure SV.21.

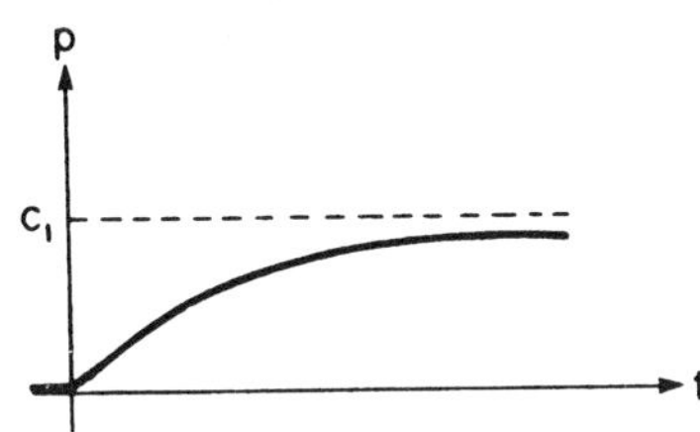

Figure SV.21

22. (a) Yes. If the number of molecules of A decreases by $\alpha(t)$ then the number of molecules of B increases by $\alpha(t)$. Hence $x + y = a$ for all time. Physically the number of molecules is conserved. Initially there are a molecules of A and 0 molecules of B;

(b) $\dfrac{dy}{dt} = kx - k'y$, $\dfrac{dx}{dt} = k'y - kx$, where $y(0) = 0$, $x(0) = a$. $y(t) = \dfrac{ak}{k+k'}[1 - e^{-(k+k')t}]$;

(c) $\lim_{t \to \infty} y(t) = \dfrac{ak}{k+k'}$, $\lim_{t \to \infty} x(t) = \dfrac{ak'}{k+k'}$;

(d) $\beta = \dfrac{ak}{k+k'}$. This value is never reached since $\dfrac{dy}{dt} > 0$ if $t > 0$;

(e) half-life is $\tau = \dfrac{\log 2}{k+k'}$, $y = \dfrac{3\beta}{4}$ when $t = \dfrac{\log 4}{k+k'}$.

23. The critical concentration is reached at 3:05 a.m.

24. (a) $\dfrac{dc}{dt} = k(c_s - c) - \dfrac{\Theta}{V}$; (c) $c_e = c_s - \dfrac{\Theta}{kV}$; (f) plot $\log(c - c_e)$ *vs.* t.

25. $\dfrac{3}{40}$ N/m^2 per second.

26. Total reaction is $\frac{1}{2}$.

Chapter VI

1. $10+\frac{1}{150}\simeq 10.0067$.

2. The approximate value is $\frac{1}{\sqrt{2}}\left[1-\frac{\pi}{180}\right]\simeq 0.6948$. Since $f(x)=\cos x$ is such that $f''(x)<0$ if $0\leq x<\frac{\pi}{2}$ (radians), the approximation is greater than the actual value of $\cos 46°$.

3. 0.203.

4. $14-\frac{1}{28}\simeq 13.964$.

5. (a) $dy=3x^2\,dx$; (b) $5-\frac{4}{75}\simeq 4.9467$.

6. $\sqrt{82}\simeq 9+\frac{1}{18}\simeq 9.0556$, $\sin 29°\simeq\frac{1}{2}-\frac{\pi\sqrt{3}}{360}\simeq 0.4849$.

7. $10-\frac{1}{20}=9.95$.

8. $2-\frac{1}{32}\simeq 1.9688$.

9. Let $f(x)=(\sin x)^{10}$, $f'(x)=10(\sin x)^9\cos x$, $f\left(\frac{\pi}{4}+0.01\right)\simeq\frac{11}{320}\simeq 0.0344$.

10. 10,540.

11. $25+\frac{11}{48}\simeq 25.229$.

12. $\frac{27\pi}{4}\simeq 21.206$ ounces.

13. $8\pi\simeq 25.13$.

14. ± 0.0125 m^2.

15. (a) $\frac{\pi}{4}$; (b) $\frac{\pi}{4}+0.005\simeq 0.7904$.

16. 2.426.

17. 0.618.

18. (b) $x_1=0$, $\hat{x}_2=\frac{1}{2}$.

19. (b) $n=3$; (c) $\hat{x}_2=\frac{8}{3}$.

20. There is only one real root $r=0.49$.

21. (b) $x_1=\frac{\pi}{3}$.

22. $r=1.96885\pm 0.00005$.

24. (a) $T\left(\frac{1}{4}\right) \simeq 0.7489$; (c) $T\left(\frac{1}{4}\right) < \frac{\pi}{4} < T\left(\frac{1}{4}\right) + \frac{1}{8}$; $\pi = 3.2456 \pm 0.25$.

25. (a) $T\left(\frac{1}{2}\right) = 0.697023\ldots$; (b) $-\frac{1}{96} < E_T\left(\frac{1}{2}\right) < -\frac{1}{768}$;
(c) If the error is not centred, $n = 13$. If the error is centred, $n = 9$.

26. (a) $S\left(\frac{1}{4}\right) = 0.39269906\ldots$; (b) $J = \frac{\pi}{8} = 0.3926990817\ldots \Rightarrow$ error is $+1.91 \times 10^{-8}$.

28. (a) $S\left(\frac{1}{2}\right) = 4.64593$; (b) $\left|E_s\left(\frac{1}{2}\right)\right| < \frac{1}{15}$. (Show that the exact answer is $\frac{26}{3} - \frac{5}{2}\log 5 = 4.64307\ldots$)

29. (a) $T(\frac{1}{4}) = 0.697023\ldots$; (b) $S(\frac{1}{4}) = 0.693253\ldots$.
(c) A polygonal approximation to $f(x) = \frac{1}{x}$ leads to an overestimation of $\log 2$ since $f''(x) > 0$. A parabolic interpolation has the correct concavity.

30. (a) If the error is not centred, 183. If the error is centred, 151;
(b) 11.

31. $T(1) = S(1) = 10$.

32. 3.87.

33. 2.08.

34. $1 - \frac{1}{2} + \frac{1}{8} - \frac{1}{48} \simeq 0.6042$.

35. $1 - \frac{1}{200} + \frac{1}{240000} \simeq 0.995004$.

36. $-\frac{1}{5}$.

37. (a) $1 - \frac{1}{2}x + \frac{3}{8}x^2 - \frac{5}{16}x^3$;
(b) $\frac{14}{10}\left[1 + \frac{1}{100} + \frac{3}{20000} + \frac{5}{2000000} + \frac{35}{800000000}\right] \simeq 1.4142135336$
after the error is centred.

38. $\frac{1}{50} - \frac{1}{5^6} \simeq 0.019936$.

39. $\frac{2}{5} + \frac{2}{625} = 0.4032$.

40. 4.

41. 0.1.

42. (a) 0.8978; (b) 0.8974.

43. (a) $x-\frac{x^3}{6}+\frac{x^5}{120}$; (b) $\sin 1 \simeq 1-\frac{1}{6}+\frac{1}{120} \simeq 0.84167$; (c) $I=1-\frac{1}{30}+\frac{1}{1080}+\text{error}$ where $-\frac{1}{(13)(7!)}<\text{error}<0$. Hence $I=0.96758496 \pm 0.000008$.

44. (a) $\cos 2x=\sum_{k=0}^{\infty}\frac{(-1)^k 2^{2k} x^{2k}}{(2k)!}$ with radius of convergence ∞;

(b) $I \simeq \frac{1}{3}\left(\frac{\pi}{4}\right)^3-\frac{1}{15}\left(\frac{\pi}{4}\right)^5 \simeq 0.14157$.

45. 0.09992.

46. (a) $\left(\sum_{k=1}^{9}\frac{(-1)^{k+1}}{k^4}\right)-\frac{1}{(2)(10^4)}$;

(b) *Method 1*: $S=S_n+R_n$,

$$0<R_n<\int_n^{\infty}\frac{dx}{x^4}=\frac{1}{3n^3}.\text{ Hence } S=S_{19}+R_{19} \quad \text{where}$$

$$S_{19}=\sum_{k=1}^{19}\frac{1}{k^4}, \qquad 0<R_{19}<5\times 10^{-5}.$$

Method 2: the error of Method 1 is centred. Hence

$$S=\left(\sum_{k=1}^{10}\frac{1}{k^4}\right)+\frac{1}{6}\left[\frac{1}{10^3}+\frac{1}{11^3}\right]+E_{10},$$

$$|E_{10}|<5\times 10^{-5}.$$

Method 3: $S=S_n+R_n$

where

$$\int_{n+1}^{\infty}\frac{dx}{x^4}<R_n<a_{n+1}+\int_{n+1}^{\infty}\frac{dx}{x^4};$$

$$a_{n+1}=\frac{1}{(n+1)^4}.\text{ Thus } R_n=\frac{1}{3(n+1)^3}+\frac{1}{2(n+1)^4}+E_n.$$

Hence

$$S=\sum_{k=1}^{9}\frac{1}{k^4}+\frac{1}{(3)(10)^3}+\frac{1}{(2)(10)^4}+E_9, \qquad |E_9|<5\times 10^{-5}.$$

47. $\left(\sum_{k=3}^{22}\frac{1}{k^3}\right)+\frac{1}{(4)(22)^2} \simeq 0.07659$ after the error is centred.

48. $n!>n^n e^{-n+1}$.

49. $\frac{1}{4}''$.

50. $I = 0.09966867 \pm 5 \times 10^{-8}$.

51. (a) no $\left(\dfrac{F(1.2) - F(1.1)}{1.2 - 1.1} < 0 \text{ and use Mean Value Theorem}\right)$;
(b) 0.6404;
(c) 2.2680;
(d) -0.609, using the secant line joining data points at 1.2 and 1.4;
(e) (ii).

Chapter VII

2. (b) $f(x)$ is discontinuous at $x = 1$. Redefine $f(1) = 2$.

3. (a) $x = -1$ and $x = 1$ are discontinuities of $f(x)$, $f(x)$ is continuous at $x = 0$; (b) $x = -1$ is an essential discontinuity, $x = 1$ is a removable discontinuity.

5. 1.

6. 2.

7. $A = -\dfrac{9}{4}$, $B = \dfrac{3}{4}$, $C = \dfrac{41}{4}$, $D = \dfrac{13}{4}$.

8. (a) yes; (b) no.

9. $f(x)$ is continuous for all x; $f(x)$ is differentiable for all x except $x = 3$.

10. Yes.

11. If $x \neq 0, 1, -1$, $f'(x) = \left(\dfrac{|x^3 - x|}{x^3 - x}\right)(3x^2 - 1)$. The chain rule does not apply when $u = x^3 - x = 0$.

12. $f(x) = g(x) + 1$.

13. $f(x) = x$.

14. $f(x) = |x|^{3/2}$, $a = 0$.

15. $h'(0) = g(0)$.

18. No, since $f'(0)$ does not exist.

19. (a) First show that $|f'(x)| \leq K$ for some constant K. (b) The converse is false (Let $f(x) = |x|$ on $[-1, 1]$.); (c) $f(x) = |x|^{1/2}$ on $[0, 1]$.

20. $xg(x) - F(g(x))$.

21. $x^x > e^x > x > \log x$.

22. $\log x^5 < x^3 < e^x < 10^x < e^{20x} < e^{x^2}$.

23. No.

25. (a) If $f(x)$ is continuous on $[a, b]$ and $f(a) < k < f(b)$, then there is some $c \in (a, b)$ such that $f(c) = k$; (b) $p(0) = 1$, $p(1) = -1$, $p(2) = 3$. Now apply (a).

26. $\log 2$.

27. (a) at $x = \hat{x}$ where $f'(\hat{x}) = 0$; (b) $y'' > 0 \leftrightarrow$ concave upward, $y'' < 0 \leftrightarrow$ concave downward; (c) $f''(a) > 0 \leftrightarrow$ relative minimum, $f''(a) < 0 \leftrightarrow$ relative maximum, $f''(a) = 0 \leftrightarrow$ neither; (d) $H(b) - H(a)$.

28. Let $v = \dfrac{1}{t}$.

29. $a = 0$.

34. (b) $0 \le S_1 \le 1$.

35. (c) $y = \sinh x$ is 1 : 1; (d) $\sinh^{-1} x = \log(x + \sqrt{x^2 + 1})$.

36. (a) $a_2 < 0$, $c_2 < 0$; (b) $a_1 = 4$, $a_2 = 0$, $c_1 = 4$, $c_2 = 0$; (c) $a_2 = c_2 = 0$, $a_1 = b = c_1$.

37. $f(x) = \begin{cases} 0 & \text{if } x < -1, \quad x > 1. \\ 1 & \text{if } -1 \le x < 0, \quad 0 < x \le 1. \\ 2 & \text{if } x = 0. \end{cases}$

38. $f(x) = \begin{cases} \dfrac{1}{q} & \text{if } x = \dfrac{p}{q} \text{ where } p, q \text{ are integers with no common divisors.} \\ 0 & \text{if } x \text{ is irrational.} \end{cases}$

39. (a) If $\alpha > \dfrac{2}{3\sqrt{3}}$, $r < -\dfrac{2}{\sqrt{3}}$; if $\alpha < \dfrac{2}{3\sqrt{3}}$, $r > \dfrac{1}{\sqrt{3}}$. $|\alpha| > \dfrac{2}{3\sqrt{3}} \leftrightarrow$ polynomial having one real root, $|\alpha| < \dfrac{2}{3\sqrt{3}} \leftrightarrow$ polynomial having three real roots;

(b) $\dfrac{dr}{d\alpha} = \dfrac{1}{1 - 3r^2}$. See Figure SVII.39.

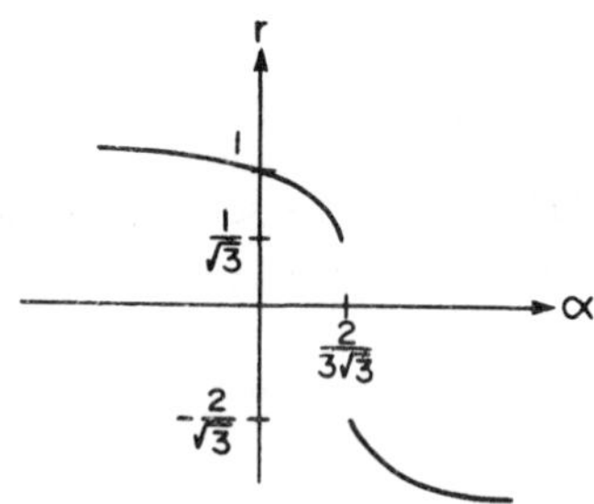

Figure SVII.39

Chapter VIII

1.1. (a) no solution for x; (b) $3 \le x \le 7$; (c) $x = 2, 8$.

1.2. $2 \le x \le 3$.

1.3. $1 \le x \le \frac{5}{3}$.

1.4. $x \le \frac{3}{2}$.

1.5. $-\frac{3}{2} \le x \le \frac{1}{4}$.

1.6. $|x| < 1$, $x \ne 0$.

1.7. $x < -2$, $x \ge -1$.

1.8. $x < -2$, $-1 < x < 4$.

1.9. $x < -2$, $-\sqrt{2} \le x < -1$, $x \ge \sqrt{2}$.

1.10. $y = -\sqrt{3}\,x - 2 - \sqrt{3}$.

1.11. $k = 29$.

1.12. centre is $(-3, 4)$, radius is 2.

1.13. $(x-2)^2 + (y+1)^2 = 34$.

1.14. The curve is the ellipse $\dfrac{(x-1)^2}{2^2} + \dfrac{(y+2)^2}{4^2} = 1$. The foci are located at $(1, -2 \pm 2\sqrt{3})$.

1.15. $2\sqrt{5}$.

1.16. $k = -1, 2 \leftrightarrow$ parabola; $-1 < k < 2 \leftrightarrow$ ellipse; $k = \frac{1}{2} \leftrightarrow$ circle; $k < -1$, $k > 2 \leftrightarrow$ hyperbola.

1.17. $y = 4\sin\left(x + \dfrac{\pi}{3}\right)$: period $= 2\pi$, amplitude $= 4$, phase displacement $= -\dfrac{\pi}{3}$.

1.18. $\dfrac{24}{25}$.

1.19. $x = \frac{1}{2}, e^{2/3}$.

1.20. (a) $\frac{1}{4}$; (b) 127; (c) $f^{-1}(x) = \left|\dfrac{x+1}{2}\right|^{1/3}$ if $x \ge -1$, $f^{-1}(x) = -\left|\dfrac{x+1}{2}\right|^{1/3}$ if $x < -1$.

1.21. (a) $f^{-1}(x) = -\sqrt{x-1}$, $1 < x \le 2$; (b) $f^{-1}(x) = \dfrac{\log x}{2}$, $x > 0$.

1.22. domain of $f = \mathbb{R}$, range of $f = (0, \infty)$, $f^{-1}(x) = |\log x + 8|^{1/3}$ if $x \geq e^{-8}$, $f^{-1}(x) = -|\log x + 8|^{1/3}$ if $0 < x < e^{-8}$, domain of $f^{-1} = (0, \infty)$, range of $f^{-1} = \mathbb{R}$.

1.23. $k = -1$.

2.1. $\dfrac{8x}{3(x^2+2)^{1/3}\left[(x^2+2)^{2/3}+1\right]^2}$.

2.2. $\dfrac{1}{\sqrt{x^2-1}}$.

2.3. $\left(\dfrac{3}{2}\right)\left(\dfrac{5-2x}{3+5x-x^2}\right)$.

2.4. $2|x|$.

2.5. $\dfrac{-4}{(x+1)^3}$.

2.6. $\dfrac{2x^2+2x-1}{(x-1)^4}$.

2.7. (a) $\dfrac{3\sqrt{x}}{2(\log 10)(1+x\sqrt{x})}$; (b) $x^x(\log x + 1)$.

2.8. $\sqrt{3}$.

2.9. $\frac{1}{2}$.

2.10. -1.

2.11. $\dfrac{\csc^2 x + y3^{xy}\log 3}{1 - x3^{xy}\log 3}$.

2.12. $\dfrac{-\left[2\cosh y + \sqrt{\dfrac{y}{x}}\right]}{2x\sinh y + \sqrt{\dfrac{x}{y}}}$.

2.13. $\dfrac{dy}{dx} = -\left(\dfrac{b^x}{a^y}\right)\left(\dfrac{\log b}{\log a}\right)$; $\dfrac{d^2y}{dx^2} = -\left[\dfrac{(\log b)^2}{\log a}\right]\left(\dfrac{b^x}{a^{2y}}\right)a^2b^2$.

2.14. (a) $f'(x) = 2xe^{x^2} > 0$ if $x > 0$; (b) $y = f(x)$ is monotonically increasing; $f'(x) > 0$ if $x > 0$; (c) $[1, \infty)$; (d) 1; (e) see Figure SVIII.2.14.

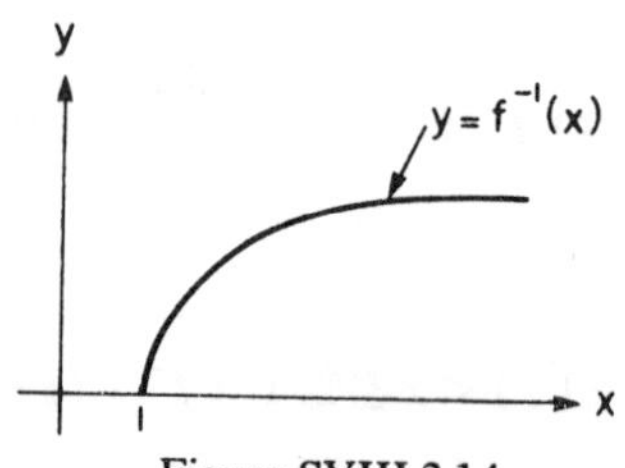

Figure SVIII.2.14

2.15. (a) $f'(x) = 2 + \dfrac{1}{x} > 0$ if $x > 0$; (b) $(-\infty, \infty)$; (c) $\dfrac{1}{3}$.

2.16. (a) $f(x)$ has an inverse; $f^{-1}(x) = (\log x)^2 - 1$, $x \geq e$;
(b) $\dfrac{\sec^2 x}{2\sqrt{\tan x}}$.

2.17. (a) (i) See Figure SVIII.2.17*a*. (ii) See Figure SVIII.2.17*b*. (iii) $(0, \infty)$; (iv) $(0, \infty)$; (v) 2;

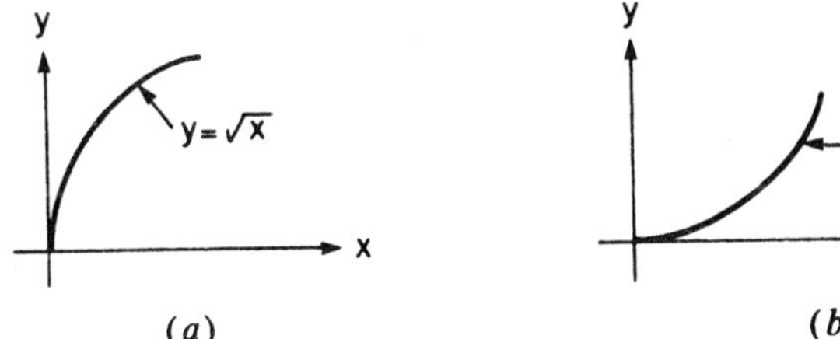

Figures SVIII.2.17*a* and *b*.

(b) (i) See Figure SVIII.2.17*c*. (ii) See Figure SVIII.2.17*d*. (iii) $\mathbb{R}$; (iv) $(0, \pi)$; (v) $-\frac{1}{2}$.

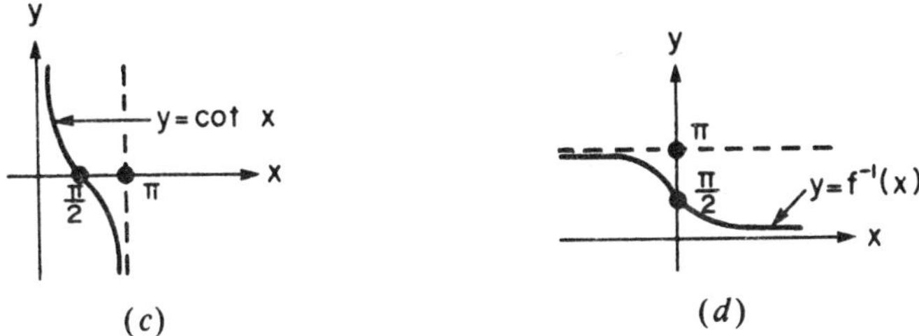

Figures SVIII.2.17*c* and *d*.

2.18. (a) See Figure SVIII.2.18*a*. (b) See Figure SVIII.2.18*b*.
(c) $\dfrac{1}{1-x^2}$.

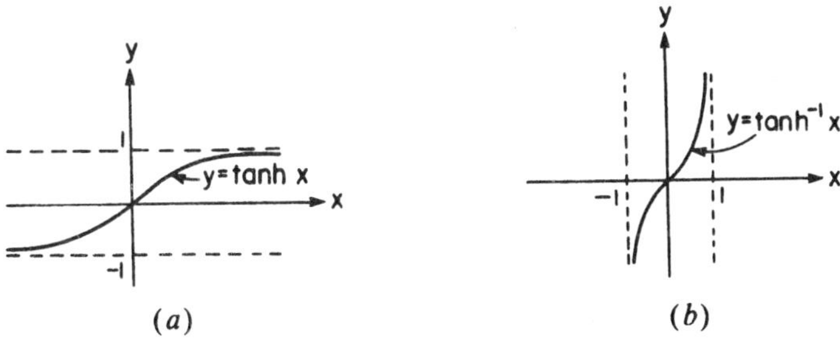

Figures SVIII.2.18*a* and *b*.

2.19. $2^x + x^2 - 3.$

2.20. $\cos x\, g(\sin x).$

2.21. $2xe^{x^2}(e^{2x^2}+1).$

2.22. $2x\sin(x^4).$

2.23. $\dfrac{2x}{(1+x^4)^3} - \dfrac{3x^2}{(1+x^6)^3}.$

2.24. $e^x\sqrt{1+e^{2x}} - \sqrt{1+x^2}.$

2.25. $f(x) = \dfrac{(1+x^2)\cos x - 2x\sin x}{(1+x^2)^2}.$

3.1. $\tan^{-1}2\sqrt{2}.$

3.2. $y = x - 1.$

3.3. $a = -\frac{3}{8},\ b = -\frac{3}{4},\ c = \frac{45}{8}.$

3.4. $a = -2,\ b = 3,\ c = d = 0.$

3.5. All such straight lines intersect the curve at least once. These straight lines are given by the equation $y = mx - 2$. Such a line intersects the given curve thrice if $m > 2$, twice if $m = 2$, once if $m < 2$.

3.6. $A = B = 1.$

3.7. $y = x \pm 4\sqrt{3}$ at $(\mp 2\sqrt{3}, \pm 2\sqrt{3}).$

3.8. $(1,0)$ and $\left(-\frac{1}{2}, -\frac{27}{8}\right).$

3.11. $x = 2.$

3.12. $(-1,0).$

3.13. $y = 2x - 1.$

3.14. $y + 2 = -\frac{1}{5}(x-1).$

3.16. $y' = \dfrac{5-3(x+y)^2}{3(x+y)^2+1} = \dfrac{8y-10x-3}{16x-2y+3}$; the tangent line is $y - \frac{5}{6} = \frac{1}{2}\left(x - \frac{1}{6}\right)$; the normal line is $y - \frac{5}{6} = -2\left(x - \frac{1}{6}\right).$

3.17. (a) $-\frac{22}{3}$; (b) $y = \frac{3}{22}x + 1.$

3.19. $y - 1 = (2-\sqrt{3})(x-1).$

3.20. The radius is 2.7742 to 5 significant figures.

3.21. $f(x)$ is increasing on $(-\infty, -4)$ and $(\frac{1}{4}, 4)$; $f(x)$ is decreasing on $(-4, \frac{1}{4})$ and $(4, \infty)$; $x = -4 \leftrightarrow$ local maximum; $x = \frac{1}{4} \leftrightarrow$ local minimum.

3.22. 16.

3.23. 33.

3.24. -5.

3.25. -3.

3.26. $\dfrac{7}{16}$.

3.27. absolute maximum is 18; absolute minimum is -8.

3.28. minimum is -2; maximum is 2.

3.29. 5 hours per week.

3.30. maximum is $\sqrt{3}$ at $x = 0$; minimum is 0 at $x = \frac{3}{2}$.

3.31. (a) 6 at $x = \frac{1}{2}$; (b) $\frac{19}{3}$ at $x = \frac{3}{4}$; (c) 7 at $x = \frac{1}{2}$.

3.32. maximum is $\frac{3}{2}$; minimum is -3.

3.33. 2.

3.34. $\dfrac{m^m n^n}{(m+n)^{m+n}}$.

3.35. (a) $F'(u) = \dfrac{4u \log u}{1 + 4(\log u)^2}$;

(b) $f(e) = \dfrac{1}{2}, f\left(\dfrac{1}{e}\right) = -\dfrac{1}{2}$.

4.1. $\log(1 + \sqrt{x})^2 + C$.

4.2. $-\sqrt{4 - x^2} + C$.

4.3. (a) $\dfrac{\cos^7\theta}{7} - \dfrac{\cos^5\theta}{5} + C$;

(b) $\dfrac{\theta}{8} - \dfrac{\sin 4\theta}{32} + C$.

4.4. $3(x^3 + x^2 + 7)^{1/3} + C$.

4.5. $\frac{1}{4}\arcsin 2x^2 + C$.

4.6. $\dfrac{x}{2}[\sin(\log x) + \cos(\log x)] + C$.

4.7. $\log\left|\dfrac{x}{x+1}\right| + C$.

4.8. (a) $\dfrac{x}{2}\sqrt{4 - x^2} + 2\arcsin\dfrac{x}{2} + C$;

(b) $\dfrac{1}{2}\left[\log\left(\dfrac{|x|}{\sqrt{x^2+2}}\right) - \dfrac{1}{x} - \dfrac{1}{\sqrt{2}}\arctan\dfrac{x}{\sqrt{2}}\right] + C$.

4.9. (a) $2\log|x+3|+\log|x-1|+C$;
(b) $\frac{1}{2}\log(x^2+4x+8)-2\arctan\left(\frac{x+2}{2}\right)+C$.

4.10. (a) $\frac{9x^2+2x-3}{x^4-1}=\frac{2}{x-1}-\frac{1}{x+1}+\frac{6-x}{x^2+1}$;
(b) $\log\left[\frac{(x-1)^2}{|x+1|\sqrt{x^2+1}}\right]+6\arctan x+C$.

4.11. $\log\left[\frac{(x+2)^2}{x^2+4x+5}\right]+\arctan(x+2)+C$.

4.12. $\log\left[|x^2-1|\sqrt{x^2+1}\right]+\arctan x+C$.

4.13. $\frac{1}{2}\left[\log\left[\frac{\sqrt{x^2+1}}{|x-1|}\right]-\frac{1}{x-1}\right]+C$.

4.14. $\log(x^2+4)+3\log|x-1|+\frac{1}{x-1}+\frac{3}{2}\arctan\frac{x}{2}+C$.

4.15. $6\log|x+4|-5\log|x+3|+C$.

4.16. $\frac{1}{5}\log\left|\frac{2+\tan x}{\sec x}\right|+\frac{2x}{5}+C$.

4.17. $\log\left|1+\tan\frac{x}{2}\right|+C$.

4.18. $\log|\csc x-\cot x|-\frac{1}{1+\cos x}+C$.

5.1. $\frac{1}{2}$.

5.2. $\frac{9\pi}{2}$.

5.3. $\frac{8}{15}$.

5.4. $\frac{8}{15}$.

5.5. $\frac{23}{3}$.

5.6. $\frac{2\sqrt{2}}{3}$.

5.7. $\frac{-11}{3}$.

5.8. (a) $\frac{1}{\sqrt{2}}+\frac{\log 2}{2\sqrt{2}}-1$; (b) $\log 2-\frac{3}{8}$; (c) $\frac{\sqrt{3}}{12}$.

5.9. $1+\log\frac{5}{8}+\frac{3}{2}\arctan\frac{1}{2}-\frac{3\pi}{8}$.

5.10. 7.41.

5.11. $2\sqrt{3}$.

5.12. $\frac{\sqrt{2}}{2}$.

5.14. (b) $\frac{384}{945}$.

6.1. I converges.

6.2. $\frac{1}{4}$.

6.3. (a) $\frac{32}{5}\log 2 - \frac{31}{25}$; (b) diverges.

6.4. I is divergent.

6.5. (a) diverges; (b) converges.

6.6. $I = J + K$ where $J = \lim_{a \to 0^-} \int_{-1}^{a} \frac{dx}{x^2} = +\infty$, $K = \lim_{b \to 0^+} \int_{b}^{2} \frac{dx}{x^2} = +\infty$. Hence I diverges.

6.7. $\frac{\pi}{4}$.

6.8. $\frac{\pi}{8} - \frac{1}{4}$.

6.9. (a) converges; (b) diverges.

7.1. $A = 8$.

7.2. $\frac{37}{12}$.

7.3. $a = \frac{8}{9}$.

7.4. $A = 4\sqrt{3} - \frac{20}{3}$.

7.5. $k = 2^{-1/3}$.

7.6. $A = \frac{3}{2} - 2\log 2$.

7.7. $\frac{9}{2}$.

7.8. $\frac{9}{2}$.

7.9. $A = 2\sqrt{3} - \frac{2\pi}{3}$.

7.10. $\frac{a^2}{6} - ab + \frac{4}{3}a^{1/2}b^{3/2} - \frac{b^2}{2}$.

7.11. $\frac{1}{2} + 5\arcsin\frac{3}{\sqrt{10}}$.

7.12. $A = \frac{3}{4}$.

7.13. $\sqrt{2}-1$.

7.14. $2\sqrt{2}$.

7.15. 12.

7.16. $e-2$.

7.17. $n=3$.

7.19. (a) $\sqrt{\dfrac{2}{e}}$.

7.20. $\dfrac{8\pi}{3}$.

7.21. 16π.

7.22. $V=8\pi$. (Use cylindrical shells.)

7.23. $V=\dfrac{\pi}{2}(e^4-e^2)$.

7.24. $\dfrac{544\pi}{15}$.

7.25. (a) $\pi\left(\dfrac{\sqrt{3}}{2}+\dfrac{\pi}{3}\right)$;
(b) $\dfrac{4\pi}{5}(4^{5/4}-3^{5/4})$.

7.26. $V=12\pi$.

7.27. (a) $\frac{4}{3}\pi a^2 b$; (b) $\frac{4}{3}\pi ab^2$.

7.28. $\dfrac{32\pi}{3}$.

7.29. $\dfrac{7\pi}{6}$.

7.30. $V=\dfrac{\pi}{2}(e^4-5e^2+4e+2)$.

7.32. $\frac{1}{3}\pi r^2 h$.

7.33. (b) $\dfrac{7}{3}$; (c) $\dfrac{76\pi}{15}$.

7.34. $V=\pi\displaystyle\int_0^M xe^{-x}\,dx=\pi[1-(M+1)e^{-M}]$; $\displaystyle\lim_{M\to\infty} V=\pi$.

7.35. (a) $A=1$; (b) yes; $V\le 3\pi A$.

7.36. (a) 128π; (b) $2\sqrt{2}$.

7.37. $\dfrac{14}{3}$.

7.38. $4\sqrt{3}$.

7.39. $40(\cosh\frac{3}{4})(\sinh\frac{1}{4})$ metres.

7.40. $\dfrac{32\pi}{3}$.

7.41. (a) $\pi\int_0^2 (4-y^2)^2\,dy$;

(b) $2\pi\int_1^4 x\sqrt{4-x}\,dx$;

(c) $\int_0^2 \sqrt{1+4y^2}\,dy$.

7.42. (a) $\int_1^3 [g(x)-f(x)]\,dx$;

(b) $3+f(1)+g(4)+\int_1^3 \sqrt{1+[f'(x)]^2}\,dx+\int_3^4 \sqrt{1+[g'(x)]^2}\,dx$;

(c) $\pi\int_3^4 [f^2(x)-g^2(x)]dx$;

(d) $\pi\Big\{[f(1)+3]^2+[g(4)+3]^2+2\int_1^3 [f(x)+3]\sqrt{1+[f'(x)]^2}\,dx$
$+2\int_3^4 [g(x)+3]\sqrt{1+[g'(x)]^2}\,dx\Big\}$.

7.43. (a) $\dfrac{8\pi}{3}(2-\sqrt{3})$; (b) $\pi(10-4\sqrt{3})$.

8.1. $x=a\cos t,\ y=b\sin t,\ 0\le t<2\pi$.

8.2. $x=a(1+\sqrt{2}\cos t),\ y=b(1+\sqrt{2}\sin t),\ 0\le t<2\pi$.

8.3. See Figure SVIII.8.3.

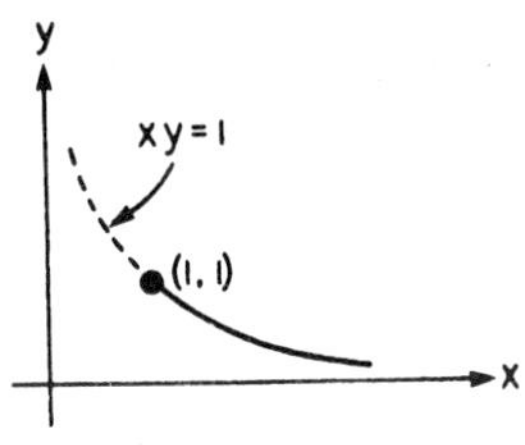

Figure SVIII.8.3

8.4. $-\sqrt{3}$.

8.5. $\dfrac{e^t+2t}{1+\cos t}$.

8.6. (a) $(-11,-4)$ and $(13,4)$; (b) no; (c) $\dfrac{4\sqrt{6}}{9}$.

8.7. $\dfrac{2(t^2+t+1)}{9(t^2-1)^3}$.

8.8. $e^t(t^2+t)$.

8.9. $\dfrac{dy}{dx}=\dfrac{\cos t}{2t+\sin t},\ \dfrac{d^2y}{dx^2}=-\dfrac{(2t\sin t+2\cos t+1)}{(2t+\sin t)^3}$.

8.10. $x = a\sin^2 t + R\cos t,\ y = -a\sin t\cos t + R\sin t.$

8.11. $\dfrac{8\sqrt{2}}{3}a^2.$

8.12. (b) The area of the region bounded by the line joining the origin to $P = (\cos T, \sin T)$, the circle $x^2 + y^2 = 1$ and the positive x-axis is $\dfrac{T}{2}$.

8.13. $\dfrac{8[10^{3/2} - 1]}{27}.$

8.14. (a) $\displaystyle\int_{t_0}^{t_1} \sqrt{\left(\frac{dx}{dt}\right)^2 + \left(\frac{dy}{dt}\right)^2}\, dt$; (b) 72.

8.15. $\sqrt{2}\,[e^{\pi} - 1].$

8.16. (a) $\dfrac{dy}{dx} = \dfrac{3}{2}t,\ \dfrac{d^2y}{dx^2} = \dfrac{3}{4t}$; the tangent line is $2y = 3x - 7$;
(b) $\dfrac{13\sqrt{13} - 8}{27}.$

8.17. (a) 8; (b) 3π.

9.1. (a) $\sqrt{x^2 + y^2} = 3 + \dfrac{2y}{\sqrt{x^2 + y^2}}$; (b) no, $(0,0) \leftrightarrow r = 0$ but $3 + 2\sin\theta \geq 1$.

9.2. Graph $\leftrightarrow r = \sin\left(\theta + \dfrac{\pi}{4}\right) \leftrightarrow$ circle of radius $\dfrac{1}{2}$ centred at $\left(\dfrac{1}{2\sqrt{2}}, \dfrac{1}{2\sqrt{2}}\right)$; see Figure SVIII.9.2.

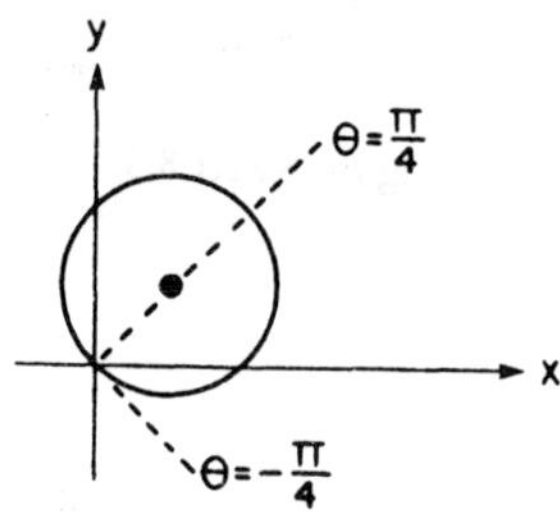

Figure SVIII.9.2

9.3. $r = \dfrac{p}{1 - e\cos\theta}$ where $p = \dfrac{1}{2},\ e = \dfrac{1}{2} \leftrightarrow$ ellipse $\dfrac{\left(x - \frac{1}{3}\right)^2}{\left(\frac{2}{3}\right)^2} + \dfrac{y^2}{\left(\dfrac{1}{\sqrt{3}}\right)^2} = 1$;

see Figure SVIII.9.3.

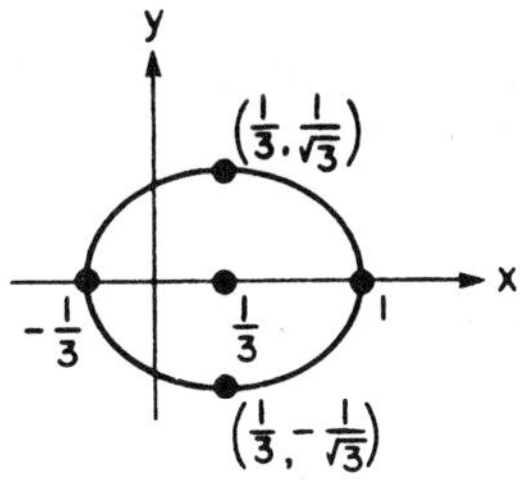

Figure SVIII.9.3

9.4. $\dfrac{dy}{dx} = \dfrac{\cos\theta}{1-\sin\theta}$; horizontal tangent when $\theta = \dfrac{3\pi}{2}$; vertical tangent when $\theta = \dfrac{\pi}{2}$.

9.5. 1.

9.6. 6π.

9.7. 11π.

9.8. $-\displaystyle\int_{\pi}^{2\pi}\left[\sin\theta + \frac{\sin^2\theta}{2}\right]d\theta$.

9.9. π.

9.10. (a) $x^2+(y-1)^2=1$, $(x-1)^2+y^2=1$; (b) $\dfrac{\pi}{2}-1$.

9.11. $\dfrac{11\pi}{12}$.

9.12. See Figure SVIII.9.12; $A = \dfrac{3\pi}{8}$.

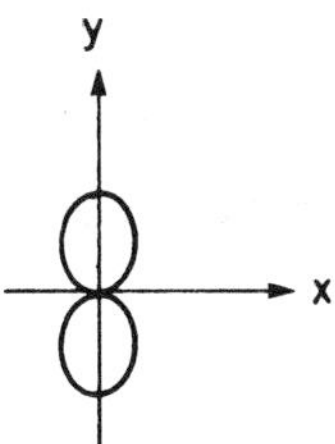

Figure SVIII.9.12

9.13. $a^2\left[\dfrac{\sqrt{3}}{2} - \dfrac{\pi}{6}\right]$.

9.14. $\frac{1}{4}[\pi - 1]$.

9.15. $8+\pi$.

9.16. (a) $r = \dfrac{ab}{\sqrt{a^2\sin^2\theta + b^2\cos^2\theta}}$ where $a > 0$, $b > 0$;

(b) $\theta = \pm\dfrac{\pi}{4}, \pm\dfrac{3\pi}{4}$;

(c) $4ab\tan^{-1}\left(\dfrac{b}{a}\right)$ if $a \geq b$; $4ab\tan^{-1}\left(\dfrac{a}{b}\right)$ if $a \leq b$.

9.17. $4\sqrt{2}\,\pi a^2$.

10.1. converges absolutely (use integral test).

10.2. (a), (b) both converge by alternating series test.

10.3. (a) converges (integral test);

(b) converges $\left(\text{comparison with } \sum_{n=1}^{\infty} \dfrac{1}{n^2}\right)$;

(c) diverges $\left(\lim_{n\to\infty} a_n = 1 \text{ with } a_n = n\sin\dfrac{1}{n}\right)$;

(d) converges (alternating series test);

(e) converges (ratio test);

(f) diverges ($S_n = \log(n+1)$).

10.4. diverges for $k = \pm\dfrac{\pi}{2} + 2\pi n$ for any integer n, converges for all other values of k.

10.5. 2.

10.6. (a) $S = S_n + R_n$ where $S_n = \sum_{k=1}^{n} \dfrac{1}{k^6}$ and $\int_{n+1}^{\infty} \dfrac{dx}{x^6} < \mathrm{R_n} < \int_{n}^{\infty} \dfrac{dx}{x^6} \Rightarrow S = S_n + \dfrac{1}{10}\left[\dfrac{1}{n^5} + \dfrac{1}{(n+1)^5}\right] + E_n$ where $|E_n| < \dfrac{1}{2n^6}$; (b) 1.0175.

10.7. $\lim_{n\to\infty} \sum_{i=0}^{n} \dfrac{1}{n+i} = \int_1^2 \dfrac{dx}{x} = \log 2$.

10.8. (a) $a_1 - a_{n+1}$; (b) $a_1 - A$; (c) $a_1 + a_2 - 2A$.

10.9. Let a_n be the nth term of the sequence:

(a) $a_n = n$;

(b) $a_n = (-1)^n$;

(c) $a_n = \dfrac{1}{n}$, $\sum_{n=1}^{\infty} \dfrac{1}{n}$ diverges;

(d) $a_n = \dfrac{1}{n^2}$, $\sum_{n=1}^{\infty} \dfrac{1}{n^2}$ converges.

10.10. (a) $a_n = \dfrac{1}{2}(-1)^n$;

(b) $a_n = \dfrac{1}{n}$, $b_n = -\dfrac{1}{2n}$;

(c) $a_n = -\dfrac{20}{3}\left(\dfrac{-3}{7}\right)^n$.

10.11. converges for all x.

10.12. $1+x-\frac{x^2}{2}$.

10.13. $1+5(x-1)+10(x-1)^2+10(x-1)^3+5(x-1)^4+(x-1)^5$.

10.14. (a) $f(x)=f(1)+f'(1)(x-1)+\frac{f''(1)}{2!}(x-1)^2+\cdots$;
(b) $1+2(x-1)+\frac{1}{4}(x-1)^2$; (c) $\frac{33}{16}$; (d) $\frac{-1}{64}<\text{error}<\frac{-9}{1225}$.

10.15. $e^2+3e(x-e)+\frac{5}{2}(x-e)^2$.

10.16. $f(x)=x-x^4+\frac{x^7}{2!}-\frac{x^{10}}{3!}+\frac{x^{13}}{4!}-\cdots$; $f^{(4)}(0)=-24, f^{(5)}(0)=0$.

10.17. (a) $x-\frac{x^3}{6}$; (b) $x^2-\frac{x^6}{6}$; (c) $\frac{x^3}{3}-\frac{x^7}{42}$.

10.18. $1+x-2x^2-\frac{x^3}{6}+\frac{2x^4}{3}+\frac{x^5}{120}$.

10.19. For $|x|<1$, note that $\frac{d}{dx}\left(\frac{1}{1-x}\right)=\frac{1}{(1-x)^2}$. Hence $\frac{1}{(1-x)^2}$
$=\sum_{n=1}^{\infty}nx^{n-1}=\sum_{n=0}^{\infty}(n+1)x^n$.

10.20. $\frac{1}{6}\sum_{n=20}^{\infty}(n-17)(n-18)(n-19)x^n$.

10.21. (a) $\sum_{n=0}^{\infty}(-1)^n x^{2n+1}$, $|x|<1$;
(b) $\frac{2}{\sqrt{\pi}}\sum_{n=0}^{\infty}\frac{(-1)^n x^{2n}}{(2n+1)n!}$, all x;
(c) $\sum_{n=0}^{\infty}\left[\frac{e^2+(-1)^{n+1}e^{-2}}{2}\right]\frac{(x-2)^n}{n!}$, all x.

10.22. $\frac{a}{(1-r)^2}$.

10.23. $-(31!)$.

10.24. (a) $\sum_{n=0}^{\infty}\frac{x^n}{2^{n+1}}$; (b) $\frac{x}{1+x-2x^2}=\frac{-x}{(2x+1)(x-1)}=$
$-\frac{1}{3}\left[\frac{1}{x-1}+\frac{1}{2x+1}\right]=\sum_{n=0}^{\infty}\left[\frac{1+(-1)^{n+1}2^n}{3}\right]x^n$, $|x|<\frac{1}{2}$.

11.1. 2.

11.2. (a) $\frac{11}{9}$; (b) $\frac{1}{2}$.

11.3. 3.

11.4. 0.

11.5. $\frac{2}{3}$.

11.6. $f(k)=3$ if $k \neq 2$; $f(2)=-1$.

11.7. -1.

11.8. (a) limit does not exist; (b) $\frac{1}{2}$.

11.9. 0.

11.10. 0.

11.11. 1.

11.12. 2.

11.13. 0.

11.15. $\dfrac{f(0)}{f(4)}$.

11.16. e.

11.17. $\dfrac{-\cos 1}{\pi}$.

11.18. 0.

11.19. $\frac{1}{3}$.

11.20. $\frac{1}{2}$.

11.22. 0.

11.23. $e^{-1/2}$.

11.24. limit is $+\infty$ if $x \to +\infty$; limit is 0 if $x \to -\infty$.

11.25. e^3; $[(e^x+2x)^{1/x}=e(1+2xe^{-x})^{1/x}]$.

11.26. (a) e^{15}; (b) 1; (c) $-\infty$; (d) $e^{25}\log 5$.

11.27. (a) $\frac{2}{9}$; (b) 1; (c) 0.

11.28. (a) 2; (b) ∞ (limit is $-\infty$ if $x \to 0^+$, limit is $+\infty$ if $x \to 0^-$); (c) $e^{-1/2}$.

11.29. (a) $\sin(2x)=2x-\dfrac{4x^3}{3}+\dfrac{4x^5}{15}-\cdots$, $\cos(3x)=1-\dfrac{9x^2}{2}+\dfrac{27x^4}{8}-\cdots$, $e^{-x^2}=1-x^2+\dfrac{x^4}{2}-\cdots$; (b) (i) 0.0997, (ii) $-\frac{7}{4}$.

12.1. π.

12.2. 3.

12.3. $\dfrac{\pi}{3}$.

12.4. (a) $\dfrac{dR}{dS} = \dfrac{k}{S}$, k = constant; (b) $R = k\log\dfrac{S}{S_0}$.

12.5. $y^2 = 4x(\log x - 1) + 8$.

12.6. $3y^4 = 4x^3 - 60$.

12.7. $y = \frac{2}{3}x^3 + \frac{3}{2}x^2 + C_1x + C_2$ where C_1 and C_2 are arbitrary constants.

12.8. $12y = 6x^2 - x^4 - 20x + 27$.

13.1. $y = -\left[\dfrac{1}{x-1} + \dfrac{1}{x+1}\right]$, $y^{(n)}(x) = (-1)^{n+1}n!\left[\dfrac{1}{(x-1)^{n+1}} + \dfrac{1}{(x+1)^{n+1}}\right]$.

13.2. Note that $f(x) = -1 + \dfrac{2}{1+\sqrt{1+x}}$.

(a) $Df = \dfrac{-1}{\sqrt{1+x}\,(1+\sqrt{1+x})^2}$, $\lim_{x\to\infty} f(x) = -1$;
(b) domain of f: $x \geq -1$, domain of f^{-1}: $-1 < x \leq 1$;
(c) -4;
(d) see Figures SVIII.13.2(a) and (b).

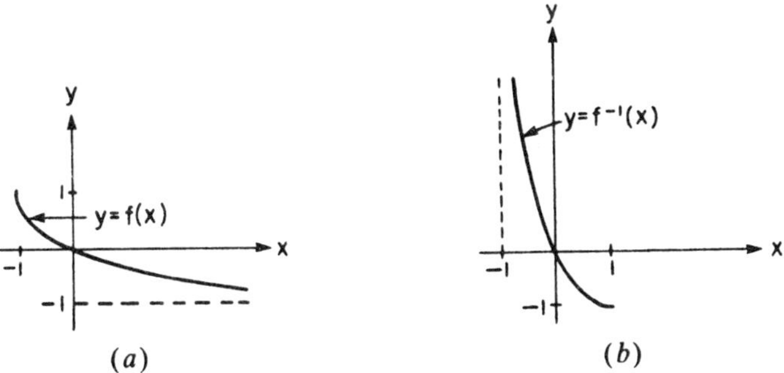

Figures SVIII.13.2a and b.

13.3. $k_1 = -\frac{1}{8}$, $k_2 = 6$.

13.4. (a) See Figures SVIII.13.4(a), (b), and (c);

(b) (i) $0 < k < 2$: $\left(\dfrac{1}{2^{1/k}}, \dfrac{1}{2^{1/k}}\right)$; $k = 2$: all points; $k > 2$: $(0,1)$ and $(1,0)$;

(ii) $0 < k < 2$: $(0,1)$ and $(1,0)$; $k = 2$: all points; $k > 2$: $\left(\dfrac{1}{2^{1/k}}, \dfrac{1}{2^{1/k}}\right)$;

(c) $k = \dfrac{2}{3}$: $\dfrac{3}{2}$; $k = 1$: $\sqrt{2}$; $k = 2$: $\dfrac{\pi}{2}$; as $k \to \infty$, arc length $\to 2$; (d) the tangent line through (u, v) is $y - v = -\left(\dfrac{u}{v}\right)^{k-1}(x - u)$; length of segment is $\sqrt{u^{2-2k} + v^{2-2k}}$; length is independent of (u, v) if $k = 1, \dfrac{2}{3}$.

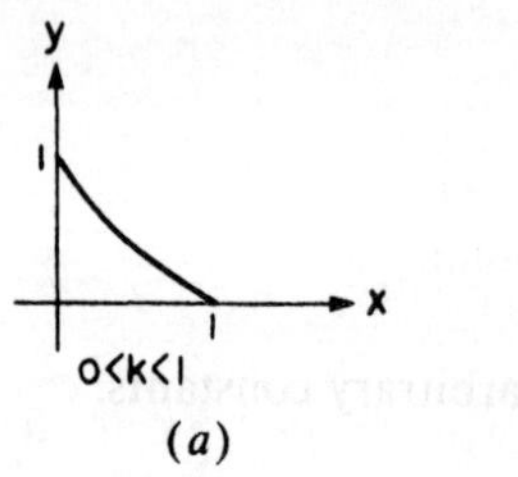

(a)

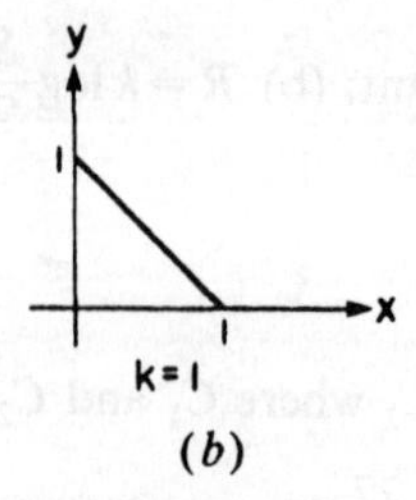

(b)

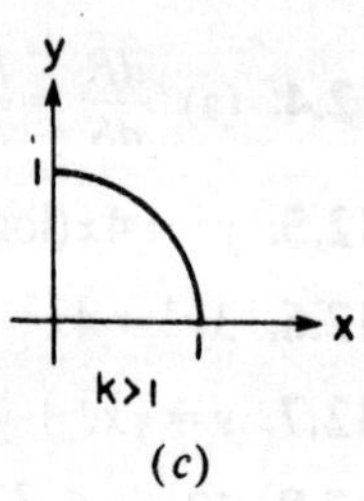

(c)

Figures SVIII.13.4$a-c$.

13.5. (a) (i) $(x^2+1)\tan^{-1}x - x + C$, (ii) $\log|1+\tan x| + C$;
(b) $y = (x^2+1)\tan^{-1}x - x + \log|1+\tan x| + 2$.

13.7. possibly (b), (d), (g).

Problem Sources

1. Solved Problems

Acadia University: I.8; II.10, 25; VIII.3.8, 3.13, 7.3, 7.12, 12.4.

Carleton University: I.12; II.12, 14, 18; III.12, 15, 19, 21; V.1, 2; VII.13, 16; VIII.3.1, 3.5, 3.10, 3.12, 4.10, 5.3, 7.6, 7.10.

Concordia University: VIII.3.6, 3.11, 10.3, 10.16.

Dalhousie University: III.30; VI.9, 15; VIII.3.7, 5.4, 8.4, 8.5, 8.9, 8.11, 11.5.

École Polytechnique: VIII.6.3, 8.2, 9.7, 9.8.

Lakehead University (W. Eames): V.7, 11.

McGill University: II.20, 34; III.2, 22, 24; VIII.2.2, 7.7.

Queen's University: I.4, 6, 16; II.28; III.27; IV.2, 7, 8, 16; V.5; VI.1, 6, 10, 11, 13, 24; VII.12, 14; VIII.2.7, 4.4, 5.8, 6.1, 6.5, 7.1, 7.14, 10.1, 10.2, 10.7, 10.11 to 10.13, 10.15, 10.17, 11.6, 11.8, 11.9, 12.3.

Royal Roads Military College: II.13, 16, 22, 32, 40; IV.1; VIII.1.5, 1.9, 4.13, 7.15, 8.1, 11.1, 11.11.

Université de Sherbrooke: V.8 to 10.

University of British Columbia: II.3, 23, 30, 31, 35, 38; III.9, 13; IV.11, 14; VI.17; VII.19; VIII.1.4, 1.6, 1.8, 6.6, 7.5, 8.6, 9.5, 10.5, 10.18, 12.6.

University of Lethbridge: I.2; III.6; IV.4; VI.21 (D. Connolly); VIII.2.4, 3.3, 7.2.

University of Manitoba: III.8, 18, 20, 23, 28; VIII.3.2, 9.3, 12.2.

University of New Brunswick: I.1; II.8; VII.5; VIII.1.2, 1.10, 4.6, 4.12, 5.9, 8.8, 9.1, 11.10, 12.5.

University of Ottawa: I.11; II.36; VII.6, 17, 18; VIII.2.9, 4.1, 11.2.

University of Saskatchewan: I.5, 13; II.7, 9, 15, 19, 21, 27; III.3, 11, 14, 25, 26, 31; IV.5, 13, 15; VI.3, 4, 7; VII.2, 11; VIII.7.4.

University of Toronto: I.15, 17; II.37; III.4, 29; IV.3, 9, 10; V.6; VI.19; VIII.4.2, 4.3, 4.5, 5.6, 7.13, 10.8, 10.14, 11.12.

University of Victoria: II.1, 6; III.7; IV.6; VIII.3.15, 3.16, 4.9, 6.7, 7.9, 7.16, 8.3, 9.2, 9.6.

University of Waterloo: I.3; II.4, 11, 17, 33, 39; III.16, 17; VI.2, 16; VII.1, 3, 4, 10; VIII.1.1, 1.7, 2.3, 2.5, 3.4, 3.14, 4.7, 6.4, 10.6, 10.9, 10.10, 11.13.

University of Western Ontario: I.7, 9; II.2, 5, 29; III.1, 10, 32; VI.12; VIII.2.8, 5.1, 5.2, 7.8, 7.11, 8.10, 11.7.

University of Winnipeg: VII.8, 9; VIII.8.7, 12.1.

Wilfrid Laurier University: II.24; IV.12, 17 to 20; V.3; VIII.4.11, 5.5.

York University: II.26; III.5; VIII.1.3, 2.1, 4.8, 6.2.

2. Supplementary Problems

Acadia University: II.6, 27, 38, 46, 52, 62, 70, 71, 77, 100; III.37, 54; IV.19; VII.10, 27; VIII.3.10, 5.13, 7.2, 8.9, 10.7.

Carleton University: I.17; II.16, 22, 28, 33; III.15, 18, 44, 58; IV.6, 9 to 11, 26, 27; V.4; VII.11, 14, 25; VIII.2.8, 3.16, 3.17, 3.21, 3.24, 4.2, 4.3, 4.9, 5.3, 5.10, 7.15, 7.16, 7.18, 7.25, 7.26, 7.31, 11.8, 12.1 to 12.3, 12.9, 12.10, 13.5.

Concordia University: II.41, 50, 73; V.2; VI.7, 14; VIII.3.27, 12.7, 12.8.

Dalhousie University: II.45, 61, 64; IV.12, 14; V.25; VI.1, 25, 36; VIII.1.20, 2.10, 3.23, 3.25, 4.5, 5.1, 7.32, 8.6, 8.13, 8.16, 9.1, 9.4, 9.5, 9.7, 9.10, 10.3, 10.13, 10.19.

École Polytechnique: VIII.6.4, 7.10, 7.11, 8.1, 8.2, 8.10, 8.11, 9.13 to 9.16.

Lakehead University (W. Eames): V.14 to 19.

McGill University: II.5, 14, 26, 58, 67, 72, 90, 93, 95; III.3, 42, 46, 48, 57; V.26; VI.11 to 13; VIII.1.12, 1.13, 1.15, 7.29.

Queen's University: I.3, 7, 15, 23, 24, 26; II.23, 30, 31, 35, 79, 94; III.19 and 20 (J. Ursell), 22; IV.15, 22, 23, 30, 38; V.12; VI.24, 31, 32, 38, 49 (P. Taylor), 51; VII.5, 39 (P. Taylor); VIII.2.2, 2.9, 2.11, 2.12, 2.14, 2.19, 2.21, 2.24, 3.5, 4.7, 4.8, 4.11 to 4.13, 5.2, 5.5, 5.7, 5.8, 6.1, 6.3, 6.8, 7.23, 7.30, 10.1, 10.2, 10.4, 10.6, 10.9, 10.11, 10.12, 10.14, 10.16, 11.9, 11.11 to 11.13, 12.5.

Royal Roads Military College: II.53, 85; III.12, 24, 25, 33, 40, 47; VII.1, 26; VIII.1.19, 2.1, 2.3, 2.6, 2.13, 3.1, 3.22, 3.28, 3.32, 4.4, 5.11, 6.6, 7.24, 7.35, 7.40, 8.7, 9.11, 11.24.

Université de Sherbrooke: V.21 to 24.

University of British Columbia: I.20, 21 (R. Riddell); II.18, 34, 42, 47, 55, 59, 66, 82, 88, 91, 97, 98, 99; III.6, 14, 16 (K. Lam), 23 (R.A. Adams), 51, 52 (K. Lam); IV.13, 33, 42; V.6, 9; VI.2, 34, 35, 37, 39, 40, 43; VII.16 and 23 (R. Riddell), 28, 31; VIII.1.9 (R. Riddell), 1.10, 1.11, 1.16 (R. Riddell), 1.17, 2.20, 3.9 and 3.13 (R. Riddell), 3.18 to 3.20 (R. Riddell), 3.31, 3.33, 3.34, 5.15, 7.6, 7.9, 7.13, 7.27, 7.43, 8.3, 8.4, 9.2, 9.3, 9.6, 9.17, 10.18, 10.21, 11.3, 11.14, 11.29, 13.1.

University of Lethbridge: I.4, 10; II.11, 25, 43, 65; III.4, 8, 10, 17; IV.5; VI.4 to 6; VIII.1.5, 1.8, 7.7, 13.7.

University of Manitoba: II.8, 39, 44, 80, 83, 86; III.21, 28, 31, 34, 35, 38; IV.3; VII.30, 35.

University of New Brunswick: I.5, 11, 12, 25; II.32, 49, 60, 81; III.5, 13; IV.4; VII.6; VIII.1.6, 1.23, 2.7, 2.16, 3.11, 3.12, 3.14, 3.15, 4.1, 4.10, 5.12, 6.5, 7.5, 7.22, 7.28, 8.5, 8.14, 8.17, 9.8, 11.2, 11.23.

University of Ottawa: II.4, 92; III.29; VI.19, 26 to 28; VIII.1.22, 3.30, 4.17, 4.18, 5.16, 9.12, 11.4, 11.6, 13.2.

University of Saskatchewan: I.6, 9, 16; II.10, 17, 48, 54, 74, 76, 87; III.9, 11, 36, 41, 45, 50; IV.1, 2, 35; V.13; VI.3, 22, 29; VII.2, 8, 9, 13, 15; VIII.1.2 to 1.4, 1.7, 2.22, 3.7, 3.8, 3.29, 5.4, 7.1, 7.4, 7.33, 12.6.

University of Toronto: I.22; II.2, 3, 13, 37, 96; III.2, 7, 30, 55; IV.7, 18, 20, 24, 28, 31, 36; V.8, 10, 11; VI.23, 44; VII.36 to 38; VIII.1.1, 4.14, 4.16, 7.3, 7.42, 10.10, 10.22, 11.5, 11.15, 11.16, 11.19, 11.21, 11.27, 13.3, 13.4.

University of Victoria: I.1, 8; II.9, 24, 29, 63, 75, 78; III.1, 32, 39, 43; IV.8, 16; V.1, 3, 7; VIII.1.14, 2.5, 3.2, 7.14, 7.19, 8.15.

University of Waterloo: I.19; II.15, 20, 21, 36, 40, 89; III.27; IV.32; V.20; VII.4, 7, 12, 17, 20, 32, 34; VIII.1.18, 2.4, 2.15, 2.26, 3.6, 3.35, 5.9, 5.14, 6.2, 7.17, 7.36, 7.38, 7.39, 8.12, 10.5, 10.8, 11.17, 11.30, 13.6.

University of Western Ontario: I.13, 14; II.7, 12, 19, 56; III.26, 53, 59; IV.46; VI.9, 16, 18, 21; VII.18, 21, 22; VIII.3.3, 3.4, 4.6, 6.9, 7.12, 7.21, 7.34, 7.37, 7.41, 9.9, 10.15, 10.17, 11.1, 11.10, 11.22, 12.4.

University of Winnipeg: I.18; II.1, 51, 68, 69; VI.8; VIII.2.25, 3.26, 8.8.

Wilfrid Laurier University: II.84; IV.17, 25, 29, 34, 37, 39 to 41, 43 to 45; VI.17; VII.3; VIII.4.15, 5.6, 7.8, 11.7, 11.25.

York University: I.2; II.57; V.5; VI.10; VIII.1.21, 7.20.

Index

Each entry has three possible references:

1. Page listing(s) to introductory chapter material and/or key problems.
2. Solved Problems (in parentheses).
3. Supplementary Problems [in brackets].

Index of Symbols

Symbol	Meaning	Page
$f'(x)$	the derivative of f with respect to x	1
$f''(x)$	the second derivative of f with respect to x	1
$x \to \pm\infty$	x approaches plus or minus infinity	1
$(f \circ g)(x)$	the composition function f of g of x	2
$\leftrightarrow$	corresponds to; is equivalent to	3
$a \equiv b$	a is identically equal to b	3
$\simeq$	approximately equals	6
$y(\pm 1) = \mp 2$	$y(1) = -2$ and $y(-1) = 2$	6
$\Rightarrow$	implies that	6
$\sim$	is asymptotic to; behaves like	8
$y \to 0^{\mp}$	y approaches 0 from the left $(-)$ or right $(+)$	8
$\log y$	the natural logarithm of y	13
$\cap$	intersection	21
$x \neq 0$	x does not equal 0	31
$D_x f$	the derivative of f with respect to x	33
ΔOPQ	triangle OPQ	40
$\angle OQP$	angle OQP	54
rpm	revolutions per minute	71
dA	differential element of area	97
$\cup$	union	99
$F(x)\|_a^b$	$F(b) - F(a)$	112
dV	differential element of volume	119
$\exp[f]$	e^f	153
$a \ll b$	a is very much less than b	194
$x \in [a, b]$	x belongs to the closed interval $a \leq x \leq b$	195
$r = a \pm \varepsilon$	$a - \varepsilon < r < a + \varepsilon$	196
$\Leftrightarrow$	if and only if	222
$\mathbb{R}$	the real line	234
1:1	one-to-one	239
r.h.s.	right hand side	251
l.h.s.	left hand side	251